Advances in Structural Analysis and Rehabilitation for Existing Structures

Advances in Structural Analysis and Rehabilitation for Existing Structures

Guest Editor

Alessio Cascardi

MDPI

Basel • Beijing • Wuhan • Barcelona • Belgrade • Novi Sad • Cluj • Manchester

Guest Editor
Alessio Cascardi
Civil Engineering
University of Calabria
Arcavacata di Rende
Italy

Editorial Office
MDPI AG
Grosspeteranlage 5
4052 Basel, Switzerland

This is a reprint of articles from the Special Issue published online in the open access journal *Eng* (ISSN 2673-4117) (available at: www.mdpi.com/journal/eng/special_issues/KN03J54453).

For citation purposes, cite each article independently as indicated on the article page online and using the guide below:

Lastname, A.A.; Lastname, B.B. Article Title. *Journal Name* **Year**, *Volume Number*, Page Range.

ISBN 978-3-7258-1962-1 (Hbk)
ISBN 978-3-7258-1961-4 (PDF)
https://doi.org/10.3390/books978-3-7258-1961-4

Contents

About the Editor

Alessio Cascardi

Alessio Cascardi is an associate professor at the University of Calabria. He is a multi-graduate in Building Systems Engineering and Civil Engineering with a Ph.D. in Materials and Structural Engineering and is mainly interested in the field of structural and energy reinforcement and the consolidation of existing built heritage in masonry and/or reinforced concrete with traditional and innovative composite materials. He currently focuses his research on the experimental, analytical, and numerical aspects aimed at defining new predictive formulas for the design and sizing of reinforcement. This goal is pursued through traditional data processing techniques, such as simple and multiple linear regressions, and innovative ones such as artificial neural networks and data mining. He is the author of more than 70 scientific articles published in national and international journals. He is the holder of two patents relating to the S.S.D. ICAR/09 – Construction Technique.

eng

MDPI

Editorial

Special Issue: Advances in Structural Analysis and Rehabilitation for Existing Structures

Alessio Cascardi

Department of Civil Engineering, University of Calabria, Via P. Bucci Cube 45B, 87036 Arcavacata di Rende, Italy; alessio.cascardi@unical.it

Citation: Cascardi, A. Special Issue: Advances in Structural Analysis and Rehabilitation for Existing Structures. *Eng* **2024**, *5*, 1804–1810. https://doi.org/10.3390/eng5030095

Received: 1 August 2024
Accepted: 3 August 2024
Published: 6 August 2024

1. Introduction

In the dynamic realm of civil engineering, the principles of structural analysis and rehabilitation are pivotal in extending the lifespan and enhancing the performance of existing structures. The following Special Issue, entitled "*Advances in Structural Analysis and Rehabilitation for Existing Structures*", delves into the forefront of research and innovation in the field. It explores the evolving methodologies, technologies, and strategies that engineers and researchers worldwide are employing to address the challenges posed by aging infrastructure. As populations grow and urbanization intensifies, the demand for sustainable infrastructure becomes increasingly urgent. Existing structures, whether historic landmarks or essential infrastructure, require continuous assessment and maintenance to ensure safety, functionality, and longevity. This collection of contributions aims to illuminate the latest advancements that enable engineers to analyze, diagnose, and rehabilitate structures with precision and efficiency. From advanced computational models and non-destructive testing techniques to innovative materials and retrofitting strategies, the chapters in this Special Issue encompass a broad spectrum of approaches. Each article offers insights into how these advancements are reshaping the practice of structural engineering, paving the way for more resilient and sustainable built environments.

Moreover, the following Special Issue serves as a platform for collaboration and knowledge exchange among researchers, practitioners, and academicians. By sharing cutting-edge research and practical applications, we aim to inspire further innovation and foster a deeper understanding of the complexities involved in maintaining and upgrading our infrastructure. Through this compilation, we invite readers to explore the frontiers of structural analysis and rehabilitation, where theory meets practice and where the future of sustainable infrastructure is being defined. Whether you are a seasoned professional, a budding researcher, or simply curious about the evolving landscape of civil engineering, we hope that this collection sparks new ideas and insights that contribute to the ongoing evolution of our built environment.

Welcome to "*Advances in Structural Analysis and Rehabilitation for Existing Structures*"—a testament to the ingenuity and dedication driving the advancement of structural engineering worldwide.

2. Overview of the Published Articles

This Special Issue contains eleven papers, including eight reviews, published by a number of authors interested in cutting-edge developments in the field of engineering. The authors hail from 12 countries, including Germany, France, Portugal, Canada, the UK, the USA, Singapore, Brazil, China, Malaysia, Iran, and Iraq.

The authors of the paper "Carbon Fiber-Reinforced Polymer Composites Integrated Beam–Column Joints with Improved Strength Performance against Seismic Events: Numerical Model Simulation" explore the use of carbon fiber-reinforced polymer (CFRP) composites in beam–column joints to enhance their strength and resilience during seismic events. The study authors utilize numerical simulations to investigate how CFRP can effectively reinforce these critical structural connections, offering insights into its performance and potential benefits in seismic-resistant design. Key findings highlight the improved strength and durability of CFRP-integrated joints, suggesting promising applications for enhancing structural safety in earthquake-prone regions.

The authors of the paper "Performance Evaluation of Self-Compacting Glass Fiber Concrete Incorporating Silica Fume at Elevated Temperatures" examine the behavior of self-compacting concrete (SCC) reinforced with glass fibers and silica fume when subjected to high temperatures. Through a series of experimental tests, the study authors evaluate the concrete's mechanical properties, including its strength, workability, and durability under elevated temperature conditions. The results indicate that the addition of glass fibers and silica fume enhances the thermal resistance and overall performance of SCC, making it a viable option for construction applications where high-temperature exposure is a concern.

The authors of the paper "Minimum Shear Reinforcement for Reactive Powder Concrete Beams" investigate the requirements for shear reinforcement in reactive powder concrete (RPC) beams. Through experimental testing and analysis, the study authors determine the optimal amount of shear reinforcement needed to ensure structural integrity and prevent failure. Findings suggest that RPC beams, known for their high strength and durability, require less shear reinforcement compared to conventional concrete beams, due to their superior material properties. The research provides guidelines for the design of RPC beams, aiming to improve efficiency and safety in structural engineering applications.

The authors of the paper "Numerical Investigation of the Seismic Performance of an Innovative Type of Buckling-Restrained Brace (BRB)" explore the effectiveness of a newly designed buckling-restrained brace in enhancing the seismic performance of structures. Using numerical simulations, the study assesses the brace's ability to withstand seismic loads and prevent buckling under stress. Their study results indicate that the innovative BRB design significantly improves energy dissipation and overall structural resilience during earthquakes. The findings highlight the potential of this new BRB type to provide superior seismic protection, offering valuable insights for improving earthquake-resistant building design.

The authors of the paper "Assessment of Groundwater Contamination in the Southeastern Coast of Brazil: A Potential Threat to Human Health in Marica Municipality" investigate the quality of groundwater in Marica Municipality, Brazil, and its implications for human health. Through comprehensive water quality testing and analysis, the study authors identify various contaminants present in the groundwater, including heavy metals and microbial pollutants. The findings reveal significant levels of contamination, posing potential health risks to the local population. The research underscores the urgent need for effective groundwater management and pollution mitigation strategies to protect public health in the region.

The authors of the paper "A Parametric Study on the LDB Strength of Steel-Concrete Composite Beams" examine the lateral–distortional buckling (LDB) strength of steel-concrete composite beams through a series of parametric analyses. By varying key parameters such as beam geometry, material properties, and loading conditions, the study authors evaluate their impact on the LDB behavior and strength of these composite structures. The results provide a deeper understanding of the factors influencing LDB strength and offer guidelines for optimizing the design of steel-concrete composite beams to enhance their stability and performance in structural applications.

The authors of the paper "Insight into the Optimization of Implementation Time in Cob Construction: Field Test and Compressive Strength Versus Drying Kinetics" explore the optimal timing for implementing cob construction to balance compressive strength and drying kinetics. Through field tests, the study authors examine how different implementation times affect the drying process and the resulting compressive strength of cob materials. Findings indicate that optimizing the implementation time is crucial for achieving desirable strength characteristics while ensuring efficient drying. The research provides practical recommendations for enhancing the durability and performance of cob structures, contributing to more effective and sustainable building practices.

The authors of the paper "Seismic Resilience and Design Factors of Inline Seismic Friction Dampers (ISFDs)" investigate the effectiveness of ISFDs in enhancing the seismic resilience of structures. By analyzing various design factors, the study authors evaluate how these dampers perform under seismic loading conditions. The results show that ISFDs significantly improve energy dissipation and structural stability during earthquakes. Key design factors such as damper configuration, material properties, and installation methods are found to influence the overall performance of ISFDs. The research offers valuable insights and guidelines for incorporating ISFDs into building designs to achieve superior earthquake resistance.

The authors of the paper "Improved Structural Health Monitoring Using Mode Shapes: An Enhanced Framework for Damage Detection in 2D and 3D Structures" present a novel approach to structural health monitoring that leverages mode shapes for more accurate damage detection. The enhanced framework is designed to identify and locate damage in both two-dimensional and three-dimensional structures. Through numerical simulations and experimental validation, the study authors demonstrate that using mode shapes improves the sensitivity and reliability of damage detection methods. The findings suggest that this advanced monitoring technique can significantly contribute to maintaining the safety and integrity of various structural systems.

The authors of the paper "Chemo-Thermo-Mechanical FEA as a Support Tool for Damage Diagnostic of a Cracked Concrete Arch Dam: A Case Study" explore the use of finite element analysis (FEA) incorporating chemical, thermal, and mechanical factors to diagnose damage in a cracked concrete arch dam. The case study demonstrates how this comprehensive FEA approach can effectively simulate the complex interactions and stressors affecting the dam's integrity. Their study results highlight the method's ability to accurately identify damage locations and assess the severity of cracks. The research supports the use of chemo-thermo-mechanical FEA as a valuable tool for enhancing the precision of structural diagnostics and maintenance strategies for concrete arch dams.

The authors of the paper "Surface Waterproofing Techniques: A Case Study in Nova Lima, Brazil" investigate various surface waterproofing methods applied to structures in Nova Lima, Brazil. The study authors evaluate the effectiveness of different techniques in preventing water infiltration and mitigating related damage. Through field experiments and performance assessments, the authors identify which waterproofing strategies provide the best protection under local environmental conditions. The study findings suggest that certain techniques significantly enhance the durability and lifespan of the structures. The study offers practical insights and recommendations for selecting and implementing surface waterproofing solutions in similar climatic regions.

3. Closing Remarks

In editing the eleven papers comprising the herein-presented Special Issue, several key insights and conclusions emerge, highlighting the evolving landscape of structural engineering and materials science.

The integration of advanced materials and innovative design approaches plays a crucial role in improving structural resilience. For instance, carbon fiber-reinforced polymer (CFRP) composites have been shown to enhance the strength and seismic performance of beam–column joints, offering a promising solution for earthquake-resistant structures. Similarly, the use of glass fibers and silica fume in self-compacting concrete has demonstrated significant improvements in thermal performance and mechanical properties, indicating their potential for high-temperature applications.

Recent research on reactive powder concrete (RPC) beams reveals that their superior properties allow for reduced shear reinforcement compared to traditional concrete, optimizing material use without compromising structural integrity. The above aligns with findings on buckling-restrained braces (BRBs), where innovative designs have been found to significantly improve energy dissipation and stability during seismic events, underscoring the importance of advanced seismic protection technologies.

Groundwater contamination studies in Nova Lima, Brazil, highlight a critical public health concern, emphasizing the need for effective water management and pollution mitigation strategies. This concern extends to the structural domain as well, where optimizing design parameters for steel–concrete composite beams can enhance their lateral–distortional buckling strength, contributing to more stable and efficient construction practices.

In the realm of construction techniques, the optimization of implementation time in cob construction has been shown to affect both compressive strength and drying efficiency, illustrating the importance of precise timing for material performance. This principle of optimization extends to the use of inline seismic friction dampers (ISFDs), which offer improved seismic resilience through enhanced energy dissipation.

Structural health monitoring benefits from the use of mode shapes, which provide more accurate damage detection in both two-dimensional and three-dimensional structures. This method's reliability is further supported by advanced chemo-thermo-mechanical finite element analysis (FEA) techniques, which have proven effective in diagnosing damage in concrete arch dams by considering a comprehensive range of factors.

Lastly, the evaluation of surface waterproofing techniques in Nova Lima, Brazil, underscores the variability in effectiveness based on environmental conditions, stressing the importance of selecting appropriate methods to prevent water damage and extend the lifespan of structures.

Together, these studies reflect a growing emphasis on integrating advanced materials, innovative designs, and precise monitoring techniques to enhance the performance, safety, and longevity of structures across diverse applications and environments.

In addition, a valuable short references list is reported herein in order to make the present section as comprehensive as possible. In their study, Cascardi et al. (2024) [1] explore the use of carbon fabric-reinforced cementitious mortar (CFRCM) for enhancing the confinement of concrete cylinders. The study authors focus on evaluating the effectiveness of multi-ply wrapping configurations in improving the mechanical properties and structural performance of concrete under different loading conditions. Key findings from the research highlight the influence of the mortar matrix in CFRCM on overall performance, particularly in terms of strength enhancement and ductility. The authors experimentally investigate various parameters such as ply orientation, number of layers, and mortar composition to determine their impact on confinement efficiency. The results suggest that the matrix composition significantly affects the confinement effectiveness of CFRCM, with implications for optimizing design and application in structural engineering. The study provides valuable insights into the mechanics of composite materials and offers practical guidelines for utilizing CFRCM in enhancing the resilience and durability of concrete structures. Overall, Cascardi et al. contribute to the understanding of CFRCM technology by emphasizing the importance of mortar matrix properties in achieving enhanced performance of reinforced concrete elements through multi-ply wrapping techniques.

The authors of [2] investigate the application of fiber-reinforced polymers (FRP) as a removable method for confining and reinforcing heritage masonry columns. The study aims to address the need for reversible strengthening techniques in the preservation of historical structures. The authors propose a novel approach that allows FRP materials to be used temporarily, ensuring that the original state of the masonry columns can be restored when necessary. This method balances the preservation requirements of heritage structures with the need for structural enhancement to improve their load-bearing capacity and stability. Key aspects of the research include experimental tests on masonry columns confined with removable FRP wraps. The results demonstrate that the FRP confinement significantly enhances the mechanical performance of the columns, increasing their strength and ductility. Additionally, the reversibility of the FRP application is confirmed, making it a viable solution for temporary strengthening. The study provides detailed insights into the behavior of FRP-confined masonry columns, including the effects of various confinement configurations and materials. The authors also discuss the practical implications of their findings for the conservation of heritage buildings, emphasizing the potential for FRP technology to offer both structural benefits and compliance with preservation standards.

A further contribution is [3], whose authors address the critical issue of evaluating the seismic performance of existing reinforced concrete (RC) structures. The study authors provide a comprehensive framework for assessing the vulnerability of such structures to seismic events and propose methodologies for their evaluation and retrofitting. Key elements of the research include:

- The authors outline systematic procedures for the seismic assessment of existing RC structures. These procedures incorporate both qualitative and quantitative evaluations, considering factors such as structural design, construction quality, material properties, and the current state of the structure.
- Ghobarah discusses various performance criteria that should be met to ensure the safety and functionality of RC structures during and after seismic events. These criteria include maintaining structural integrity, minimizing damage, and ensuring the safety of occupants.
- The authors review different methods for evaluating seismic performance, including linear and nonlinear analysis techniques. Ghobarah emphasizes the importance of accurate modeling and analysis to predict the behavior of RC structures under seismic loads.
- The authors also explore various retrofitting strategies to enhance the seismic performance of RC structures. These strategies range from local strengthening techniques, such as the addition of shear walls and steel bracings, to global approaches like base isolation and energy dissipation devices.
- To illustrate the practical application of the proposed assessment and retrofitting methodologies, the paper includes case studies and examples of existing RC structures that have undergone seismic evaluation and strengthening.

In conclusion, Ghobarah's research offers a valuable contribution to the field of structural engineering by providing a detailed and practical framework for the seismic assessment and retrofitting of existing RC structures. The study emphasizes the importance of a thorough evaluation to ensure that these structures can withstand seismic events, thereby safeguarding lives and reducing economic losses.

Lastly, in the paper by Mesquita et al. (2016) [4], the authors provide a comprehensive review of the advancements in structural health monitoring (SHM) technologies. The study focuses on the development and implementation of SHM platforms to enhance the safety and longevity of civil infrastructure. Key aspects of the research include:

- The paper begins with an overview of the current state of SHM technologies, highlighting their importance in detecting damage, assessing structural integrity, and ensuring the safety of civil structures.
- The authors discuss various sensor technologies used in SHM, including traditional sensors such as strain gauges and accelerometers, as well as advanced optical fiber sensors and wireless sensor networks. These technologies are evaluated based on their accuracy, reliability, and ease of deployment.
- The authors examine the methods for data acquisition and processing, emphasizing the role of real-time monitoring and data analytics in providing accurate and actionable information about structural health. The use of machine learning and artificial intelligence in SHM data processing is also explored.
- The authors review the integration of SHM systems with communication platforms, enabling remote monitoring and analysis. The authors also highlight the importance of robust and secure communication channels for the effective implementation of SHM systems.
- To demonstrate the practical applications of SHM technologies, the study includes several case studies showcasing successful implementations in bridges, buildings, and

other civil structures. These examples illustrate the benefits of SHM in extending the service life of structures and reducing maintenance costs.

- ➢ The authors identify the challenges faced in the field of SHM, such as the need for standardized protocols, the high cost of advanced sensor technologies, and the integration of diverse data sources. They also discuss future directions, including the development of more cost-effective solutions and the incorporation of emerging technologies such as the Internet of Things (IoT) and cloud computing.

In conclusion, Mesquita et al. provide a detailed and insightful overview of the advancements in SHM platforms, highlighting their critical role in the maintenance and safety of civil infrastructure. Their paper underscores the need for continuous innovation and integration of new technologies to address the challenges in SHM and to enhance the effectiveness of monitoring systems.

Acknowledgments: I extend my personal heartfelt gratitude to all of the authors who contributed to this Special Issue. Your dedication, expertise, and rigorous research have significantly advanced our understanding of critical topics in structural engineering and materials science. I am particularly thankful for the insightful papers provided, which explore a wide range of innovative solutions—from advanced materials such as CFRP composites and glass fiber-reinforced concrete to cutting-edge techniques in structural health monitoring and seismic resilience. Each contribution has enriched this collection with valuable knowledge and practical applications. My appreciation also goes to the reviewers and editorial team whose constructive feedback and support ensured the high quality of the presented research. Your efforts in refining and shaping these papers have been instrumental in bringing this Special Issue to fruition. Thank you all for your commitment to advancing the field and for your valuable contributions to this important collection of research.

Conflicts of Interest: The author declares no conflicts of interest.

List of Contributions

1. Leitão, N.; Castilho, E. Chemo-Thermo-Mechanical FEA as a Support Tool for Damage Diagnostic of a Cracked Concrete Arch Dam: A Case Study. *Eng* **2023**, *4*, 1265–1289. https://doi.org/10.3390/eng4020074.
2. Zamani Kouhpangi, M.; Yaghoubi, S.; Torabipour, A. Improved Structural Health Monitoring Using Mode Shapes: An Enhanced Framework for Damage Detection in 2D and 3D Structures. *Eng* **2023**, *4*, 1742–1760. https://doi.org/10.3390/eng4020099.
3. Gomes, L.; Gomes, H.; Reis, E. Surface Waterproofing Techniques: A Case Study in Nova Lima, Brazil. *Eng* **2023**, *4*, 1871–1890. https://doi.org/10.3390/eng4030106.
4. Naghshineh, A.; Bagchi, A.; Tehrani, F. Seismic Resilience and Design Factors of Inline Seismic Friction Dampers (ISFDs). *Eng* **2023**, *4*, 2015–2033. https://doi.org/10.3390/eng4030114.
5. Touati, K.; Al Sahmarany, B.; Le Guern, M.; El Mendili, Y.; Streiff, F.; Goodhew, S. Insight into the Optimization of Implementation Time in Cob Construction: Field Test and Compressive Strength Versus Drying Kinetics. *Eng* **2023**, *4*, 2075–2089. https://doi.org/10.3390/eng4030117.
6. Rossi, A.; de Carvalho, A.; de Oliveira, V.; de Souza, A.; Martins, C. A Parametric Study on the LDB Strength of Steel-Concrete Composite Beams. *Eng* **2023**, *4*, 2226–2253. https://doi.org/10.3390/eng4030128.
7. Publio, M.; Delgado, J.; Pierri, B.; Lima, L.; Gaylarde, C.; Baptista Neto, J.; Neves, C.; Fonseca, E. Assessment of Groundwater Contamination in the Southeastern Coast of Brazil: A Potential Threat to Human Health in Marica Municipality. *Eng* **2023**, *4*, 2640–2655. https://doi.org/10.3390/eng4040151.
8. Naghshineh, A.; Fischer, O.; Pathan, N.; Couch, L.; Tehrani, F. Numerical Investigation of the Seismic Performance of an Innovative Type of Buckling-Restrained Brace (BRB). *Eng* **2023**, *4*, 2978–2990. https://doi.org/10.3390/eng4040167.
9. Sultan, H.; Huseien, G. Minimum Shear Reinforcement for Reactive Powder Concrete Beams. *Eng* **2024**, *5*, 801–818. https://doi.org/10.3390/eng5020043.
10. Sultan, H.; Noor, A.; Huseien, G. Performance Evaluation of Self-Compacting Glass Fiber Concrete Incorporating Silica Fume at Elevated Temperatures. *Eng* **2024**, *5*, 1043–1066. https://doi.org/10.3390/eng5020057.

11. Aljabbri, N.; Karim, A.; Majeed, F. Carbon Fiber-Reinforced Polymer Composites Integrated Beam–Column Joints with Improved Strength Performance against Seismic Events: Numerical Model Simulation. *Eng* **2024**, *5*, 1112–1139. https://doi.org/10.3390/eng5020061.

References

1. Cascardi, A.; Verre, S.; Ombres, L.; Aiello, M.A. Carbon Fabric Reinforced Cementitious Mortar confinement of concrete cylinders: The matrix effect for multi-ply wrapping. *Compos. Struct.* **2024**, *332*, 117919. [CrossRef]
2. Micelli, F.; Cascardi, A.; Aiello, M.A. A removable use of FRP for the confinement of heritage masonry columns. *Mater. Struct.* **2023**, *56*, 184. [CrossRef]
3. Ghobarah, A. Seismic assessment of existing RC structures. *Prog. Struct. Eng. Mater.* **2000**, *2*, 60–71. [CrossRef]
4. Mesquita, E.; Antunes, P.; Coelho, F.; André, P.; Arêde, A.; Varum, H. Global overview on advances in structural health monitoring platforms. *J. Civ. Struct. Health Monit.* **2016**, *6*, 461–475. [CrossRef]

eng

MDPI

Article

Carbon Fiber-Reinforced Polymer Composites Integrated Beam–Column Joints with Improved Strength Performance against Seismic Events: Numerical Model Simulation

Noor Alhuda Sami Aljabbri *, Abdulamir Atalla Karim and Fareed Hameed Majeed

Department of Civil Engineering, Faculty of Engineering, University of Basrah, Basrah 61004, Iraq; abdulamir.karim@uobasrah.edu.iq (A.A.K.); fareed.majeed@uobasrah.edu.iq (F.H.M.)
* Correspondence: noorsami738@gmail.com; Tel.: +964-772-629-4505

Abstract: Strength enhancement of non-seismic concrete beam–column joints (NSCBCJs) via carbon fiber-reinforced polymer composites (CFRPCs) integration has become a viable strategy. However, the implementation of these NSCBCJs without transverse reinforcement shows poor performance during earthquakes in seismic locations. Thus, strengthening the anti-seismic performance of NSCBCJs to meet the acceptance criteria of ACI 374.1-05 is fundamentally significant. Yet, in addition to limited experimental results, only a few numerical studies based on the finite element model have been performed to determine the anti-seismic behavior of NSCBCJs. Consequently, the stress contribution of CFRPCs to NSCBCJs is not clearly understood. Therefore, we used a finite element model to examine the strength contribution of CFRPCs to NSCBCJs. The performance of the proposed finite element model was validated using the experimental results, demonstrating a good agreement between them. It was shown that the strength of NSCBCJs was improved due to CFRPC incorporation, thereby achieving compliance with the seismic requirements of ACI 374.1-05. In addition, CFRPCs presence could enhance the confinement, reduce the deformation of the NSCBCJs and, thus, decrease their stiffness and strength degradation, while simultaneously improving the energy dissipation.

Keywords: CFRPCs; NSCBCJs; strength performance; finite element model; plastic hinge

Citation: Aljabbri, N.A.S.; Karim, A.A.; Majeed, F.H. Carbon Fiber-Reinforced Polymer Composites Integrated Beam–Column Joints with Improved Strength Performance against Seismic Events: Numerical Model Simulation. *Eng* **2024**, *5*, 1112–1139. https://doi.org/10.3390/eng5020061

Academic Editor: Alessio Cascardi

Received: 11 May 2024
Revised: 7 June 2024
Accepted: 11 June 2024
Published: 14 June 2024

1. Introduction

In the context of construction technology, the term "strengthening" refers to the retrofit, rehabilitation, and repair of building materials. In retrofitting, the structural components are installed with the goal of improving their overall performance in terms of shear, flexural, and compressive strength [1–3]. This improvement in the structural integrity becomes more significant than the initial performance. Rehabilitation is the process of reparation and restoration of the performance and strength lost in the structures under varied situations [4–7]. In the repairing process, the structural members' quality are improved to meet the desired requirements. Over the decades, numerous experimental and analytical studies have been conducted to gain an in-depth understanding regarding the behavior of beam-to-column joints (BCJs). In the reinforced concrete structures, the critical zones are identified at the BCJ wherein the vertical and lateral loads meet and are transferred to the footing [8–10]. Consequently, the risk of failure at the joints is much higher compared to other structural components. During an earthquake, the plastic hinge mechanism (PHM) can potentially occur near the joints. Essentially, BCJs are a segment of columns shared by the beams at their intersections in reinforced concrete buildings [11–13]. Needless to say that with the increase in the frequency of earthquakes worldwide, the sustainability of building structures in the civil engineering sectors has become a major concern unless steps are taken to inhibit this damage. It has been realized that the implementation of stable and durable BCJs with high performance is essential in the construction process [14]. To determine the legitimacy of such joints, in-depth experimental, analytical, and numerical modelling and simulation studies are vital.

In recent years, many details of the BCJs have been examined to protect the BCJs from failure during seismic vibration. In this regard, impacts of several critical factors, such as the degree of steel reinforcement in various types of concretes [15], changes in the longitudinal reinforcement [16] and shear reinforcement [17–20], and the use of rebar at different diameters [21] on the BCJs were evaluated. It is established that diverse strengthening methods, like the incorporation of different carbon fiber sheets (CFRP and glass fiber-reinforced polymer (GFRP)) [22–24], steel plate jacketing [25], and external post-tension rods [26], can effectively be used to improve the anti-seismic behavior of BCJs. The results of several experimental and theoretical studies on the BCJs concerning influencing factors and strengthening methods have significantly contributed towards their anti-seismic performance.

The concept of using fiber-reinforced polymer (FRP) confinement involves bonding the FRP to the concrete surface with a bonding agent, such as epoxy resin. This bond allows the FRP and concrete to function as a composite structure, enhancing the BCJs' seismic performance when properly designed and constructed [27,28]. FRP is an anisotropic material with superior engineering properties along the direction of the reinforcing fiber, but it is weak in the transverse direction. Therefore, it is highly recommended to confine the structure in the principal fiber direction of the FRP [29–31]. The application process of FRP begins with cleaning the concrete surface, applying a layer of primer and epoxy, and then laying up the FRP sheets with epoxy until the required number of layers is achieved. A variety of FRP materials have been used to retrofit BCJs, including CFRP [32–34], GFRP [35,36], aramid fiber-reinforced polymer [37], basalt fiber-reinforced polymer (BFRP) [38], FRP hybrid composites [39], quasi-isotropic laminates [40], and sprayed FRP [41]. Among these composite materials, CFRP is the most common due to its superior mechanical properties [42].

The application of CFRP has become a recognized method for strengthening or rehabilitating existing reinforced concrete elements (RCEs). Many studies have reported the excellent efficacy of CFRP in enhancing the flexural performance of RC beams under monotonic loading [29,43,44]. The main elements of the CFRP-concrete bonded system used for external reinforcement of RC structures include CFRP dry fiber sheets, adhesive, and anchors when necessary [45]. It is generally assumed that complete composite action occurs between the concrete elements, like beams, and the CFRP materials. However, the bonding quality relies heavily on the shear stiffness and strength of the adhesive, which must be sufficient to transfer shear forces between the CFRP and the concrete substrate [46–48]. Furthermore, the bond quality is notably affected by the properties of the concrete and the preparation of the surface.

In study conducted by Laseima et al. [49], the authors have been examined the seismic behavior of exterior RC-BCJs strengthened with CFRP. From the obtained results and compared to control specimens, the authors reported that the strengthened system improved by 62%, 209%, and 61.9% in terms of strength, energy dissipation, and ductility, respectively. In [50], the authors examined the impact of using CFRP with different anchors on the seismic performance of RC-BCJs assemblies. The results showed that the ultimate load-carrying capacity increased by between 19 and 43.6%, depending on the anchoring configurations, compared to the control specimens. Additionally, all strengthened specimens demonstrated greater total energy dissipation and damping ratios than the control specimens.

In a recent study conducted by Karayannis and Golias [51], the authors investigated full-scale RC-BCJs that were externally reinforced with CFRP ropes. The joints were strengthened using X-shaped ropes, either singly or doubly, on each side of the joint, while single or double straight ropes were applied to each side of the beam. The strengthened specimens demonstrated an improved hysteretic response compared to the unstrengthened ones. Enhancements were observed in maximum loads per loading step, stiffness, and energy dissipation. The maximum strength of the reinforced samples increased by 20% to 57%, depending on the number of CFRP ropes used.

In experiments conducted by Lim et al. [52], four RC-BCJs were retrofitted utilizing two different techniques involving CFRP grids covered with engineered cementitious composite materials and high-strength mortar. The dimensions of the retrofitted specimens were preserved by removing the concrete cover of the joint and applying a cementitious matrix. The results showed that the failure mode of the specimens could be modified by redistributing the load concentration from the joint to the beam. Additionally, the overall performance of the specimens in terms of strength improved, exhibiting greater ductility and, therefore, delaying failure.

For instance, Atari et al. [53] strengthened ten BCJs through different types of fibers and evaluated their influence on the ductility, durability, strength performance, and energy dissipation capacity against earthquakes. Golias et al. [54] examined six external joints subjected to seismic loads, where three of them were strengthened with carbon fiber ropes of different shapes. In addition, Mady et al. [55] designed five T-shaped joints to withstand seismic loads wherein the effects of some vital factors, like the details of the reinforcement with carbon fiber GFRP in strengthening the performance were determined. Bsisu and Hiari et al. [56] theoretically studied the behavior and strength of CFRPG-reinforced joints to examine the feasibility of improving their performance against seismic loading. Ercan et al. [57] used four different strengthening patterns for the joints to examine their anti-seismic response. These joints were strengthened from the inside by changing the reinforcement details and from the outside with carbon fiber CFRPCs.

In early structural designs, the significance of BCJs is frequently overlooked, making the retrofitting of these joints crucial for structural stability. Over time, external confinement has proven to be one of the most effective retrofit methods. Nevertheless, its practical application is hindered by several factors, such as the absence of comprehensive guidelines for analysis, design, and confinement schemes. Few studies have examined the seismic performance of reinforced concrete structures enhanced with carbon fiber-reinforced polymers.

Considering the basic significance of and applied interest in high-performance CBCJs, we performed a numerical study to determine the possibility of strengthening NSCBCJs by integrating CFRPCs. The results obtained from the numerical model simulation were compared with the state-of-the-art experimental findings available in the literature. In addition, more details about the selected specimens and evaluation methods are presented. The results revealed NSCBCJ strength enhancement because of CFRPC integration, fulfilling the compliance with the seismic requirements standard ACI 374.1-05.

2. Experimental Work

Following the recommendation of Trung et al. [58], this work selected four concrete specimens to examine the anti-seismic behavior of the proposed NSCBCJ-integrated CBCJs. The first specimen was designed using non-seismic details. The second specimen was made based on the first specimen details of CFRPC to strengthen the joint, beam area, and volume adjacent to the joint. Those specimens had identical dimensions and degrees of reinforcement. Figure 1 illustrates the reinforcement details of specimens for the non-seismic (NS) and seismic design (SD). The designed specimens (NS and SD) had identical dimensions and the same number of longitudinal reinforcements. The NS specimen differed from the SD specimen in that it lacked transverse reinforcements in the beam–column joint region, had relatively large stirrup spacing in the beam, and featured down-bending anchorage of beam bars away from the joints. In contrast, the SD specimen was designed to meet the requirements for an intermediate moment frame as specified by the ACI 318-02 code [59], which included confinement reinforcement in the beam–column joint region and relatively small stirrup spacing in the beam. Comparing the behaviors of these two specimens under cyclic loading can reveal the impact of adequate transverse reinforcement on the shear capacity and ductility of the joints. Figure 2a shows the experimental design for testing the specimens. The joint supports were fixed to the columns and beams at both ends to indicate the inflection points in the element's central span. Next, the joint support was attached under the column to the strong floor. The roller support was connected at the end of the beam to enable rotation and longitudinal movements only. In contrast, the column base was connected to a hinge support. The load on the specimens (at the end of the column) was applied using a 500 kN actuator. The cyclic load test was conducted under displacement control, as described by Trung [58] in the figure below (Figure 2b). A displacement-controlled quasi-static analysis was performed in ABAQUS using the dynamic implicit method, which is optimal for simulating the nonlinear behavior of beam–column joints. The default full Newton solving algorithm was utilized. The mesh was refined to increase node points, ensuring a reasonable mesh density in the joint areas with an acceptable aspect ratio of the elements. The two materials shared the same nodes because the steel reinforcing truss element was connected between the nodes of each concrete element to create an embedded bond. The embedded bond assumption used in the structural modeling did not result in significant errors in the predicted load and displacement response. In the CFRP specimen, full contact was simulated at the interface between the CFRP shell elements and the solid concrete elements.

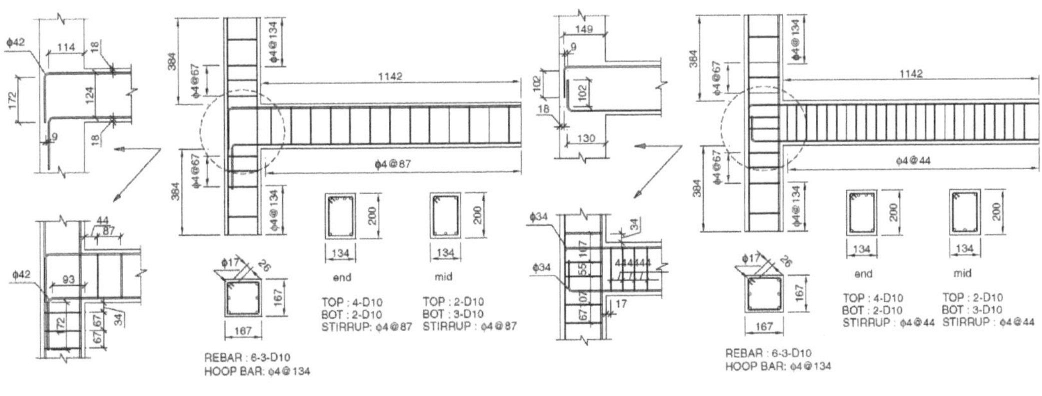

a. Non-seismic detail b. Seismic detail

Figure 1. Details of designed non-seismic beam–column joint [58].

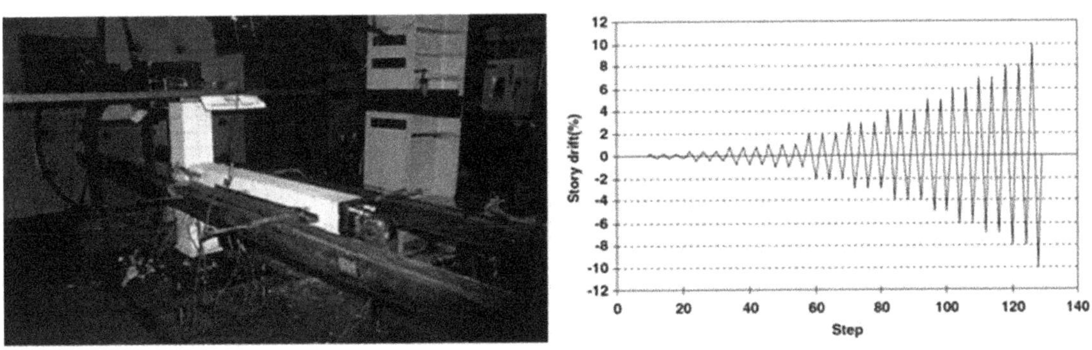

(a) (b)

Figure 2. (**a**) The proposed experimental setup; (**b**) lateral displacement applied to the model [58].

3. Model Simulation Using Abacus

3.1. Element Type

The steel reinforcing bar and concrete are the two main components of any reinforced CBCJs. Moreover, the key components of the retrofitted CBCJs are the concrete, steel reinforcement bars, and external CFRP sheet. To introduce a realistic model under cyclic force, it is necessary to accurately replicate the real material characteristics of every component. The ABAQUS 19 library (computer software, user's manual, Providence, RI, Rising Sun Mills) offers effective material models for three-dimensional (3D) continuum elements that can simulate the actual behavior of each component with acceptable accuracy. Using three different element types, including solid, shell, and link, various geometries were modeled. We used a 3D eight-node linear brick element with reduced integration and hourglass control (C3D8R) for the solid section, like concrete (Figure 3a). The linear brick components were chosen because they can be utilized with contact, unlike the quadratic brick elements that require more time to compute the constant nodal loads over the surface. The truss elements T3D2 (Figure 3b) used for the steel bars were embedded in the concrete. The discrete reinforcement bars in the CBCJ are modeled as T3D2 elements. Four-node, doubly curved, linear, three-dimensional general-purpose shell elements with decreased integration and hourglass control (S4R5) were used in the CFRP plate (Figure 3c). Since the nodes of a conventional shell element are situated in a well-defined planar dimension, the thickness of this shell element can be determined by the section properties. The conventional shell elements are regarded to be more accurate in contact modelling than continuum shell elements because they can measure strain or slip without affecting the thickness of the CFRP composite plates. Furthermore, the conventional shell elements have exceptional computational efficiency.

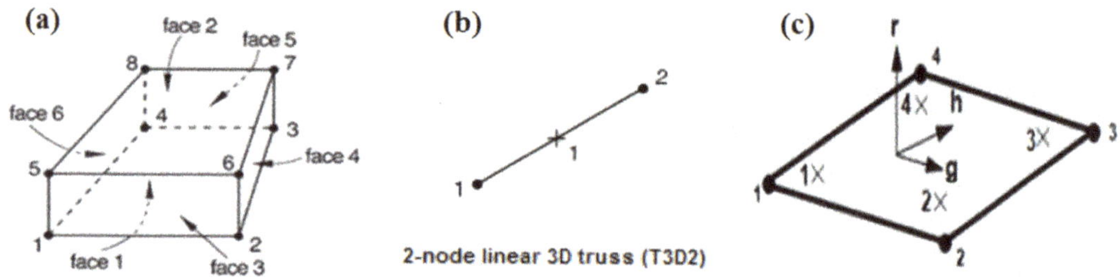

Figure 3. Elements used in ABAQUS to model: (**a**) C3D8R for concrete, (**b**) T3D2 for longitudinal and transverse reinforcement, and (**c**) the S4R5 four-node shell for FRP.

3.2. Material Properties

In this study, the material properties used to prepare the specimens, such as concrete compressive strength and splitting tensile strength, longitudinal reinforcement and stirrups' yield strength, as well as the tensile strength and the elastic modulus of CRFP and epoxy, are summarized in Table 1. Concrete is known to exhibit a high compressive strength (CS) and low tensile strength (TS), and it is prone to fracture when subjected to tensile loads. This is mainly due to the susceptibility of aggregates in concrete that can endure compressive stress and, thus, allow it to carry compressive loads. However, cracks that separate the cement particles in the aggregates are generated due to tensile stresses. The separation of cement particles causes the entire structure to fail as cracks start propagating. The issue with concrete can be addressed via the reinforcing of elements, such as metallic bars or fibers. These components serve as the framework of the entire construction and can support the aggregates when subjected to tensile forces. This process is referred to as concrete reinforcing. The term "brittle material" might apply to concrete. The difference in behavior between concrete and ductile materials, like steel, is due to their distinct load responses. Brittle materials tend to develop tensile fracture perpendicular to the direction of the largest tensile strain. Thus, the cracks tend to develop parallel to the maximum compressive stress when a concrete specimen is subjected to a uniaxial compressive load. In compression, the stress–strain curve of concrete being elastic can gradually reach up to the maximum CS followed by a decrease into a softening region and eventually crushing failure at an ultimate strain. Tao et al. [60] and Chen et al. [61] modeled the normal concrete stress–strain relationship (Figure 4). The stress–strain characteristics of concrete can be expressed as follows:

$$f_c = \frac{E_0\, \varepsilon_c}{1 + \left(\frac{E_0\, \varepsilon_c}{f_p} - 2\right)\left(\frac{\varepsilon_c}{\varepsilon_p}\right) + \left(\frac{\varepsilon_c}{\varepsilon_p}\right)^2} \quad (1)$$

where f_c, ε_c, f_p, ε_p, and E_0 are the concrete's corresponding stress (in MPa), strain, maximum experimental stress (in MPa), strain at maximum experimental stress (taken as 0.002), and elastic modulus (in MPa), $E_0 = 4700\sqrt{f_c}$, respectively.

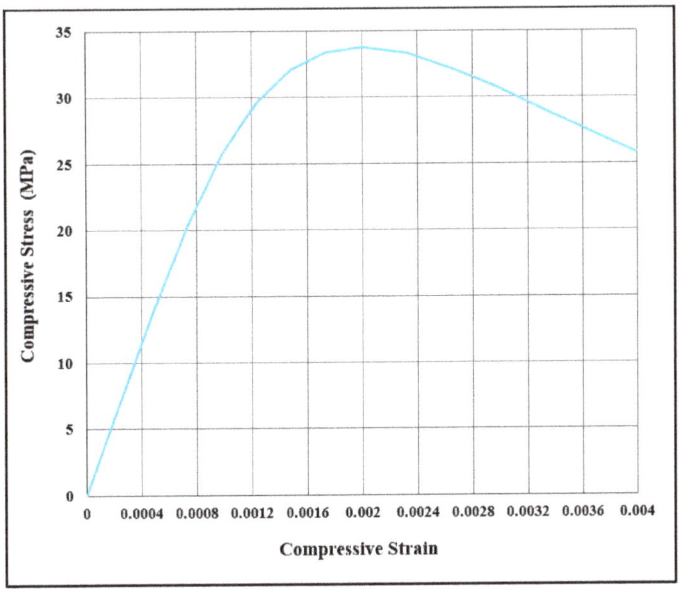

Figure 4. Uniaxial stress–strain curve of concrete.

The stress–strain curve of concrete under tension is approximately linearly elastic up to the maximum TS. After this point, the cracks and strength of concrete decrease gradually to zero [26,27]. The uniaxial TS of concrete can be modeled via the following equation:

$$f_t = 1.4\left(\frac{f_p - 8}{10}\right)^{2/3} \quad (2)$$

The damaged plasticity model in ABAQUS is characterized by the degradation of the compression and tension. When plasticizers are introduced, the elastic stiffness of the material decreases due to its deterioration, making it unable to restore the original elastic stiffness. This is crucial for cyclic loading, as the two damage parameters are considered to be dependent on the plastic strains, indicating a reduction in the elastic stiffness. Figure 5 displays the stress–strain curve for tension and compression modeled using the damage plasticity model.

Table 1. Concrete, steel bars, CFRP and epoxy properties.

Properties of Used Materials			
Concrete	Compressive strength, MPa	Water curing	33.8
		Air curing	36.5
	Tensile strength, MPa	Water curing	4.0
		Air curing	3.8
Steel bars	Yield strength, MPa	D10	324
		Ø4	459
CFRP (CF720)	Tensile strength 0°, MPa		4965.8
	Elastic modulus, GP		240.5
	Elastic modulus 90°, GPa		10
	In-plane shear modulus, GPa		5
	Tensile strength 90°, MPa		50
	Compressive strength 0°, MPa		1200
	Compressive strength 90°, MPa		250
	In-plane shear strength, MPa		70
	Sheet thickness, mm		0.33
	Major Poisson's ratio		0.3
	Density, g/cm^3		1.6
Epoxy topcoat (CLR67)	Tensile strength, MPa		59.5
	Elastic modulus, GP		3.7
Epoxy primer (CLR67)	Tensile strength, MPa		56.8
	Elastic modulus, GP		3.7

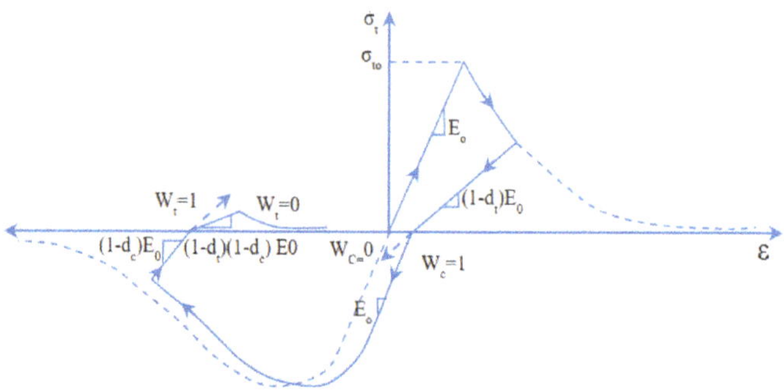

Figure 5. Stiffness recovery curve of concrete under tension and compression.

The element develops cracks and eventually damages the material when it experiences a tension that surpasses its TS. This behavior can be expressed by a variable called the damage parameter (dt). Furthermore, the behavior of the element under CS is represented by the variable damage parameters (dc). Generally, the developed cracks do not affect the stiffness in compression unless the degradation and compression stiffness reach the same level as the stiffness in tension, in which the dc becomes zero. Following the experimental results of Trung [58], this study used (in ABAQUS for the concrete material) a CS value of 33.8 MPa, a Young's modulus of 27,324 MPa, and a Poisson's ratio of 0.2. Five parameters were considered to describe the Drucker–Prager flow potential yield surface that was first developed by Lubliner et al. [62] and then upgraded by Lee et al. [63]. To accurately determine the values of the selected parameters, several tests were conducted in the model wherein the default parameters of ABAQUS were used. These parameters included (1) the dilation angle equal to 31 that denoted the volume change to shear strain ratio, (2) the eccentricity equal to 0 when the flow potential approaches a straight line or 0.1, (3) the initial uniaxial to biaxial CS ratio of 1.16 (Figure 6), (4) the viscosity that represented the relaxation time of the visco-plastic system that was assumed to be zero because the BCJ model caused a severe convergence difficulty, enabling us to improve the convergence of the model in the softening region, and (5) the yield surface on the deviatoric plane defined by the ratio of the second stress invariant on the tensile to the compressive meridian was equal to 2/3.

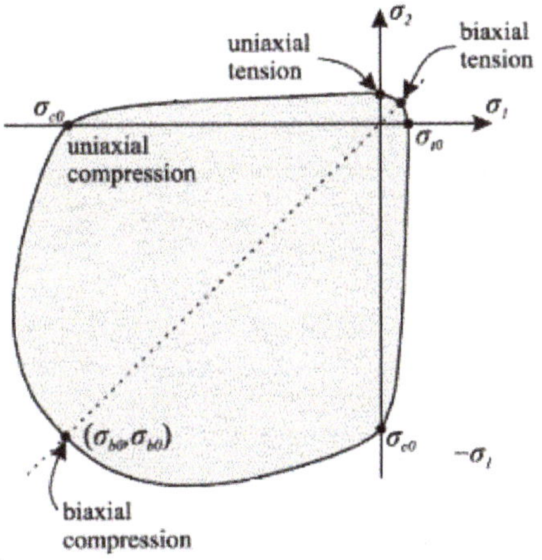

Figure 6. Model representing the failure surface of concrete [64].

The linear isotropic and bilinear kinematic model was applied to simulate the elastic and inelastic behavior of the steel components in the reinforcement bar. The material model in ABAQUS used the Young's modulus value of 200,000 MPa and Poisson's ratio of 0.3 to accurately represent the behavior of steel. Following the tensile test conducted by Trung [58], the steel material parameters utilized in the finite element models were derived. These properties were then incorporated into the model (Table 2 and Figure 7). An orthotropic elastic material was used to represent CFRPCs. Following the report of Narayanan [65], the material properties were determined. The required material properties for the 3D analysis of CFRP are displayed in Table 1. The model assumed full bonding between CFRP and concrete.

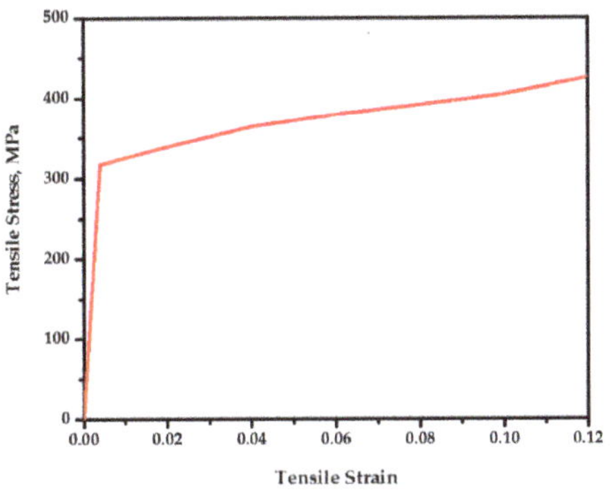

Figure 7. Stress–strain curve of steel.

Table 2. Model properties of reinforcement material.

Model Number	Materials	Material Properties		
1	Deformed steel bar (10 mm)	Linear isotropic	Young's modulus, MPa	200,000
			Poisson's ratio	0.3
		Bilinear kinematic	Strain–stress, MPa	0—0
				0.00162—324
				0.12—420
2	Deformed steel bar (4 mm)	Linear isotropic	Young's modulus, MPa	200000
			Poisson's ratio	0.3
		Bilinear kinematic	Strain–stress, MPa	0—0
				0.0022—459
				0.008—620

3.3. Geometry

As mentioned previously, to accurately represent the geometrical configuration and dimensions of the test specimens', 3D numerical models were developed. The mesh for the beam and columns was chosen so that the solid elements' node points could line up with the real positions of the reinforcement. The mesh was subdivided to obtain more node points, resulting in a reasonable mesh density in the joint areas with the acceptable aspect ratio of the elements. The two materials shared the same nodes because the steel reinforcing truss element was attached between the nodes of each concrete element to create a perfect bond. It is worth noting that the perfect bond assumption used in the structural model did not cause a significant error in the predicted load-displacement response. In the CFRPC, the full contact was simulated at the interface between the CFRPC shell and the concrete elements. Values for the cross-sectional area and initial strain of reinforcement as well as the CFRP were entered. A value of zero was entered for the initial strain due to the absence of initial stress on the reinforcement. Figure 8 illustrates the finite element mesh and geometry of the joints. Four joints were modeled including non-seismic detail, seismic detail, CFRPC to strengthen the NSCBCJs, and two layers of CFRPC specimens for strengthening the NSCBCJs.

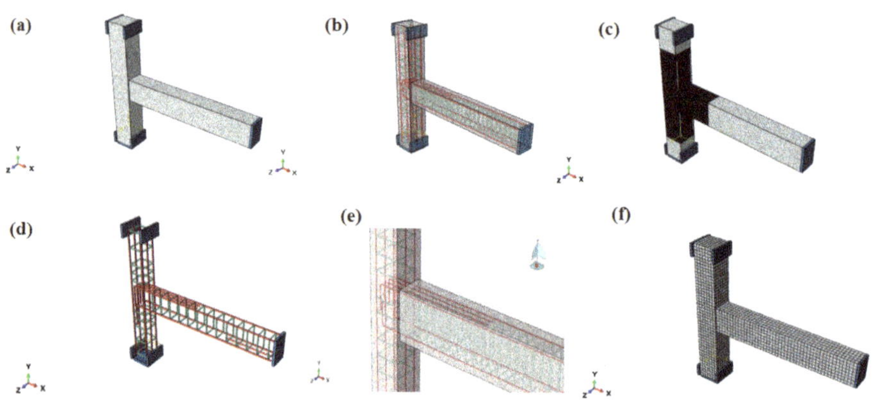

Figure 8. 3D finite element model for the analysis of NSCBCJ- and CFRPC-integrated NSCBCJs. (**a**) NSCBCJ specimen, (**b**) seismic specimen, (**c**) one and two layers of the CFRPC specimen, (**d**) longitudinal and transverse reinforcement of the NSCBCJ specimen, (**e**) longitudinal and transverse reinforcement of the seismic specimen, and (**f**) mesh of all models.

3.4. Boundary Condition

The loads and limitations applied in the finite element models were identical to the conducted test. In addition, the lateral cyclic loading was carried out at each drift ratio in the finite element models. Herein, U_X, U_Y, and U_Z are the displacement in the direction of x (beam axis), y (column axis), and z (displacement perpendicular to the x and y axis) axis, respectively. The z-direction displacements were completely restrained at the top of the column ($Uz = 0$) and the lateral displacements were applied in the x-axis direction. To simulate the hinge support, the displacements in the column bottom at each direction were restrained ($Ux = Uy = Uz = 0$). Additionally, the displacements in the y and z directions were restricted ($Uy = Uz = 0$) at the end of the beam to indicate roller support. Figure 9 shows the boundary condition used in finite element analysis.

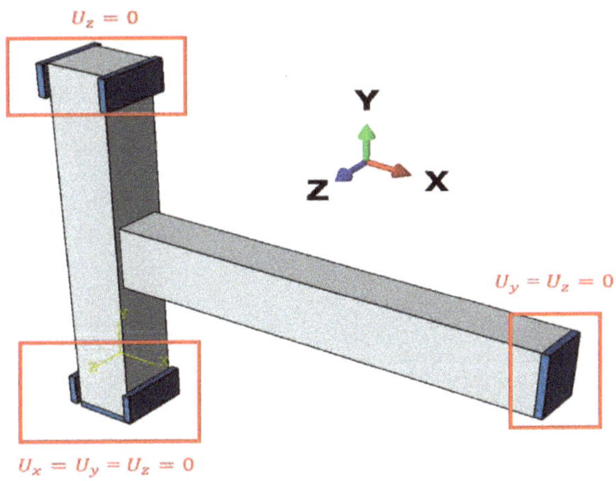

Figure 9. The boundary conditions used in the model simulation.

4. Results and Discussion

4.1. Verification of the Finite Element Models with Experiments

4.1.1. Force versus Displacement Curves

To confirm the viability of the simulated finite element models, the force–displacement curves of four specimens obtained in the current study were compared with the experimental findings of Trung [58] as shown in Figure 10. The finite element results were observed to be consistent with the experimental ones (Table 3). It was asserted that the simulated finite element model is able to predict the inelastic response of the test specimens. The observed slight difference between the finite element model simulations and experimental results may be due to (1) unexpected occurrences in the experimental work, (2) variables, like the concrete's tensile and compressive properties selection, (3) the refinement of meshes, (4) the use of idealized boundary conditions in the finite element models, and (5) nonlinearity in the materials' behavior (materials' properties modeled in ABAQUS).

Instead of comparing the results of every load cycle, the backbone curve should be developed from the experimental and numerical data for each direction of loading with unique behavior. The curve should be plotted in a single quadrant (positive force versus positive displacement or negative force versus negative displacement; consequently, the backbone curve created by connecting the first cycle peak points with the related displacement is shown in Figure 11. The results indicated that the backbone curve (envelope curve) obtained from the finite element analysis was in good agreement with the experimental results, indicating a considerable correlation at all stages of the lateral cycling loading.

The results of the finite element studies showed that a total of 30 cycles at 12 different drift levels varied from 0 to 10% applied to the specimens. The compression axial load on the column was not considered based on the experimental work. The total deformations that occurred near the top end of the columns were measured with a Linear Variable Differential Transformer as mentioned by Trung [58]. The hysteretic curves were initially linear and elastic. With the increase in displacement, the specimens were converted into an inelastic response due to the occurrence of pinch effects. The experimental and finite element results proved that the NSCBJ specimens were strengthened due to the integration of CFRPCs, displaying a good seismic behavior with endurance strength at the test end. In short, the specimens showed maximum lateral strength and no crushing of the concrete occurred. In addition, the observed difference in the main reinforcement bar that caused the force–displacement curves was different in the positive and negative direction of loading therefore it differed in the lateral resistance of the peaks.

The ideal bilinear drawing demonstrated by Paulay and Priestley and described in Section 3.4 was used to obtain the first yield and ultimate strengths, respectively, from the experimental (Q_{ye}) and finite element models (Q_{yFEM}). The ultimate strengths from the finite element and experimental model are Q_{ue} and Q_{uFEM}, respectively. The experimental and numerical results showed that the capacity of the seismic design specimen is improved by 32% compared to the non-seismic specimen. Also, the strengthened specimen with one layer of CFRPC showed a maximum ultimate strength of specimen greater than the non-seismic one because of the rise in the strength capacity of the BCJ by 28%. Finally, the specimen strengthened with two layers of CFRPC revealed better performance when the ultimate load reached to 11.2 kN, with an increase in ultimate load capacity by 41% compared to the NS specimen.

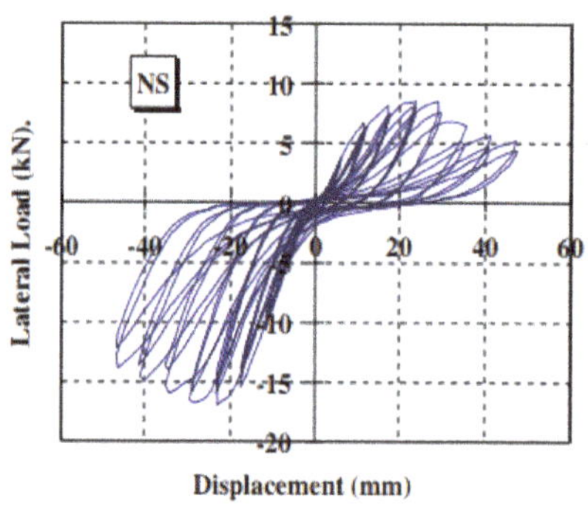

(**a**)

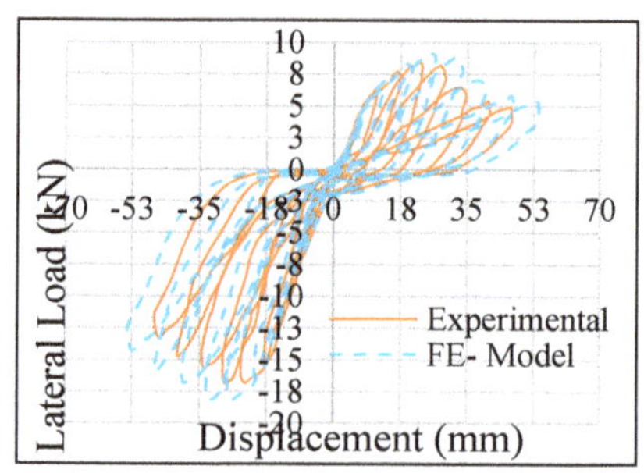

(**b**)

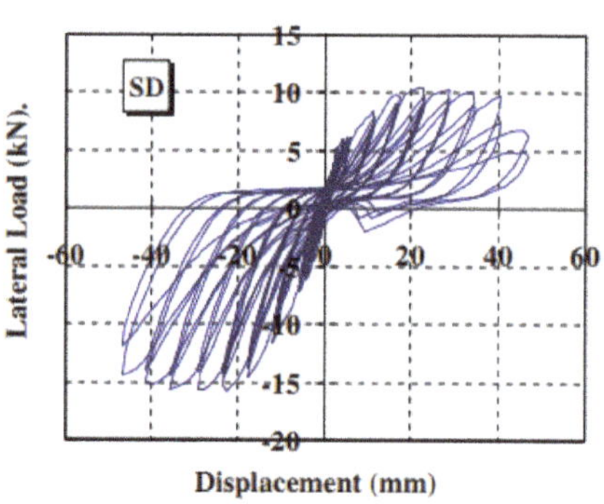

(**c**)

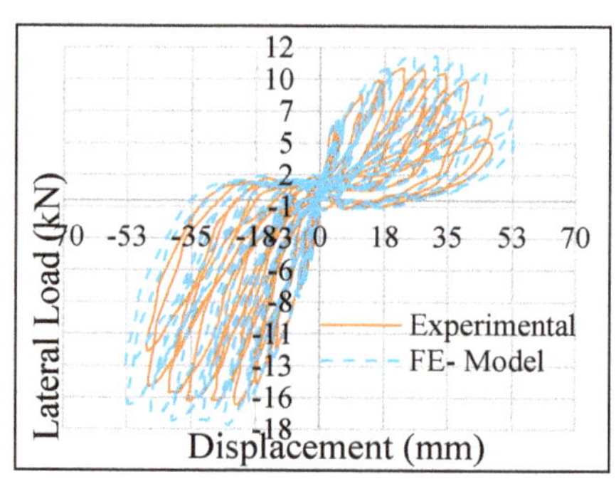

(**d**)

Figure 10. *Cont.*

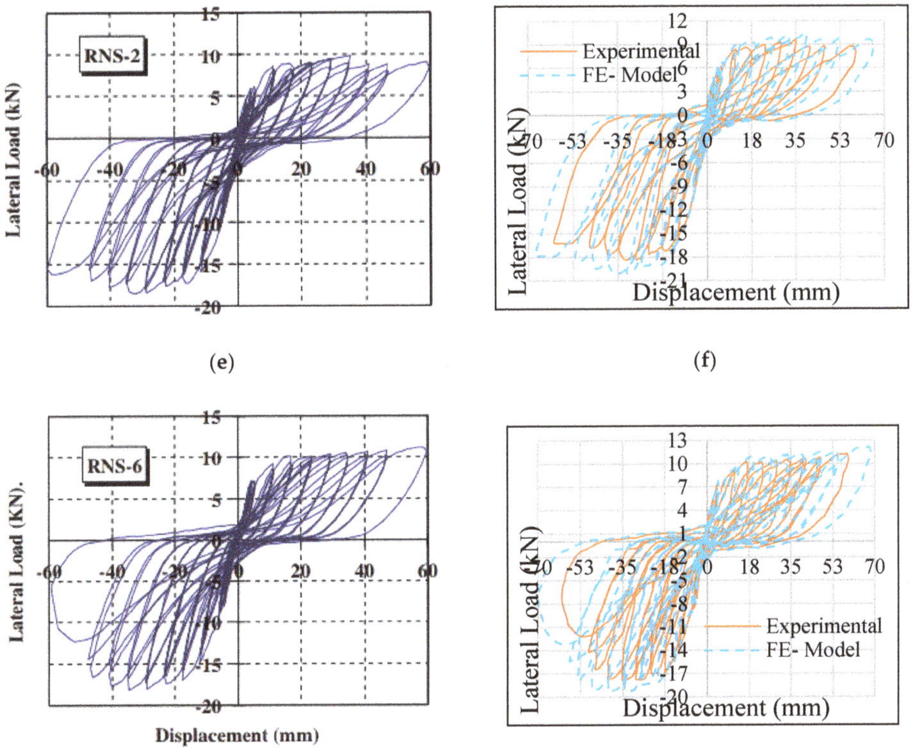

(e) (f)

(g) (h)

Figure 10. Comparison force–displacement curves for the non-seismic specimen (NS), seismic specimen (SD), strengthened one-layer CFRPC specimen, and strengthened two-layer CFRPC specimen. (**a**) Experimental results of Trung [58] on NSCBCJ; (**b**) comparison of the proposed model results with experiments with NSCBCJ; (**c**) experimental results of Trung [58] with SD; (**d**) comparison of the proposed model results with experiments with SD; (**e**) experimental results of Trung [58] with one-layer CFRPC; (**f**) comparison of the proposed model results with experiments with one-layer CFRPC; (**g**) experimental results of Trung [58] on two-layer CFRPC; (**h**) comparison of the proposed model results with experiments of two-layer CFRPC.

Table 3. Comparison between experimental and finite element results.

Specimen	Loading Direction	Yield Lateral Resistance (kN)			Ultimate Lateral Resistance (kN)		
		Q_{ye}	Q_{yFEM}	Q_{yFEM}/Q_{ye}	Q_{ue}	Q_{uFEM}	Q_{uFEM}/Q_{ue}
Non-seismic (NS)	Positive	5.40	5.925	1.09	7.3	7.9	1.08
	Negative	11.70	12.9	1.10	15.76	17.2	1.09
Seismic design (SD)	Positive	7.30	7.82	1.07	9.7	10.75	1.10
	Negative	13.15	13.85	1.05	16.15	18.23	1.13
Strengthened by CFRP (one layer)	Positive	6.75	7.42	1.09	9.10	10.15	1.12
	Negative	12.95	14.25	1.10	17.23	19.07	1.11
Strengthened by CFRP (two layers)	Positive	7.95	8.4	1.05	10.6	11.20	1.06
	Negative	13.80	14.35	1.04	16.80	17.80	1.06

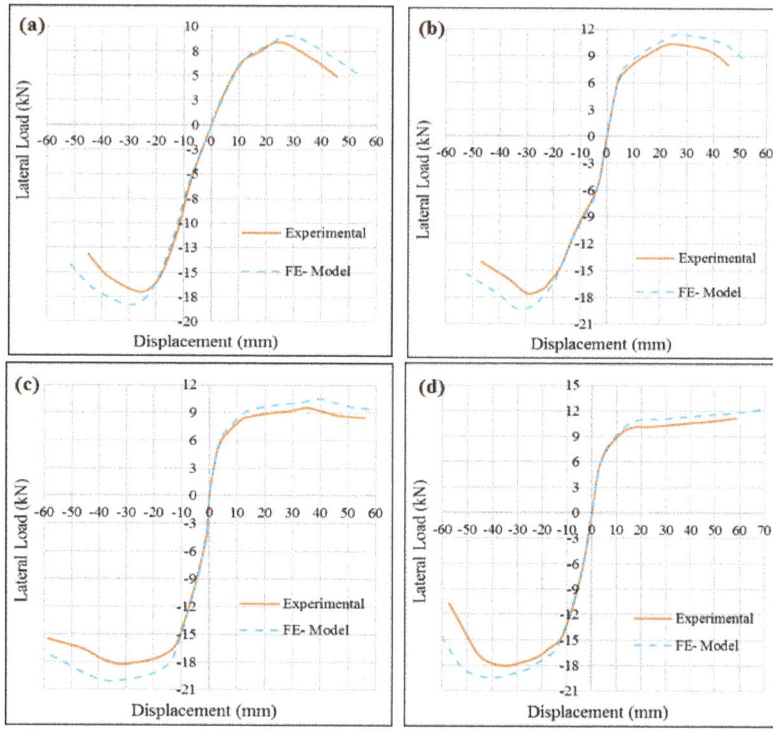

Figure 11. Comparison of backbone curves. (**a**) NSCBCJ, (**b**) SD, (**c**) strengthened one-layer CFRPC specimen, and (**d**) strengthened two-layer CFRPC specimen.

4.1.2. Failure Modes

To further verify the finite element models, the failure modes of four joint specimens simulated by ABAQUS were validated by the experimental work (Figures 12–15). The model simulation results of the failure modes were in good agreement with the experimental findings, indicating the precision of the established model calculation. According to the experimental results of Trung [58], the failure mode of the NS specimen was the joint shear failure with the beam and column bending cracks. Shear failure of the joint happened because of the transverse reinforcement deficiency in the joint. The specimen designed according to the seismic details showed the beam flexural cracks near the joint. Compared to the non-seismic specimen, the seismic specimen developed more surface cracks in the beam generated from the bending. The shear cracks at the joint for the specimen made with the seismic detail were fewer, due to sufficient joint transverse reinforcement. Thus, it can be concluded that the transverse reinforcement restricted the joint shear failure in the beam. Rao [66] acknowledged that the failure mode in the specimen strengthened by CFRPC is mainly due to the following factors: (1) concrete crushing in compression before the yield of the longitudinal steel bar, (2) the steel yield in tension with the debonding of the CFRPC sheet, (3) the steel yield in tension with the concrete crushing, and (4) the spalling of the concrete cover.

The single-layer CFRPC-integrated specimen showed the improved strength of the beam, column, and joint. In addition, experimental results of the CFRPC-strengthened specimen exhibited fewer cracks because the restraint provided by the two strips [U-shaped] reduced the T-shaped sheets' debonding in the column. Finally, the presence of CFRPC could debond at a large lateral displacement, classifying the failure mode as steel yield (Figure 14c) in tension with debonding of the CFRPC sheet. The behavior of the specimen made with two layers of CFRPC integration was same as the single-layer specimen, except that there was an improvement in the performance. This was mainly due to the increased confinement of the joint and the reduced shear cracks of the joint, meaning that the specimen attained larger lateral strength compared to the others. To understand the joint performance more precisely, the failure mechanisms of the specimens were assessed while taking the parameters, like concrete tensile and compressive damage, Von-Mises's stress of the reinforcement, and maximum principal plastic strain. In brief, the finite element models simulated the joint nonlinear behavior appropriately. Figure 12a clearly indicated that the NS beam–column developed joint, column, and beam tensile damage. From the joint to the beam, the concrete exhibited tensile damage. The longitudinal bars in the beam were found to be in tension, while yielding of the transverse joint bars occurred. Concrete damage developed in the joint. The NS specimen failed due to extensive cracking of the concrete at the joint as its ultimate strength was attained. Figure 12b shows the concrete damage under compression. The concrete damage that originated from the repeated loading was cumulative, leading to the strength and stiffness degradation. The principal plastic strain value served as an indicator for the occurrence of cracks in the concrete damage model. The cracks formed for the positive value of the principal plastic strain. Figure 12d shows that the joint failed in the concrete diagonal crack region, which refers to the same crack pattern obtained by the finite element model (Figure 12e).

Figure 13a revealed that the SD beam–column suffered from beam tensile damage wherein the concrete's tensile damage mostly spread in the beam. The yielding of the longitudinal bars occurred without yielding of the transverse joint bars (Figure 13c). In addition, the concrete's damage developed in the joint. The failure of this specimen occurred due to cracking of the concrete beam. The principal plastic strain showed the formation of cracks in the beam which is very similar to the crack pattern obtained by the finite element model (Figure 14d).

Figures 14 and 15 elucidate the finite element output and comparison between the failure mode at ultimate strength for the specimens strengthened by one and two layers of CFRPC. The stress contribution of the CFRPC sheet to the concrete was observed, damaging the concrete under tension. The compression value of the CFRPC specimen was lower compared to the NS specimen. This finding indicated that the cracks in the CFRPC-integrated specimen were lower than the NS specimen, as mentioned above. In addition, the maximum plastic strain distribution for the specimen was mainly concentrated at the beam end and joint, indicating the appearance of same crack pattern as observed in the experiment.

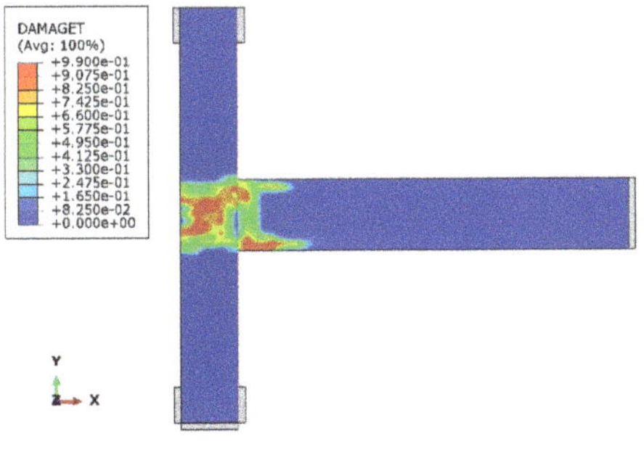

(a)

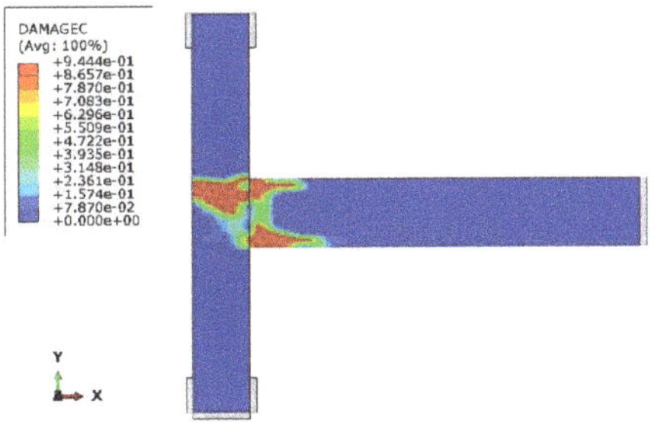

(b)

Figure 12. *Cont.*

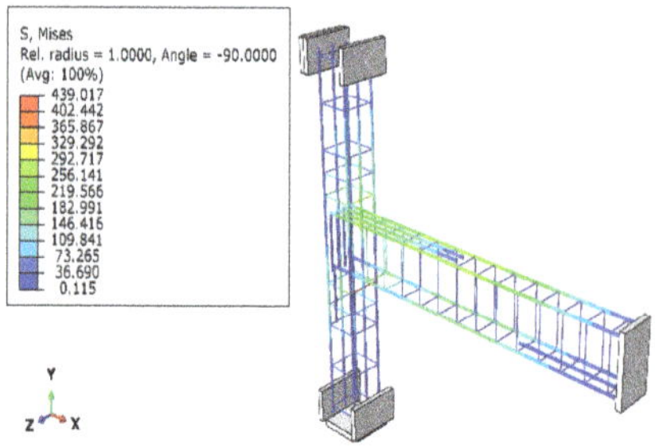

(c)

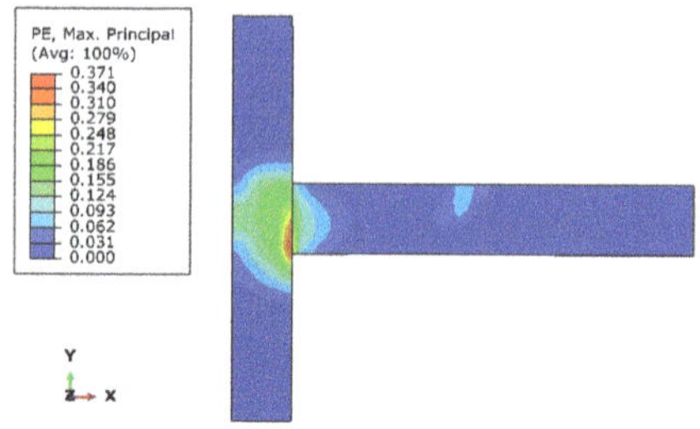

(d)

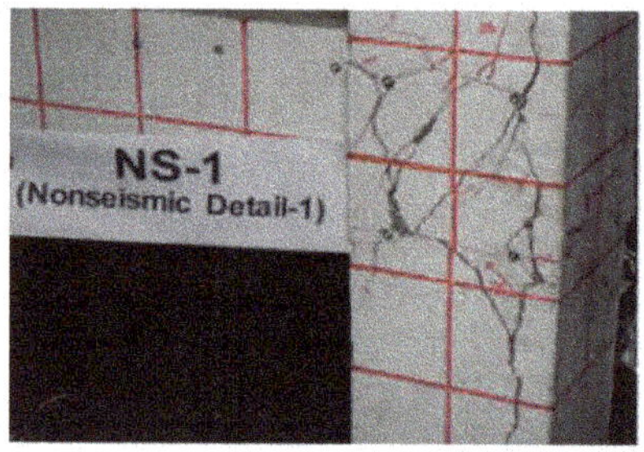

(e)

Figure 12. Finite element output of the NS beam–column joint. (**a**) Concrete tensile damage; (**b**) concrete compressive damage; (**c**) stress in the reinforcement; (**d**) FE crack pattern; (**e**) experimental crack pattern [58].

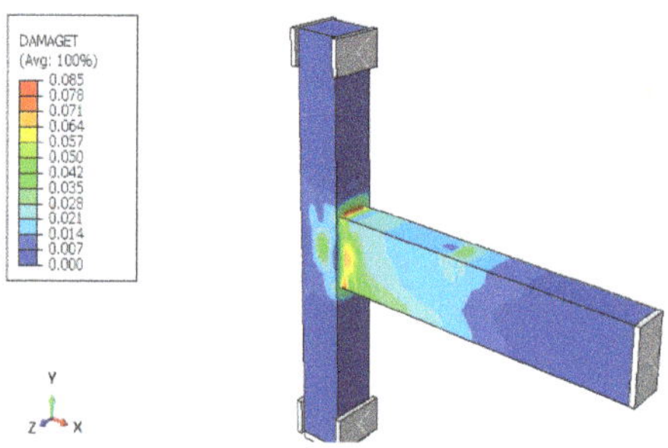

(a)

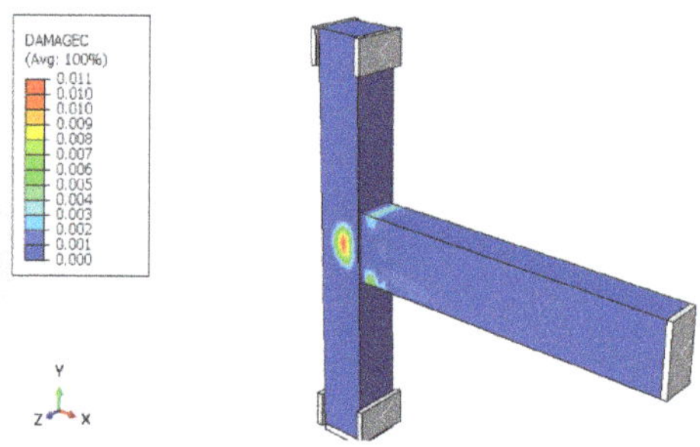

(b)

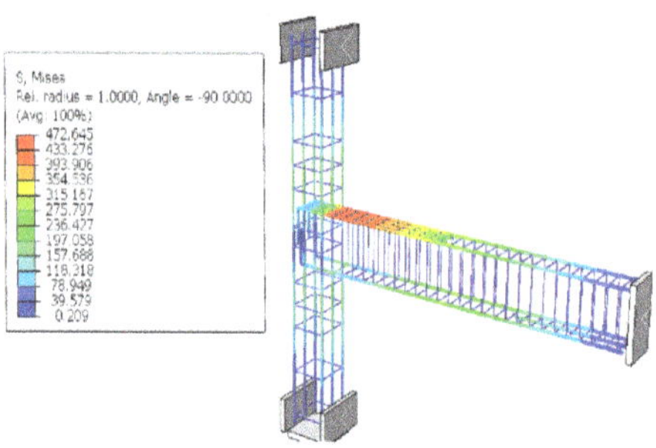

(c)

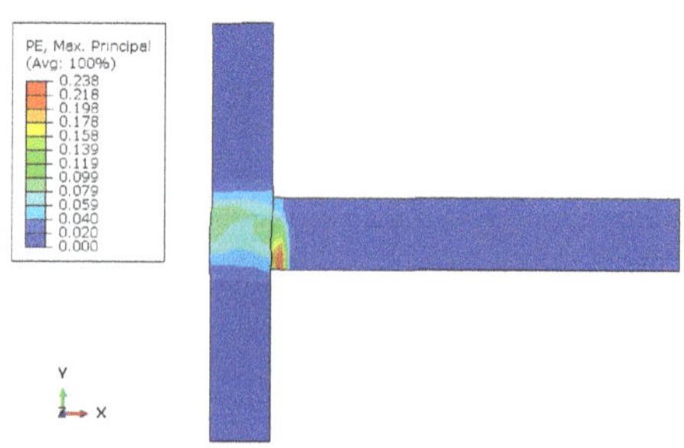

(d)

Figure 13. *Cont.*

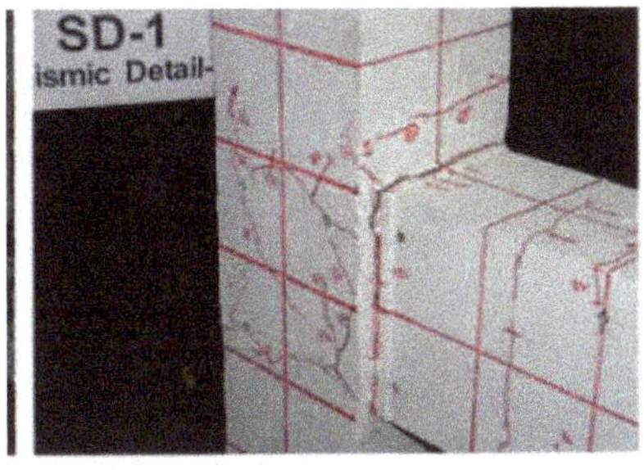

(e)

Figure 13. Finite element output of the seismic beam–column joint. (**a**) concrete tensile damage; (**b**) concrete compressive damage; (**c**) stress in the reinforcement; (**d**) FE crack pattern; (**e**) experimental crack pattern [58].

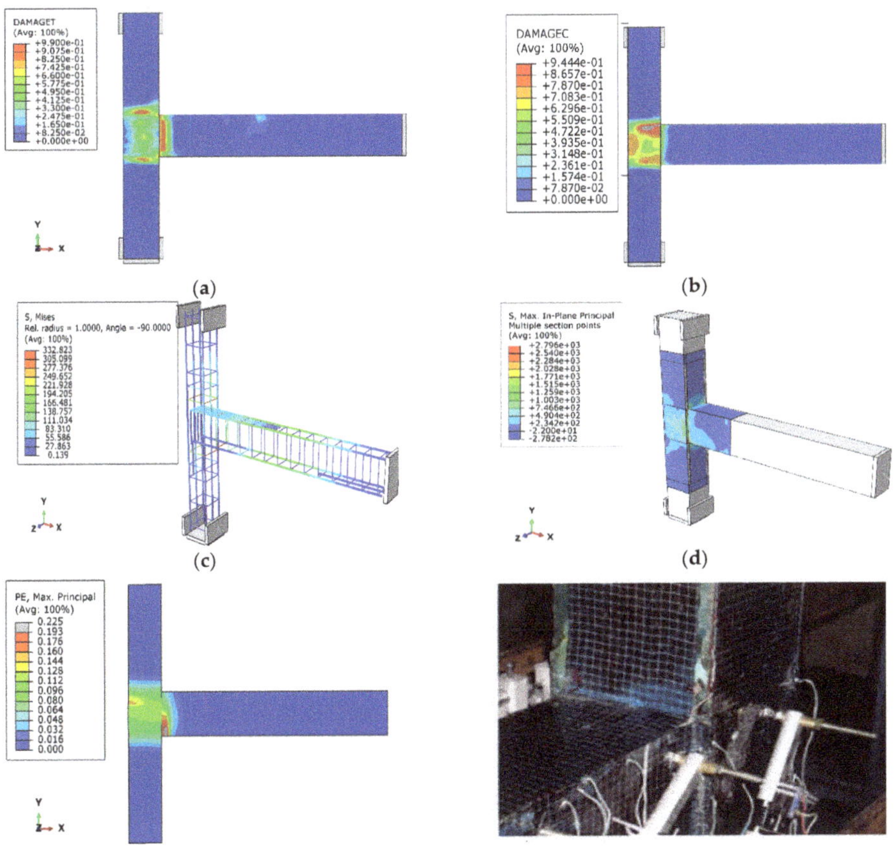

Figure 14. Finite element output of the single-layer CFRPC-strengthened beam–column joint. (**a**) Concrete tensile damage; (**b**) concrete compressive damage; (**c**) stress in the reinforcement; (**d**) stress in the CFRP; (**e**) FE crack pattern; (**f**) experimental crack pattern [58].

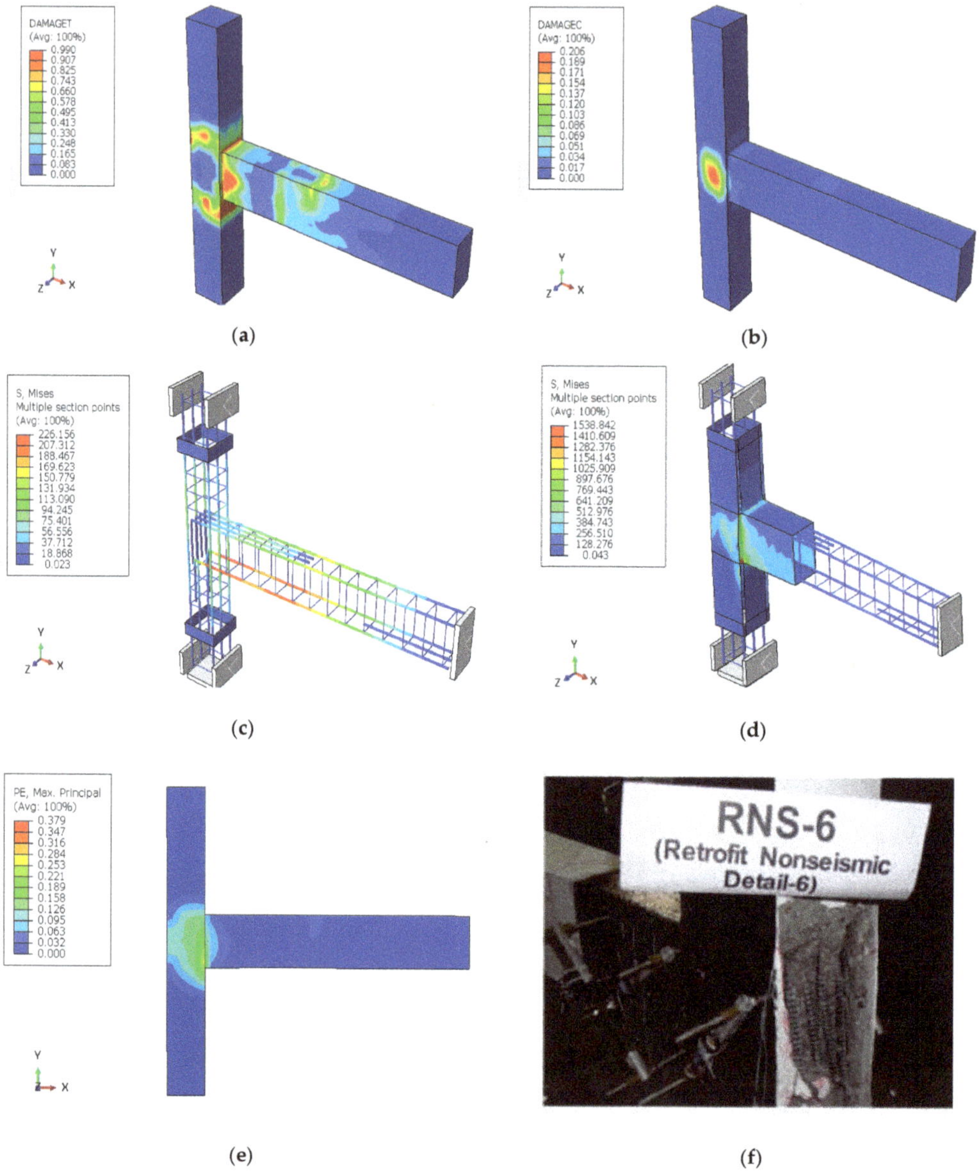

Figure 15. Finite element output of double-layer CFRPC-strengthened beam–column joint. (**a**) Concrete tensile damage; (**b**) concrete compressive damage; (**c**) stress in the reinforcement; (**d**) stress in the CFRP; (**e**) FE crack pattern; (**f**) experimental crack pattern [58].

4.2. Requirement of ACI 374.1-05

To achieve a joint that can be used in the high seismic risk area, the test results obtained by Trung were checked with the ACI 374.1-05. According to this standard, three criteria including energy dissipation, strength, and stiffness must be checked.

4.2.1. Relative Energy Dissipation

To ensure that the joints have a good damping after seismic action, relative energy dissipation at 3.5% of the drift ratio must exceed 0.125. If the beam–column joint fails to meet this condition, the structure will experience excessive oscillation after a seismic event. The dissipation energy in the cycle represents the elastic–plastic behavior. ACI 374.1-05 stipulates that the energy is calculated from the hysteric curve area at drift ratio of 3.5% (area parallelogram). The parallelogram was created by two horizontal lines and four parallel lines, with first positive stiffness on the upper parallel lines and initial negative stiffness on the bottom parallel lines. Figure 16 shows the calculated relative energy dissipation obtained from the definition of the parallelograms. Table 4 displays the specimens' relative energy dissipation ratios.

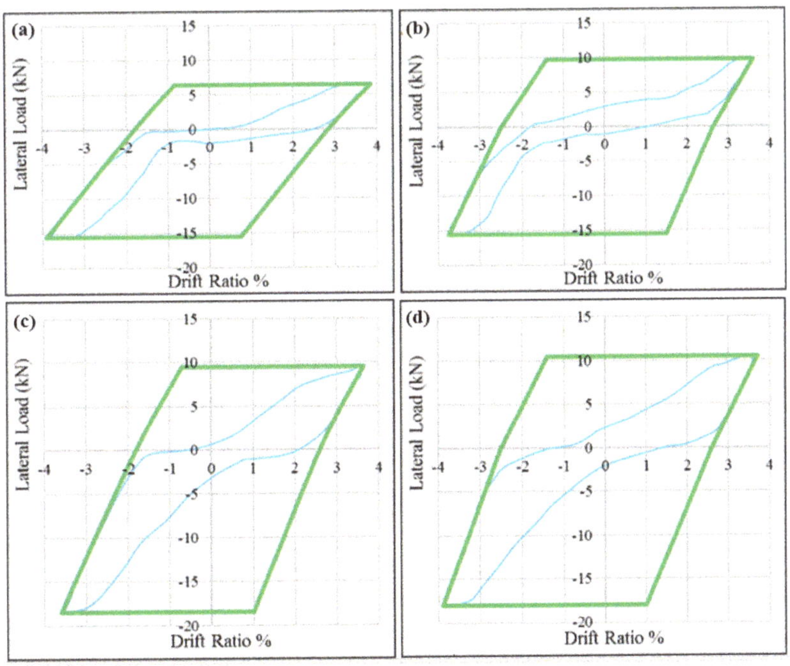

Figure 16. Relative energy dissipation parallelograms at drift ratios of 3.5% for (**a**) NS, (**b**) SD, (**c**) the one-layer CFRPC specimen, and (**d**) the two-layer CFRPC specimen.

Table 4. Validation of relative energy dissipation ratios of various specimens.

Specimen	Drift Ratio (%)	Parallelograms Area	Area of Hysteretic Loop	Relative Energy Dissipation Ratio β	Check to Meet the Criteria ≥ 0.125
NS	3.5	91.75	18.30	0.20	Ok
SD	3.5	130.33	41.20	0.32	Ok
1-layer CFRPC-integrated	3.5	124.64	43.87	0.35	Ok
2-layer CFRPC-integrated	3.5	144.08	72.60	0.5	Ok

The obtained relative energy ratio values in the table are above 0.125, which clearly shows that the specimens could meet the requirement of the standard. This can be assigned to the design of the specimens that follow the criteria of a weak beam/strong column. Consequently, plastic hinges can occur at the end of beam near joint, making the energy dissipation in plastic hinge region. In addition, transverse joint reinforcement was observed to reduce the joint shear distortion and joint cracks, resulting in higher performance of the specimens made using seismic details compared to the non-seismic specimens. The better energy dissipation obtained in the specimen strengthened by two layers of CFRPC can be ascribed to the increase in the confinement of the joint by double layers of CFRPC integration.

4.2.2. Strength Criteria

In the standard mandates, the test specimen model must adhere to the principle of a strong column and weak beam. The maximum lateral load of the model specimen (E_{max}) must be lower than λE_n (where E_n represents the nominal strength of the specimen calculated using a strength reduction factor of 1 and where λ is the percentage ratio of column strength to beam strength). Trung's test result revealed that the lateral load varies in the positive and negative loading directions, thus necessitating separate checks for E_{max} in each direction. Table 5 shows the evaluation of the strength criteria. The specimen meets this requirement since the value of E_{max} was below λE_n. The failure mode of NS is the flexural crack of the beam followed by shear failure in the joint. For the CFTPC-strengthened specimen, the crack was smaller in the retrofitted area. Thus, the obtained values of λ indicated the occurrence of beam flexural failure in the specimen.

Table 5. Validation of maximum lateral load resistance for various specimens.

Specimen	λ	E_n (kN)	λE_n (kN)	E_{max} (kN)	Remarks
NS	1.31	8.4	11.00	8.65	Ok
SD	1.31	8.4	11.00	10.42	Ok
1-layer CFRPC-integrated	1.31	12.65	16.57	9.87	Ok
2-layer CFRPC-integrated	1.31	16.02	20.98	11.27	Ok

The joints must meet the requirement of the strengths degradation as specified by the standard. The strength degradation of the joints is assessed at a drift level of 3.5% (Table 6) and should exceed $0.75E_{max}$ for the same loading direction. The results in Table 6 showed that the NSCBCJs failed to satisfy the strength criteria. Conversely, in the specimens strengthened by CFRPC and seismic details, the beam–column joint met the strength criteria. The deficiency in transverse reinforcement in the joint led to a high loss of lateral resistance capacity at a later stage. In the NS specimen, the value strength dropped dramatically by about 58% at a drift ratio of 3.5% due to the cracking of the concrete joint. The reason for an abrupt reduction in the strength value might be due to the bond failure of the longitudinal reinforcement. On the contrary, the CFRPC-strengthened specimens achieved excellent strength values due to the increased confinement provided by the CRFPC to the joint, thereby leading to a reduction in the joint cracks and bond failure.

Table 6. Assessment of joint strength degradation at a drift level of 3.5%.

Specimen	Direction	Ultimate Lateral Force in kN	3.5% (Drift Ratio)	
			Force in kN	Ratio of the Strength > 0.75
NS	Positive	8.56	6.30	0.73 *
	Negative	16.77	15.50	0.92
SD	Positive	10.42	10.01	0.96
	Negative	16.20	15.67	0.96

Table 6. *Cont.*

Specimen	Direction	Ultimate Lateral Force in kN	3.5% (Drift Ratio)	
			Force in kN	Ratio of the Strength > 0.75
1-layer CFRPC-integrated	Positive	9.87	9.58	0.97
	Negative	18.23	18.00	0.98
2-layer CFRPC-integrated	Positive	11.27	10.50	0.93
	Negative	17.86	17.98	0.99

* Less than the ratio of the strength.

4.2.3. Stiffness Criteria

After the seismic action, damage occurred in the frame structure during a large lateral displacement under a small lateral load. To overcome this effect, the standard requires that the stiffness of the joints be high (0.05) compared to the initial stiffness. The standard requires that the criterion be checked at a drift range of −0.35% to 0.35% using the secant stiffness, and it must meet the prescribed limitation. The initial positive and negative stiffnesses were calculated by dividing the load per associated displacement. Table 7 shows the results for the stiffness criterion validation of the joints, wherein the specimens were observed to satisfy this criterion. Comparing the seismic and strengthened specimens against the specimen made with non-seismic details against the requirement of ACI 374.1-05 for stiffness, it was found that the non-seismic joint had lower initial and secant stiffness due to the lack of transverse reinforcement in the joint. The specimen with seismic details demonstrated a good performance, since this specimen was designed by providing the reinforcement according to the requirement of the ACI 318-19, wherein the transverse reinforcement reduces the cracks in the joints area leading to an increase in the stiffness. This improvement was attributed to the presence of CFRPC, which enabled us to increase the strength of the beam, column, and joint. Consequently, the stiffness was increased. In short, the CFRMC-strengthened specimens demonstrated improvements in their initial and secant stiffness values.

Table 7. Validation of the secant stiffness of various specimens.

Specimen	Direction	Drift Ratio	δ (mm)	F (kN)	Secant Stiffness (kN/mm)	Initial Stiffness (kN/mm)	Ratio between Secant Stiffness and Initial Stiffness	Check to Meet the Criteria ≥ 0.05
NS	Positive	3.5	33.88	6.30	0.18	0.66	0.27	Ok
	Negative		33.88	15.50	0.45	0.76	0.59	Ok
SD	Positive	3.5	33.88	10.01	0.29	1.53	0.19	Ok
	Negative		33.88	15.67	0.46	1.83	0.25	Ok
1-layer CFRPC-integrated	Positive	3.5	33.88	9.58	0.28	1.82	0.15	Ok
	Negative		33.88	18.00	0.53	1.84	0.28	Ok
2-layer CFRPC-integrated	Positive	3.5	33.88	10.50	0.31	1.70	0.18	Ok
	Negative		33.88	17.80	0.53	1.71	0.31	Ok

The main remarks obtained from the data analysis can be described as follows: (1) the CFRPC-strengthened specimens showed an increase in lateral resisting force without any spalling of concrete; (2) compared to the non-seismic specimens, both CFRPC-strengthened specimens displayed a good amount of energy dissipation, which was increased from 40 to 57%; (3) at the end of loading, the secant stiffness of the CFRPC-strengthened specimens was higher than that of the non-seismic specimen, indicating an increase in their initial stiffness; (4) a comparative evaluation of the acceptance criteria of ACI 374-05 showed that the specimens with seismic details achieved superior seismic performances.

4.2.4. Effective Elastic Stiffness and Displacement Capacity

The backbone curve of all specimens was developed depending on the data obtained from the experiential study by Trung. The results were verified using the finite element model calculation for each loading direction, revealing a unique behavior. The curve was plotted in a single quadrant (positive force with associated positive displacement, and negative force with associated negative displacement) and the backbone curve was created by connecting the first cycle peak points from each drift level (Figure 17). The backbone curve provided the information related to the peak load (P_y) in each loading direction, effective elastic stiffness (K_e), yield displacement (Δ_y), ultimate displacement (Δ_u), and displacement ductility factor (μ). The determination of the yield displacement was challenging due to the absence of a clear yield displacement in the load-displacement curve of a joint. To begin with, the nonlinear capacity curve was simplified by representing it as a bilinear curve that included a specified yield displacement. The values of (Δ_y) for various specimens were calculated following the recommendation of Paulay and Priestley [67]. Figure 18 illustrates an idealized bilinear curve for the joints, which is then utilized to derive the ductility parameters. The values of K_e were determined using the tangent line of the secant slope from the original point and the point when the line had 0.75 of the yield load, where Δ_{yi} is the displacement corresponding to 0.75 P_y in the ascending part of the tangent line. The following relation must satisfy the intersection of the horizontal by trial and error of the line drawn in the failure part to the line drawn from the original by slope provided the Δ_y.

$$A1 + A2 = A3 + A4 \tag{3}$$

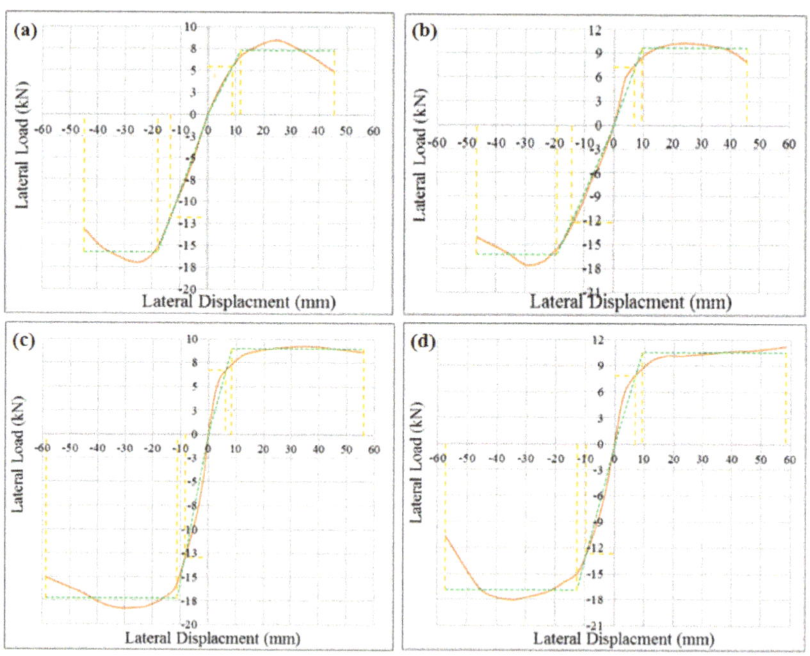

Figure 17. Normalization process of the backbone curve for (**a**) NS, (**b**) SD, (**c**) one-layer CFRPC-integrated specimen, and (**d**) two-layer CFRPC-strengthened specimen.

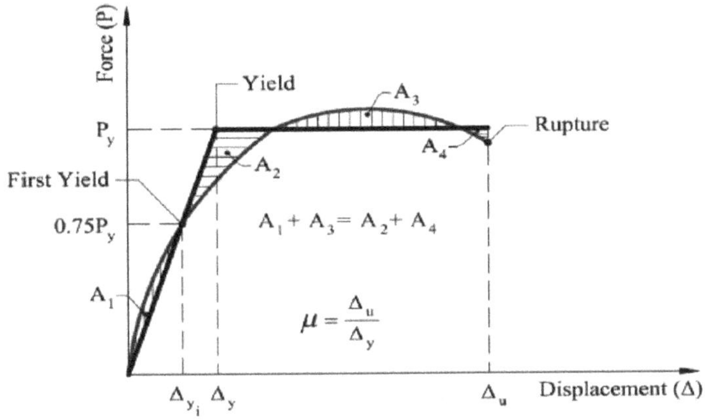

Figure 18. Ideal bilinear load–displacement behavior of the specimen.

The specimen's displacement ductility factor (μ) was calculated by dividing its ultimate displacement by the yield displacement via the following equation:

$$\mu = \frac{\Delta_u}{\Delta_y} \tag{4}$$

The bilinear diagram was used to represent the backbone curve. The effective stiffness (K_e) and the deformation capacity (Δ_y, Δ_u, and μ) in the negative and positive loading directions of the specimens are enlisted in Table 7. The results clearly indicated that the ductility factor ranged from 3.10 to 6.16, which was larger than the suggested ductility for reinforced concrete structures ($\mu \geq 2$) as obtained by Peng, Zhong, et al. [68]. It can be asserted that the ductility of the specimens was higher than the suggested ductility for the reinforced concrete type of structures. Also, the CFRPC-strengthened specimens and the specimen with seismic details had higher ductility factors compared to the non-seismic specimen. Table 8 shows the values for the effective stiffness (K_e) and the displacement capacity (Δ_y, Δ_u) in both the negative and positive loading directions for the specimen. The effective stiffness of the CFRPC-reinforced specimens was observed to be higher than that of the non-seismic specimen, which was mainly due to the presence of CFRPC in the structure, leading to an enhancement in the strength performance of the specimen. The specimen containing two layers of CFRPC showed higher effective stiffness than the specimen designed with one layer of CFRPC. The effective stiffness of the specimen designed with seismic details was higher than that of the non-seismic specimen. This strength enhancement can be ascribed to the seismic reinforcement details of the specimen.

Table 8. Effective stiffness and ductility of various specimens.

Specimen	Direction	$0.75P_y$ (kN)	Δ_{yi} (mm)	K_e (KN/m)	Avg. K_e (KN/m)	Δ_y (mm)	Δ_u (mm)	μ	Average μ
NS	Positive	5.47	8.77	623	745.5	11.70	45.72	3.9	3.10
	Negative	11.75	13.5	870		19.70	44.79	2.3	
SD	Positive	7.30	7.12	1025	973.5	9.50	45.90	4.83	3.63
	Negative	13.15	14.25	922		19.10	46.55	2.43	
1-layer CFRPC-integrated	Positive	6.75	6.375	1059	1262	8.50	56.03	6.59	5.95
	Negative	12.95	8.80	1465		11.00	58.59	5.32	
2-layer CFRPC-integrated	Positive	7.95	6.97	1140	1301	9.30	58.50	6.29	6.16
	Negative	13.80	9.30	1462		9.52	57.53	6.04	

4.2.5. Damping Ratio and Energy Dissipation

Figure 19 shows the drift ratio-dependent variation in the cumulative dissipated energy of various specimens. The energy dissipation capacity was determined using the enclosed loop area of each specimen in the hysteresis curve. The non-seismic specimen experienced pinching behavior during testing. This observation can be mainly attributed to the occurrence of cracks in the joint, leading to a decrease in the energy dissipation in the specimen. The presence of CFRPC significantly increased the energy dissipation. At the final loading, the energy dissipation capacity of the two-layer CFRPC-integrated specimen was approximately 43% higher than the one obtained for the non-seismic specimen. In addition, the one-layer CFRPC-strengthened specimen showed an increase in energy dissipation of 25% compared to the non-seismic specimen. Generally, the specimens reinforced with CFRPC exhibited a greater ability to dissipate the energy compared to the non-seismic specimen. In brief, the specimens reinforced with CFRPC displayed higher flexural strength due to the presence of CFRPC material at the beam–column junction, producing an enhanced energy dissipation. It is worth noting that the energy dissipation was directly impacted by the presence of transverse reinforcement in the joint. During the final loading stage, the seismic specimen exhibited an energy dissipation capacity approximately 20% more than that of the non-seismic specimen. Essentially, the confinement effect on the core concrete was enhanced by the transverse reinforcement, leading to a reduction in joint cracking. In contrast, the non-seismic specimen showed a smaller energy dissipation compared to the others until the end of the test.

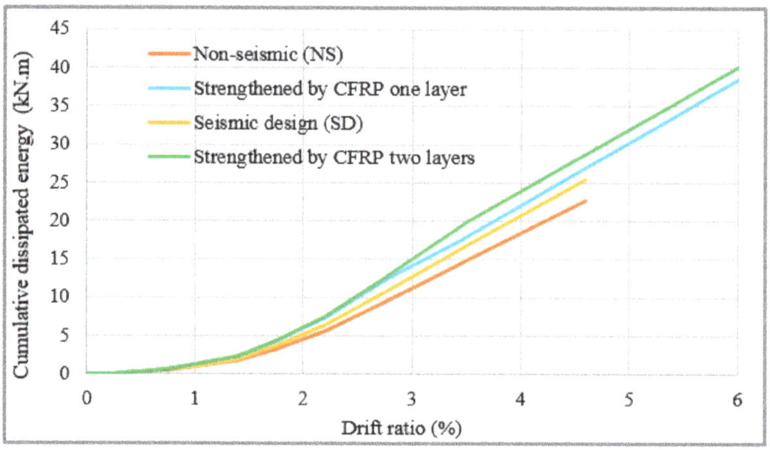

Figure 19. Drift ratio-dependent variation in the cumulative dissipated energy.

Figure 20 shows the simplified hysteresis loop that consists of the lateral load versus the displacement. The equivalent damping coefficient (h_e) of the studied specimens was calculated to evaluate the actual energy dissipation ability via the following equation:

$$h_e = \frac{1}{2\pi}\frac{S_{ABC} + S_{CDA}}{S_{OBF} + S_{ODE}} \quad (5)$$

where S_{ABC} is the area of the curve ABC (Figure 20), wherein similar definitions were used for S_{CDA}, S_{OBF}, and S_{ODE}.

Figure 21 shows the drift ratio-dependent variation in the equivalent damping coefficient. The value of (h_e) is about 0.1 for the normal reinforcement concrete joints and 0.3 for the joints with steel beam reinforcement concrete columns joints [68]. The obtained results indicated that the joints' capability for energy dissipation was within the range of reinforcement concrete joints and that it was lower than that of steel beam reinforcement concrete column joints. In the present case, the non-seismic specimen showed lower values of (h_e) compared to other specimens, while the two-layer CFRPC-strengthened sample showed the best performance.

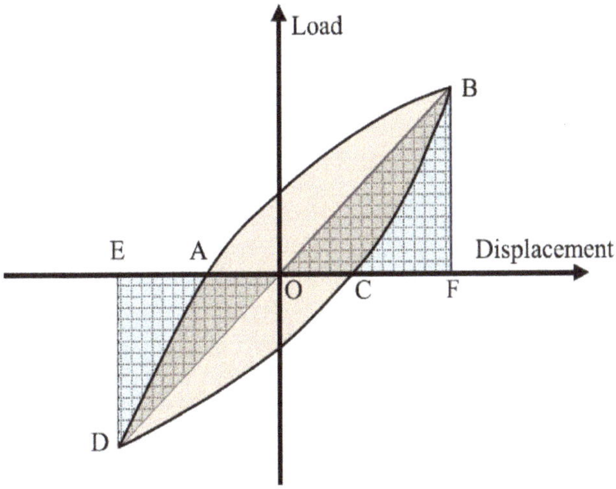

Figure 20. Simplified hysteresis loop.

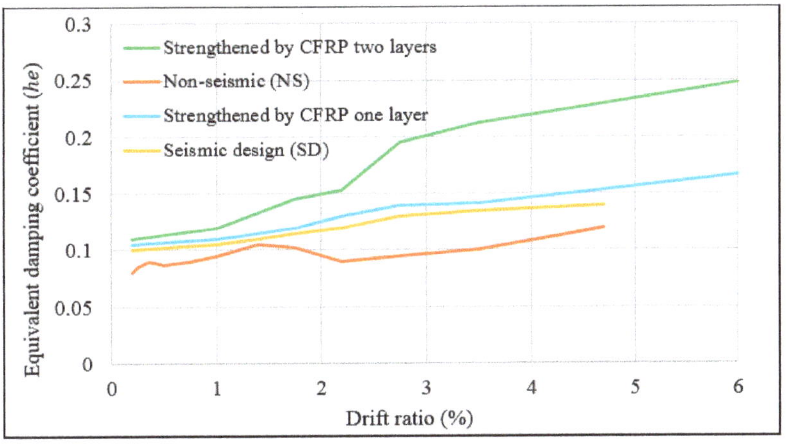

Figure 21. Drift ratio-dependent variation in the equivalent damping coefficient.

4.3. *Stiffness Degradation*

Figure 22 displays the drift ratio-dependent variation in the secant stiffnesses for all the studied specimens. The degradation of the secant stiffness for the beam–column joint in each loading cycle was computed using the following equation:

$$K_{ij} = \frac{|+F_{ij}| + |-F_{ij}|}{|+\Delta_{ij}| + |-\Delta_{ij}|} \tag{6}$$

where $+F_{ij}$, $(-F_{ij})$ is the positive (negative) peak load and $+\Delta_{ij}$, $(-\Delta_{ij})$ is the displacement associated with the positive (or negative) ultimate lateral strength [34].

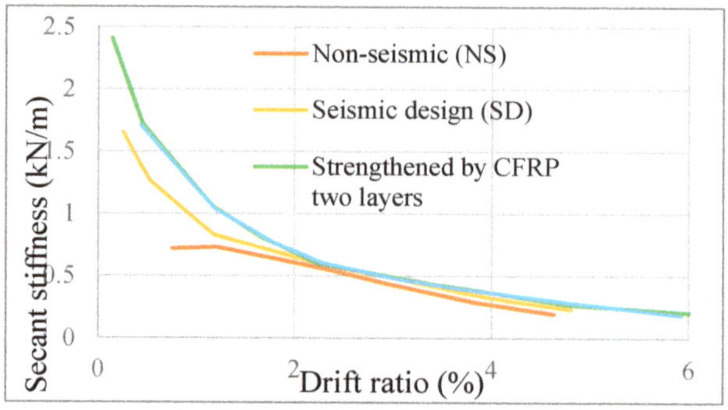

Figure 22. Drift ratio-dependent variation in the secant stiffnesses.

The degradation of the secant stiffness values for all specimens increased with the increase in drift ratio (lateral displacement) at various load steps, significantly impacting the specimens' stiffness degradation. The stiffness of the specimens was reduced with the increase in the load step. The non-seismic specimen showed lower stiffness than other specimens. Specimen strengthened by two layers of CFRPC showed higher stiffness, since all the specimens had the same value in terms of the beam reinforcement ratio. In essence, the confinement given by the presence of CFRPC to the beam–column joint could mitigate the distortion of the joint panel, hence enhancing its strength and reducing crack formation in the surrounding area. In short, CFRPC-integrated NSCBCJs showed improved performance against seismic events.

5. Conclusions

Based on the obtained results, the following conclusions can be drawn:

i The strengthening of NSCBCJs using CFRPC reinforcement enabled us to reduce crack formation in the plastic hinge location, thus leading to a reduced risk of building collapse during seismic events. The plastic hinges for the frame of the reinforced concrete building were placed near the column and beam joint.

ii The strengthening of the NSCBCJs was appropriate when assessing the behavior of non-seismic details due to the overall lateral displacement increase that produces an increase in ductility demand.

iii The NSCBCJs being non-ductile meant that they could display large lateral displacement which can be attributed to the significant cracks and deformation of the joint area. Consequently, the observed large lateral displacement can cause severe failure in the frame building during strong seismic vibration.

iv The proposed NSCBCJs showed significant joint shear failure together with the beam and column flexural failure. However, the seismic joint failure was beam flexural, while CFRPC-integrated joints showed beam flexural failure with the debonding of the CFRPC sheet. Since the studied joints had beam flexural strength, the lateral load capacity difference can be mainly ascribed to the transverse reinforcement and strengthening of the joint.

v Both seismic and CFRPC-strengthened joints revealed good strength, stiffness, ductility, and energy-dissipation capabilities.

vi Both seismic and CFRPC-strengthened joints achieved the required seismic code provisions of ACI 374.1-05. Conversely, the NSCBCJs failed to satisfy the desired requirement due to a failure in achieving the strength criteria.

vii The finite element model simulation results for the stress and strain response demonstrated the considerable impact of CFRPC integration into NSCBCJs, leading to a reasonable strength capacity enhancement in the joints, thereby reducing the joint cracks.

viii It is established that the present comprehensive numerical modeling and simulation study may contribute to the development of CFRPC-reinforced high-performance anti-seismic CBCJs, thus contributing towards the sustainable construction sector.

Author Contributions: Conceptualization, N.A.S.A. and A.A.K.; methodology, N.A.S.A.; software, F.H.M.; validation, A.A.K. and F.H.M.; formal analysis, N.A.S.A.; investigation, N.A.S.A.; resources, N.A.S.A.; data curation, N.A.S.A.; writing—original draft preparation, N.A.S.A.; writing—review and editing, A.A.K.; visualization, F.H.M.; supervision, A.A.K.; project administration, N.A.S.A.; funding acquisition, A.A.K. All authors have read and agreed to the published version of the manuscript.

Funding: This research received no external funding.

Institutional Review Board Statement: Not applicable.

Informed Consent Statement: Not applicable.

Data Availability Statement: The original contributions presented in the study are included in the article, further inquiries can be directed to the corresponding author.

Acknowledgments: The authors thank Basrah University for them support.

Conflicts of Interest: The authors declare no conflicts of interest.

References

1. Sultan, H.K.; Huseien, G.F. Minimum Shear Reinforcement for Reactive Powder Concrete Beams. *Eng* **2024**, *5*, 801–818. [CrossRef]
2. Cascardi, A.; Verre, S.; Sportillo, A.; Giorgio, G. A Multiplex Conversion of a Historical Cinema. *Adv. Civ. Eng.* **2022**, *2022*, 2191315. [CrossRef]
3. Sultan, H.K.; Noor, A.A.A.; Huseien, G.F. Performance Evaluation of Self-Compacting Glass Fiber Concrete Incorporating Silica Fume at Elevated Temperatures. *Eng* **2024**, *5*, 1043–1066. [CrossRef]
4. Saleh, S.M.; Majeed, F.H. Shear strength of headed stud connectors in self-compacting concrete with recycled coarse aggregate. *Buildings* **2022**, *12*, 505. [CrossRef]
5. Falih, Z.F.M.; Dakhi, A.J.; Saleh, S.M. Numerical Solution for Masonry Wall Using General Static Step with ABAQUS/Standard. *Math. Model. Eng. Probl.* **2023**, *10*, 2142–2148. [CrossRef]
6. Saleh, S.M.; Majeed, F.H.; Al-Salih, O.; Hussain, H.K. Torsional Behavior of Steel-Concrete-Steel Sandwich Beams with Welded Stirrups as Shear Connectors. *Civ. Eng. J.* **2023**, *9*, 208–219. [CrossRef]
7. Karim, A.A.; Abd Matooq, J.; Abdulrazzaq, O.A.; Majeed, F.H.; Saleh, S.M. The behavior of Shear Connectors in Steel-Normal Concrete Composite Structure under Repeated Loads. *Civ. Eng. J.* **2024**, *10*, 210–221. [CrossRef]
8. Deaton, J.B. Nonlinear Finite Element Analysis of Reinforced Concrete Exterior Beam-Column Joints with Nonseismic Detailing. 2013. Available online: https://repository.gatech.edu/entities/publication/78302263-8d44-44c4-bbc2-c064816bf6de (accessed on 10 May 2024).
9. Tahnat, Y.B.A.; Samaaneh, M.A.; Dwaikat, M.M.; Halahla, A.M. Simple equations for predicting the rotational ductility of fiber-reinforced-polymer strengthened reinforced concrete joints. *Structures* **2020**, *24*, 73–86. [CrossRef]
10. Gao, F.; Tang, Z.; Mei, S.; Hu, B.; Huang, S.; Chen, J. Seismic behavior of exteriorbeam–column joints with high-performance steel rebar: Experimental and numerical investigations. *Adv. Struct. Eng.* **2021**, *24*, 90–106. [CrossRef]
11. Yu, F.; Wu, C.; Xu, B.; Guan, Y.; Zhu, X.; Fang, Y. Study on shear bearing capacity of PVC-CFRP confined concrete column-RC beam exterior joint with core steel tube. *Constr. Build. Mater.* **2024**, *410*, 134197. [CrossRef]
12. Samadi, M.; Baghban, M.H.; Kubba, Z.; Faridmehr, I.; Abdul Shukor Lim, N.H.; Benjeddou, O.; Ariffin, N.F.B.; Huseien, G.F. Flexural behavior of reinforced concrete beams under instantaneous loading: Effects of recycled ceramic as cement and aggregates replacement. *Buildings* **2022**, *12*, 439. [CrossRef]
13. Kularni, S.; Patil, Y.D. A novel reinforcement pattern for exterior reinforced concrete beam-column joint. *Procedia Eng.* **2013**, *51*, 184–193. [CrossRef]
14. Abdalla, K.M.; Al-Rousan, R.; Alhassan, M.A.; Lagaros, N.D. Finite-element modelling of concrete-filled steel tube columns wrapped with CFRP. *Proc. Inst. Civ. Eng.-Struct. Build.* **2020**, *173*, 844–857. [CrossRef]
15. Kaliluthin, A.; Kothandaraman, S.; Ahamed, T.S. A review on behavior of reinforced concrete beam column joint. *Int. J. Innov. Res. Sci. Eng. Technol.* **2014**, *3*, 11299–11312.
16. Kiran, R.; Genesio, G. A case study on pre 1970s constructed concrete exterior beam-column joints. *Case Stud. Struct. Eng.* **2014**, *1*, 20–25. [CrossRef]
17. Rajagopal, S.; Prabavathy, S. Exterior beam-column joint study with non-conventional reinforcement detailing using mechanical anchorage under reversal loading. *Sadhana* **2014**, *39*, 1185–1200. [CrossRef]
18. Rajagopal, S.; Prabavathy, S. Seismic behavior of exterior beam-column joint using mechanical anchorage under reversal loading: An experimental study. *Iran. J. Sci. Technol. Trans. Civ. Eng.* **2014**, *38*, 345.

19. Fan, G.; Song, Y.; Wang, L. Experimental study on the seismic behavior of reinforced concrete beam-column joints under various strain rates. *J. Reinf. Plast. Compos.* **2014**, *33*, 601–618. [CrossRef]
20. Al-Rousan, R.Z.; Alhassan, M.A.; Al-omary, R.J. Response of interior beam-column connections integrated with various schemes of CFRP composites. *Case Stud. Constr. Mater.* **2021**, *14*, e00488. [CrossRef]
21. Patel, R.S.; Nambiyanna, B.; Prabhakara, R. An experimental study on effect of diameter of rebar on exterior beam column joint. *Int. J. Innov. Res. Sci.* **2015**, *4*, 5984–5991.
22. Rodríguez, V.; Guerrero, H.; Alcocer, S.M.; Tapia-Hernández, E. Rehabilitation of heavily damaged beam-column connections with CFRP wrapping and SFRM casing. *Soil Dyn. Earthq. Eng.* **2021**, *145*, 106721. [CrossRef]
23. Moskaleva, A.; Safonov, A.; Hernández-Montes, E. Fiber-reinforced polymers in freeform structures: A review. *Buildings* **2021**, *11*, 481. [CrossRef]
24. Al-Rousan, R.Z.; Alkhawaldeh, A. Behavior of heated damaged reinforced concrete beam-column joints strengthened with FRP. *Case Stud. Constr. Mater.* **2021**, *15*, e00584. [CrossRef]
25. Ruiz-Pinilla, J.G.; Pallarés, F.; Gimenez, E.; Calderón, P. Experimental tests on retrofitted RC beam-column joints underdesigned to seismic loads. General approach. *Eng. Struct.* **2014**, *59*, 702–714. [CrossRef]
26. Yurdakul, Ö.; Avşar, Ö. Strengthening of substandard reinforced concrete beam-column joints by external post-tension rods. *Eng. Struct.* **2016**, *107*, 9–22. [CrossRef]
27. Ong, C.; Chin, C.; Ma, C.; Tan, J.; Awang, A.; Omar, W. Seismic retrofit of reinforced concrete beam-column joints using various confinement techniques: A review. *Structures* **2022**, *42*, 221–243. [CrossRef]
28. Abdalla, J.A.; Mhanna, H.H.; Hawileh, R.A.; Sharafi, M.; Al-Marzouqi, A.; Al-Teneiji, S.; Al-Ali, K. Shear Strengthening of Reinforced Concrete T-Beams using Carbon Fiber Reinforced Polymer (CFRP) Anchored with CFRP Spikes. *Procedia Struct. Integr.* **2022**, *42*, 1223–1230. [CrossRef]
29. Fang, H.; Bai, Y.; Liu, W.; Qi, Y.; Wang, J. Connections and structural applications of fibre reinforced polymer composites for civil infrastructure in aggressive environments. *Compos. Part B Eng.* **2019**, *164*, 129–143. [CrossRef]
30. Jia, J.; Zandi, Y.; Rahimi, A.; Pourkhorshidi, S.; Khadimallah, M.A.; Ali, H.E. Numerical performance evaluation of debonding strength in fiber reinforced polymer composites using three hybrid intelligent models. *Adv. Eng. Softw.* **2022**, *173*, 103193. [CrossRef]
31. Zhang, J.-g.; Yang, G.-c.; Ma, Z.-h.; Zhao, G.-l.; Song, H.-y. A stacking-CRRL fusion model for predicting the bearing capacity of a steel-reinforced concrete column constrained by carbon fiber-reinforced polymer. *Structures* **2023**, *55*, 1793–1804. [CrossRef]
32. Vatani-Oskouei, A. Repairing of seismically damaged RC exterior beam—Column connection using CFRP. *J. Reinf. Plast. Compos.* **2010**, *29*, 3257–3274. [CrossRef]
33. Eslami, A.; Ronagh, H.R. Experimental investigation of an appropriate anchorage system for flange-bonded carbon fiber–reinforced polymers in retrofitted RC beam–column joints. *J. Compos. Constr.* **2014**, *18*, 04013056. [CrossRef]
34. Zgür Yurdakul, Ö.; Avşar, Ö. Structural repairing of damaged reinforced concrete beam-column assemblies with CFRPs. *Struct. Eng. Mech.* **2015**, *54*, 521–543. [CrossRef]
35. Sasmal, S.; Novák, B.; Ramanjaneyulu, K.; Srinivas, V.; Roehm, C.; Lakshmanan, N.; Iyer, N.R. Seismic retrofitting of damaged exterior beam–column joints using fibre reinforced plastic composite–steel plate combined technique. *Struct. Infrastruct. Eng.* **2013**, *9*, 116–129. [CrossRef]
36. Agarwal, P.; Gupta, A.; Angadi, R.G. Effect of FRP wrapping on axial behavior of concrete and cyclic behavior of external RC beam column joints. *KSCE J. Civ. Eng.* **2014**, *18*, 566–573. [CrossRef]
37. Granata, P.J.; Parvin, A. An experimental study on Kevlar strengthening of beam–column connections. *Compos. Struct.* **2001**, *53*, 163–171. [CrossRef]
38. Zhoudao, L.; Lei, S.; Jiangtao, Y. Experimental study on the seismic behaviour of strengthened concrete column-beam joints by simulated earthquake. *Procedia Eng.* **2011**, *14*, 1871–1878. [CrossRef]
39. Allam, K.; Mosallam, A.S.; Salama, M.A. Experimental evaluation of seismic performance of interior RC beam-column joints strengthened with FRP composites. *Eng. Struct.* **2019**, *196*, 109308. [CrossRef]
40. Mosallam, A. Strength and ductility of reinforced concrete moment frame connections strengthened with quasi-isotropic laminates. *Compos. Part B Eng.* **2000**, *31*, 481–497. [CrossRef]
41. Yang, Z.; Liu, Y.; Li, J. Study of seismic behavior of RC beam-column joints strengthened by sprayed FRP. *Adv. Mater. Sci. Eng.* **2018**, *2018*, 3581458. [CrossRef]
42. Mukherjee, A.; Joshi, M. FRPC reinforced concrete beam-column joints under cyclic excitation. *Compos. Struct.* **2005**, *70*, 185–199. [CrossRef]
43. Alshamrani, S.; Rasheed, H.A.; Salahat, F.H.; Borwankar, A.; Divilbiss, N. Seismic flexural behavior of CFRP strengthened reinforced concrete beams secured with fiber anchors. *Eng. Struct.* **2024**, *305*, 117728. [CrossRef]
44. Smith, S.T.; Rasheed, H.A.; Kim, S.J. Full-range load-deflection response of FRP-strengthened RC flexural members anchored with FRP anchors. *Compos. Struct.* **2017**, *167*, 207–218. [CrossRef]
45. Rasheed, H.A.; Pervaiz, S. Bond slip analysis of fiber-reinforced polymer-strengthened beams. *J. Eng. Mech.* **2002**, *128*, 78–86. [CrossRef]
46. Gao, W.; Teng, J.; Dai, J.-G. Effect of temperature variation on the full-range behavior of FRP-to-concrete bonded joints. *J. Compos. Constr.* **2012**, *16*, 671–683. [CrossRef]

47. Khodadadi, N.; Roghani, H.; Harati, E.; Mirdarsoltany, M.; De Caso, F.; Nanni, A. Fiber-reinforced polymer (FRP) in concrete: A comprehensive survey. *Constr. Build. Mater.* **2024**, *432*, 136634. [CrossRef]
48. Azuwa, S.B.; Yahaya, F.B.M. Experimental investigation and finite element analysis of reinforced concrete beams strengthened by fibre reinforced polymer composite materials: A review. *Alex. Eng. J.* **2024**, *99*, 137–167. [CrossRef]
49. Laseima, S.Y.; Mutalib, A.; Osman, S.; Hamid, N. Seismic behavior of exterior rc beam-column joints retrofitted using CFRP sheets. *Lat. Am. J. Solids Struct.* **2020**, *17*, e263. [CrossRef]
50. Saqan, E.I.; Rasheed, H.A.; Alkhrdaji, T. Seismic Behavior of Carbon Fiber-Reinforced Polymer-Strengthened Reinforced Concrete Members with Various Anchors. *ACI Struct. J.* **2020**, *117*, 3–14.
51. Karayannis, C.G.; Golias, E. Full-scale experimental testing of RC beam-column joints strengthened using CFRP ropes as external reinforcement. *Eng. Struct.* **2022**, *250*, 113305. [CrossRef]
52. Lim, C.; Jeong, Y.; Kim, J.; Kwon, M. Experimental study of reinforced concrete beam-column joint retrofitted by CFRP grid with ECC and high strength mortar. *Constr. Build. Mater.* **2022**, *340*, 127694. [CrossRef]
53. Attari, N.; Youcef, Y.S.; Amziane, S. Seismic performance of reinforced concrete beam–column joint strengthening by frp sheets. *Structures* **2019**, *20*, 353–364. [CrossRef]
54. Golias, E.; Zapris, A.G.; Kytinou, V.K.; Osman, M.; Koumtzis, M.; Siapera, D.; Chalioris, C.E.; Karayannis, C.G. Application of X-shaped CFRP ropes for structural upgrading of reinforced concrete beam–column joints under cyclic loading–experimental study. *Fibers* **2021**, *9*, 42. [CrossRef]
55. Mady, M.H.A. *Seismic Behaviour of Exterior Beam-Column Joints Reinforced with Frp Bars and Stirrups*; University of Manitoba (Canada): Winnipeg, MB, Canada, 2011.
56. Bsisu, K.; Hiari, B.O. Finite element analysis of retrofitting techniques for reinforced concrete beam-column joint. *J. Am. Sci.* **2015**, *11*, 48–56.
57. Ercan, E.; Arisoy, B.; Ertem, O.B. Experimental assessment of RC beam-column connections with internal and external strengthening techniques. *Adv. Civ. Eng.* **2019**, *2019*, 2828353. [CrossRef]
58. Le-Trung, K.; Lee, K.; Lee, J.; Lee, D.H.; Woo, S. Experimental study of RC beam–column joints strengthened using CFRP composites. *Compos. Part B Eng.* **2010**, *41*, 76–85. [CrossRef]
59. ACI Committee. Building Code Requirements For Structural Concrete (ACI 318-02) and Commentary (ACI 318R-02). Available online: https://agussugiantocom.files.wordpress.com/2016/07/aci318-02-building-code-for-structural-concrete.pdf (accessed on 10 May 2024).
60. Tao, Y.; Chen, J.-F. Concrete damage plasticity model for modeling FRP-to-concrete bond behavior. *J. Compos. Constr.* **2015**, *19*, 04014026. [CrossRef]
61. Chen, G.; Chen, J.; Teng, J. On the finite element modelling of RC beams shear-strengthened with FRP. *Constr. Build. Mater.* **2012**, *32*, 13–26. [CrossRef]
62. Lubliner, J.; Oliver, J.; Oller, S.; Oñate, E. A plastic-damage model for concrete. *Int. J. Solids Struct.* **1989**, *25*, 299–326. [CrossRef]
63. Lee, J.; Fenves, G.L. Plastic-damage model for cyclic loading of concrete structures. *J. Eng. Mech.* **1998**, *124*, 892–900. [CrossRef]
64. Neuberger, Y.M.; Andrade, M.V.; de Sousa, A.M.D.; Bandieira, M.; da Silva Júnior, E.P.; dos Santos, H.F.; Catoia, B.; Bolandim, E.A.; de Moura Aquino, V.B.; Christoforo, A.L. Numerical Analysis of Reinforced Concrete Corbels Using Concrete Damage Plasticity: Sensitivity to Material Parameters and Comparison with Analytical Models. *Buildings* **2023**, *13*, 2781. [CrossRef]
65. Narayanan, R.; James, R. Microwave nondestructive testing of galvanic corrosion and impact damage in carbon fiber reinforced polymer composites. *Int. J. Microw. Appl* **2018**, *7*, 1. [CrossRef]
66. GangaRao, H.V.; Vijay, P. Bending behavior of concrete beams wrapped with carbon fabric. *J. Struct. Eng.* **1998**, *124*, 3–10. [CrossRef]
67. Torabi, A.; Maheri, M.R. Seismic repair and retrofit of RC beam–column joints using stiffened steel plates. *Iran. J. Sci. Technol. Trans. Civ. Eng.* **2017**, *41*, 13–26. [CrossRef]
68. Peng, Z.; Dai, S.-B.; Pi, Y.-L.; Zhang, Y.-C.; Huang, J. Seismic performance of end-plate connections between T-shaped CFST columns and RC beams. *J. Constr. Steel Res.* **2018**, *145*, 167–183. [CrossRef]

eng

Article

Performance Evaluation of Self-Compacting Glass Fiber Concrete Incorporating Silica Fume at Elevated Temperatures

Hussein Kareem Sultan [1,*], Abbas Abdulhssein Abd Noor [1] and Ghasan Fahim Huseien [2,3,4,*]

1 Civil Engineering Department, Al-Muthanna University, Samawah 66001, Iraq; abbas652002@mu.edu.iq
2 Department of the Built Environment, College of Design and Engineering, National University of Singapore, Singapore 117566, Singapore
3 Guangzhou Institute of Energy Conversion, Chinese Academy of Sciences, Guangzhou 510640, China
4 Construction Materials Centre, Civil Engineering Faculty, University Technology Malaysia, Johor 81310, Malaysia
* Correspondence: hussein.ksz@mu.edu.iq (H.K.S.); fhghassan@utm.my (G.F.H.)

Abstract: In this work, the properties of self-compacting concrete (SCC) and SCC containing 0.5 and 1% glass fibers (with lengths of 6 and 13 mm) were experimentally investigated, as well as their performance at high temperatures. With a heating rate of 5 °C/min, high-temperature experiments were conducted at 200, 400, 600, and 800 °C to examine mass loss, spalling, and the remaining mechanical properties of SCC with and without glass fibers. According to the results of the flowability and passing ability tests, adding glass fibers does not affect how workable and self-compacting SCCs were. These findings also demonstrated that the mechanical properties of samples with and without glass fibers rose up to 200 °C but then decreased at 400 °C, whereas the mixture containing 0.5% glass fibers of a length of 13 mm displayed better mechanical properties. Both SCC samples with and without glass fibers remained intact at 200 °C. Some SCC samples displayed some corner and edge spalling when the temperature reached about 400 °C. Above 400 °C, a significant number of microcracks started to form. SCC samples quickly spalled and were completely destroyed between 600 and 800 °C. According to the results, glass fibers cannot stop SCC from spalling during a fire. Between 200 and 400 °C, there was no discernible mass loss. At 600 °C, mass loss starts to accelerate quickly, and it increased more than ten times beyond 200 °C. The ultrasonic pulse velocity (UPV) of SCC samples with glass fibers increased between room temperature and 200 °C, and the mixture containing 0.5% glass fibers of a length of 13 mm showed a somewhat higher UPV than other SCC mixtures until it started to decline at about 400 °C.

Keywords: self-compacting concrete (SCC); glass fibers; glass fiber-reinforced self-compacting concrete (GFR-SCC); elevated temperature

Citation: Sultan, H.K.; Noor, A.A.A.; Huseien, G.F. Performance Evaluation of Self-Compacting Glass Fiber Concrete Incorporating Silica Fume at Elevated Temperatures. *Eng* **2024**, *5*, 1043–1066. https://doi.org/10.3390/eng5020057

Academic Editor: Alessio Cascardi

Received: 24 April 2024
Revised: 24 May 2024
Accepted: 28 May 2024
Published: 1 June 2024

1. Introduction

A new generation of concrete has been produced as a result of recent developments in concrete technology, including self-compacting concrete (SCC), which has superior strength, durability characteristics, and rheology compared to freshly mixed concrete mixtures [1]. Self-compacting concrete is a type of concrete that can completely cover the formwork by shifting under its own weight, maintaining homogeneity even when reinforcing is employed, and then setting up without vibrating. Benefits of SCC include the ability to cover congested and tiny areas, shorter construction timelines, lower labor costs, and less noise pollution. Ordinary concrete has the same engineering qualities as hardened concrete; however, hardened concrete is denser and more homogeneous [2,3].

Numerous studies have been conducted on the characteristics of SCC in the literature. The majority of earlier articles tested the fresh SCC mixes using common workability tests to show the concrete's self-consolidation. Tests such as the slump cone flow, V-funnel, J-ring, and L-box were used to examine the deformability, flowability, and passing ability characteristics [4,5]. Investigations were also carried out on the rheological characteristics of SCC, including plastic viscosity and yield stress [6,7]. In addition, the mix design and mix proportions have been examined [8,9]. The performance of SCC has also been studied in relation to the impacts of mineral admixtures such fly ash, silica fume, metakaolin, powdered granulated blast furnace slag, and ladle slag [10–13], as well as chemical admixtures like superplasticizers and viscosity-modifying admixtures [14–16]. Numerous studies examined the microstructure and rate of hydration of the SCC [17–19].

The mechanical characteristics of SCC have been investigated by Domone [20]. He came to the conclusion that the information gathered was sufficient to support the overall behavior of SCC and that additional study on more specialized or corroborative evidence for specific applications was needed. The bond strength of the reinforcing steel was examined in SCC by Foroughi et al. [21]. The SCC stability test results for cracking resistance, shrinkage, and creep are also included in the literature [22–25]. A study [26] found that the shrinkage, elastic modulus, and creep of SCC were comparable to the corresponding properties of normal-strength concrete. Some of the durability tests that have been examined include those for water permeability and absorption, chloride penetration, carbonation, gas permeability, sulfate assault, electrical resistivity, frost resistance, acid attack, and scaling [18,27–29].

The features of SCC with the inclusion of steel, glass, and carbon nanotubes have also been examined by researchers [30–36]. When high temperatures are applied to concrete, the pore pressure created in thick concrete mixtures, like high-strength concrete, can be extremely detrimental to the matrix and result in concrete failure via spalling [37]. Concrete containing fibers lessens these adverse effects while enhancing the material's mechanical properties at high temperatures [38–40].

Despite the fact that SCC, high-strength concrete, and high-strength SCC are finding increasing applications, additional research is still required to establish how well these materials fare in fires and how long they will persist when exposed to high temperatures. Fire resistance is a crucial criterion for building materials. One of the most important characteristics of building materials is their capacity to prevent or postpone the spread of excessive heat or flame. The concepts and methodology of the fire test are covered in the publication [41]. The major reasons why a concrete element fails at high temperatures are spalling and strength loss [42–44]. Researchers have noted the rapid spalling of SCC in the 180–300 °C range, which results in the loss of important mechanical properties and is mostly caused by the microstructure of SCC [45,46].

Comparing SCC mixes to normal concretes, the increased cement content, superplasticizers, and addition of reactive components like silica fume have an impact on their design process [4]. SCC contains about 200 kg/m^3 more filler and a greater binder concentration than conventional concrete in addition to chemical additives such fly ash, silica fume, and ground granulated blast furnace slag (GGBFS) [47]. As a result, the permeability of SCC is reduced, which leads to an increase in internal pressure and an increased risk of spalling [48–53]. Pathak and Siddique's [54] investigation into the effects of adding class F fly ash on the mechanical properties of SCC at high temperatures (between 200 and 300 °C) revealed significant mass loss accompanied by a sharp decline in splitting tensile strength due to the escape of bound water. According to Bakhtiyari et al. [42], who investigated the fire performance of SCC containing limestone and quartz powder in the temperature range of 150–1000 °C, the temperature range of 480–650 °C is the most crucial range for spalling in SCC and normal concrete, and SCC is more prone to spalling than normal concrete. SCC showed greater mechanical property retention at a high temperature than conventional concrete.

Despite the significant uncertainty and expanding use of self-compacting concrete in a variety of applications, the temperature behavior of self-compacting concrete has not been thoroughly explored, and the various steps are not entirely understood. The purpose of this research is to evaluate the qualities of SCC created from locally available materials in both their fresh and hardened states. In this study, the effect of glass fibers on the same characteristics is also taken into account. Additionally, this study tested the compressive, flexural, and tensile strengths of self-compacting concrete samples with and without glass fibers at elevated temperatures of up to 800 °C for 1 h. The practice of keeping concrete samples in the oven for 60 min upon reaching the target temperature is essential for achieving uniform temperature distribution, complete moisture removal, stabilization of chemical reactions, adherence to standardized testing procedures, and ensuring thermal stability. This methodical approach ensures the reliability and accuracy of the test results, which are critical for evaluating the quality and performance of concrete.

2. Materials and Methods

2.1. Raw Materials

The mixture consisted of ordinary Portland cement, water, fine and coarse aggregate, superplasticizer, silica fume, and glass fibers. The proportions of the SCC mix, as well as the components and their amounts, are described in Table 1. Natural river sand passing through a 5 mm sieve and fractured dolomite angular aggregates passing thru a 9.5 mm sieve were used as fine and coarse aggregates. Figure 1 depicts the fine and coarse aggregate particle size distributions. By cement weight, 25% silica fume was added. The silica fumes had the following properties: 2.1 specific gravity and 172,000 cm^2 specific area per gram. The size distribution of silica fume particles is shown in Figure 2. In this experiment, a modified polycarboxylate aqueous solution with a specific gravity of 1.19 was utilized as a third-generation superplasticizer (Viscocrete 3425). It meets ASTM C-494 requirements [55]. The utilized glass fibers (Figure 3) were found have length of 6 to 13 mm, diameter of 13 μm, 0.91 g/cm^3 density, 150 m^2/kg specific surface area, 1700 MPa tensile strength, 72 GPa modulus of elasticity, and very high corrosion resistance. Tap water was used for mixing concrete and curing.

Table 1. The SCC mixes' ingredients (kg/m^3).

Mix No.	Cement	Silica Fume	Aggregate		Water	Super-Plasticizer	Glass Fibers	
			Coarse	Fine			6 mm	13 mm
SCC	420	105	810.5	810.5	189	4.2	0	0
SCC_{-L1}	420	105	779.1	779.1	189	4.2	0	9
SCC_{-L2}	420	105	793.5	793.5	189	4.2	0	4.5
SCC_{-S1}	420	105	779.1	779.1	189	4.2	9	0
SCC_{-S2}	420	105	793.5	793.5	189	4.2	4.5	0

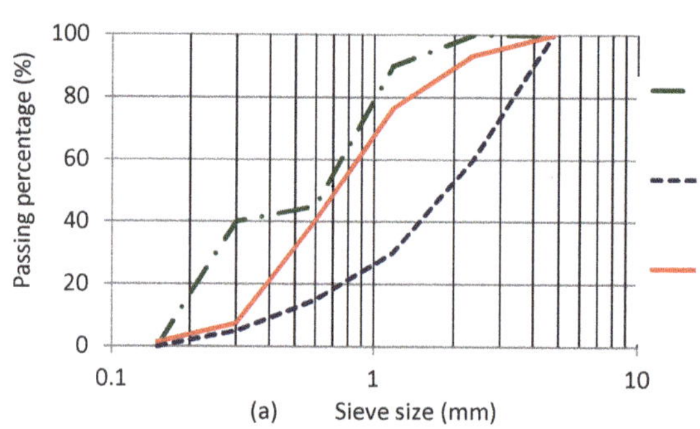

Figure 1. *Cont.*

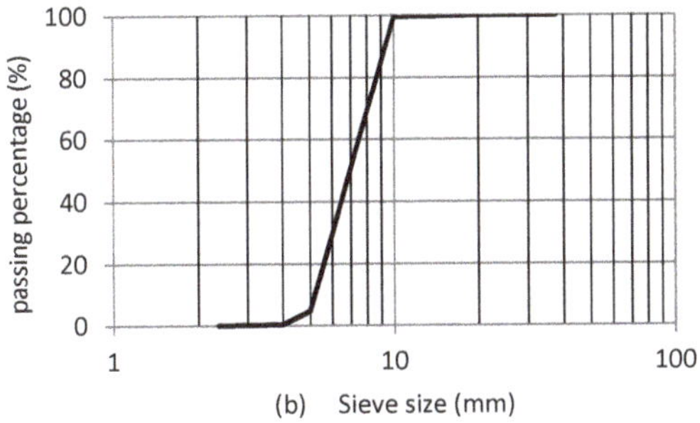

Figure 1. Aggregate particle size distribution curve, (**a**) fine aggregate, (**b**) coarse aggregate.

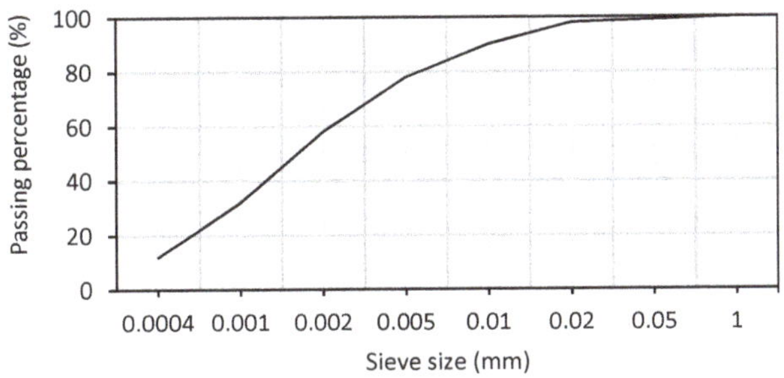

Figure 2. Silica fume particle size distribution curve (from supplier).

Figure 3. The glass fiber used in preparation of the mixtures.

Five SCC mixtures were created in this project: The first SCC mix contained no fibers (control mix), the second mix (SCC-$_{L1}$) contained 1% glass fibers with a 13 mm length, the third mix (SCC-$_{L2}$) contained 0.5% glass fibers with a 13 mm length, the fourth mix (SCC-$_{S1}$) contained 1% glass fibers with a 6 mm length, and the fifth mix (SCC-$_{S2}$) contained 0.5% glass fibers with a 6 mm length. The components for each mixture were 420 kg/m^3 ordinary Portland cement, 1% super-plasticizer, 45% water, and 25% silica fume with 0.5 or 1% glass fibers by volume.

2.2. Mixing, Moulding, and Curing

The dense packing principle was taken into consideration when creating the mix design for SCC samples. In dry conditions in a rotary mixer (Figure 4), Portland cement, silica fumes, and aggregate were mixed before super-plasticizer and water were added. Fibers were added to dry components before water was added, and then additives were added to the SCC mixtures. Compression, tension, and bending tests were conducted using a 100 mm cube, a 100 mm diameter by 200 mm high cylinder, and a 100 mm × 100 mm × 300 mm prism, respectively. To verify the self-compacting properties of mixtures, necessary experiments were run on SCC in its fresh state. This comprised L-box and GTM screen stability, as well as slump flow and flow velocity T50.

Figure 4. Dry mixing in a rotary mixer.

2.3. Tests Procedure

After the sample processing period was completed and before heating in the electric oven, the weight and speed of the ultrasonic pulse were calculated. Using an electric oven, samples were heated at 5 °C/min to 200, 400, 600, and 800 °C. After that, the temperature was maintained at the same level for an hour to ensure that all of the samples were the same temperature.

After each cycle, the temperature was cooled to room temperature. The samples were tested for compressive, tensile, and flexural strengths, as well as mass loss and ultrasonic speed after being removed from the oven. The compressive strength test was carried out according to ASTM C109-20 [56]. The tensile strength was tested according to ASTM C496 [57]. ASTM C293 [58] was used to conduct the flexural strength test. The ultrasonic pulse velocity (UPV) was determined using ASTM C597 [59]. Figure 5 shows the compressive strength test, splitting tensile strength test, flexural strength test, electrical furnace that was used for heating the samples, ultrasonic pulse velocity test, and scanning electron microscope.

(a)

(b)

(c)

Figure 5. *Cont.*

(**d**)

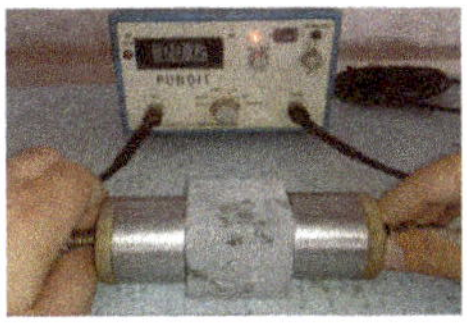

(**e**)

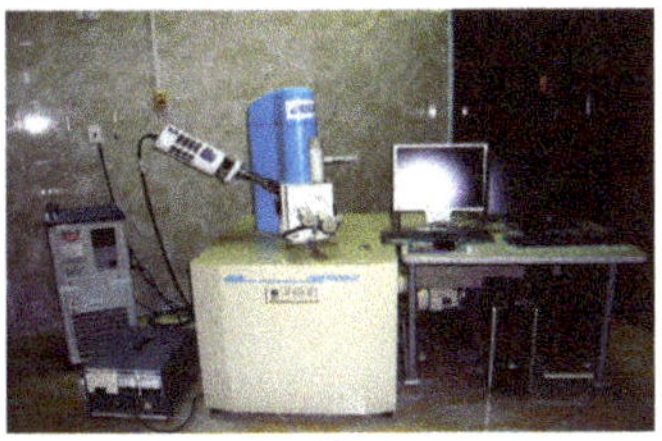

(**f**)

Figure 5. (**a**) Compressive strength test, (**b**) splitting tensile strength test, (**c**) flexural strength test, (**d**) the electrical furnace that was used for heating the samples, (**e**) ultrasonic pulse velocity test, and (**f**) scanning electron microscope.

3. Results and Discussion

3.1. Properties of Fresh SCC and Fresh SCC with Glass Fibers

SCC behavior is described, categorized, and evaluated in its early stages. The stated principles indicated in guidelines like EFNARC [59] are a method of identifying SCC behavior and determining whether it meets certain precise conditions for its use or not. Table 2 lists the slump flow, T50, L-box, and screen stability grid test method (GTM) test findings for the five SCC mixtures, as well as the EFNARC limits. Figures 6–8 show the slump flow test of SCC, L-box test of SCC, and screen stability (GTM) test of SCC, respectively. In Figures 9–12, the values of (D) reflect the largest spread diameter, T50 values represent the time needed for the concrete flowing to achieve a diameter of 0.5 m, and (H_2/H_1) is a symbol for the blocking ratio. Figures 3 and 4 show D and (H_2/H_1) in ascending order, whereas Figure 4 shows T50 in ascending order.

Table 2. Fresh properties of SCC mixtures with and without glass fibers.

Mix No.	Slump-Flow mm	T_{50} cm s	L-Box (H_2/H_1)	GTM (%)
SCC	790	2.5	0.98	11
SCC-$_{L1}$	750	4.3	0.9	8
SCC-$_{L2}$	650	5	0.8	6
SCC-$_{S1}$	760	3.4	0.93	8.7
SCC-$_{S2}$	670	4	0.85	8.3
Limit of EFNARC (2005) [60]	650–800	2–5	0.8–1.0	≤15

Table 2 contains the workability test results for SCC samples. The flow test revealed a decline in workability with the inclusion of glass fibers. Even when calculating T50 cm time, it is clear that extra time is needed for laying fresh concrete. As a result, fiber addition lowers workability and stiffens concrete. The flow value in the L-box test demonstrated a decrease in workability with the addition of fibers.

Figure 6. (a) Slump cone, (b) slump flow test of SCC with glass fibers, and (c) slump flow test of SCC without glass fibers.

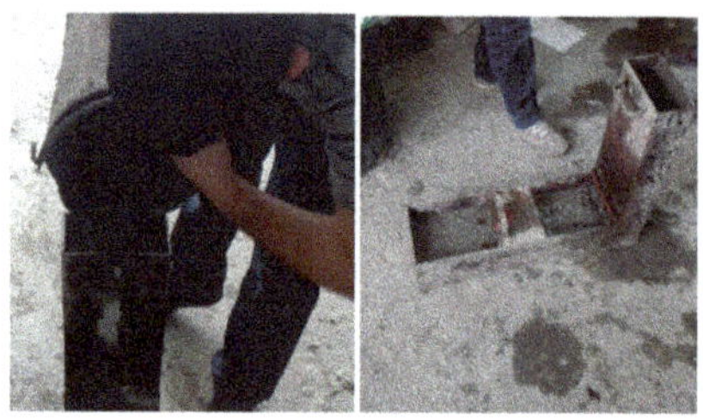

Figure 7. L-box test of SCC.

Figure 8. Screen stability (GTM) test of SCC.

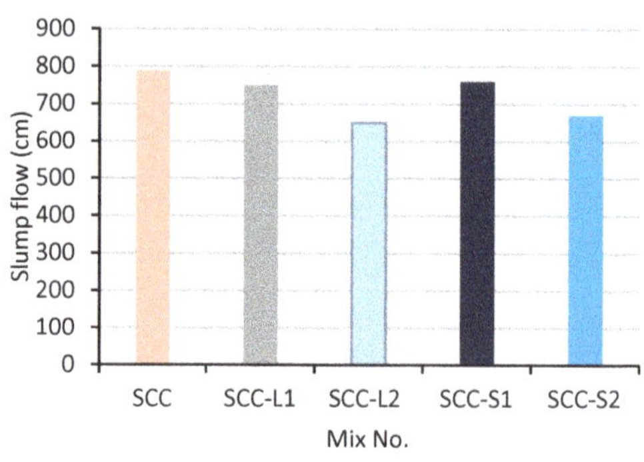

Figure 9. Diameter of slump flow (mm).

Figures 9–12 show that the SCC criteria mentioned in the EFNARC specifications are met by all of the mixtures. As a result, in terms of filling and passing ability, all of the mixtures are considered to have great homogeneity and workability. When the flow is running or stopped, there is no segmentation or bleeding.

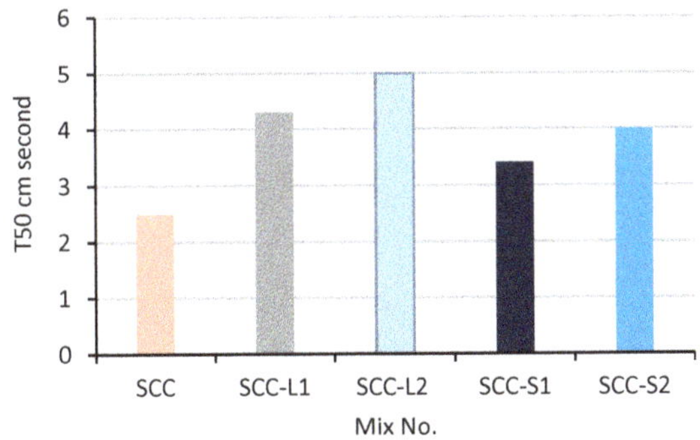

Figure 10. Circle (50 cm dia.). Time required (T_{50}).

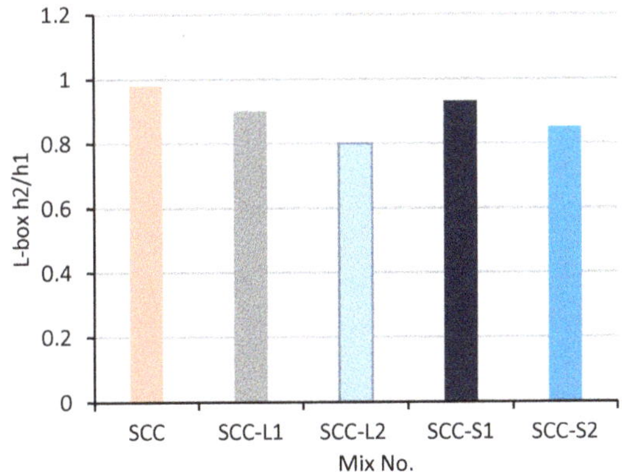

Figure 11. The effect of fiber content on the blocking ratio.

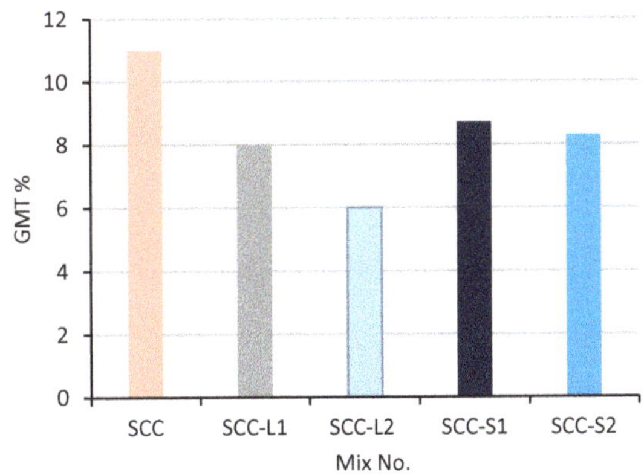

Figure 12. The effect of fiber content on the segregation ratio.

The plain SCC mixture is clearly a low-viscosity concrete, whereas the mixture with glass fibers is a high-viscosity concrete. This property is due to higher friction among aggregate grains and fibers, as well as higher viscosity due to the addition of fiber [61].

Table 2 and Figures 9–12 show that introducing 1% glass fibers with lengths of 13 and 6 mm reduces slump flow by 5.1 and 17.7%, increases T50 by 72 and 100%, reduces L-box by 8.2 and 18.4%, and reduces GTM by 27.3 and 45.5%, respectively. Introducing 0.5%

glass fibers with lengths of 13 and 6 mm reduces slump flow by 3.8 and 15.2%, increases T50 by 36 and 60%, reduces L-box by 5.1 and 13.3%, and reduces GTM by 21 and 24.5%, respectively. This reduction in workability results from the addition of glass fibers, which act as a buffer against mixture flow. The SCC's characteristics are listed in Table 2, and it can only be classified and presented as an SCC if it satisfies all the requirements. These findings are consistent with the findings of other investigators [62,63].

3.2. *Compressive Strength*

The average values of three 100 mm cubes were used to calculate the compressive strength. The compressive strength test was carried out according to ASTM C109-20 [56]. Three cubes were used to conduct this test of each mix at each step of heating. Results for the 28-day compressive strength of SCC with and without glass fibers are shown in Table 3 and Figures 13 and 14.

Table 3. All SCC mixes' compressive strength at 28 days of curing.

Mix No.	Compressive Strength at High Temp (MPa)					SD	COV
	25 °C	200 °C	400 °C	600 °C	800 °C		
SCC	50.2 (1)	50.5	30.6	17	4.6	20.255	410.282
SCC-$_{L1}$	51.8 (1)	52.3	37.3	18	6.4	20.475	419.213
SCC-$_{L2}$	52.4 (1)	56.5	45	19.4	7.2	21.652	468.79
SCC-$_{S1}$	50.4 (1)	51.1	36.6	19.3	6.2	19.678	387.217
SCC-$_{S2}$	51.0 (1)	54	41.5	19	5.8	21.007	441.288
SD	0.932	2.425	5.437	1.024	0.953		
COV	0.868	5.882	29.565	1.048	0.908		

Note: The relative compressive strength of the various SCC mixtures is shown in brackets. SD is the standard deviation, and COV is the coefficient of variation.

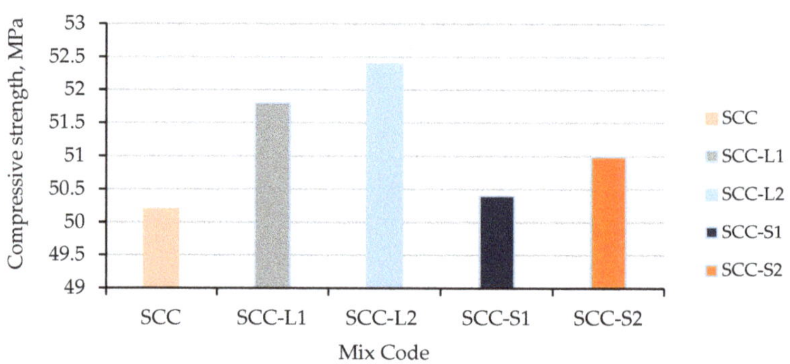

Figure 13. Compressive strength of SCC mixes is affected by fiber dose at room temperatures.

The compressive strength values for SCC specimens with and without glass fibers at room temperature are shown in Figure 13. SCC without fibers at room temperature achieved a compressive strength value of 50.2 MPa after 28 days. Samples of SCC with glass fibers showed an increase in compressive strength at room temperature. In comparison to control mixes, the compressive strength of SCC containing 1% glass fibers with lengths of 13 and 6 mm increased by about 3.2 and 4.4%, respectively, while that of SCC containing 0.5% glass fibers with lengths of 13 and 6 mm increased by about 0.4 and 1.6%, respectively. The bridging of a pre-existing crack in the interfacial transition zone (ITZ) and the addition of fiber, which improves the interface between the hardened paste and the aggregate, could both be responsible for this increase in strength [30]. The findings of this study are consistent with those of previous studies [30,64,65]. Progressive failure occurred in SCC samples with fibers, and the fibers linked the segments. In SCC samples without fibers, failure occurs quickly, and the cube breaks apart into several pieces. The failure pattern is shown in Figure 15.

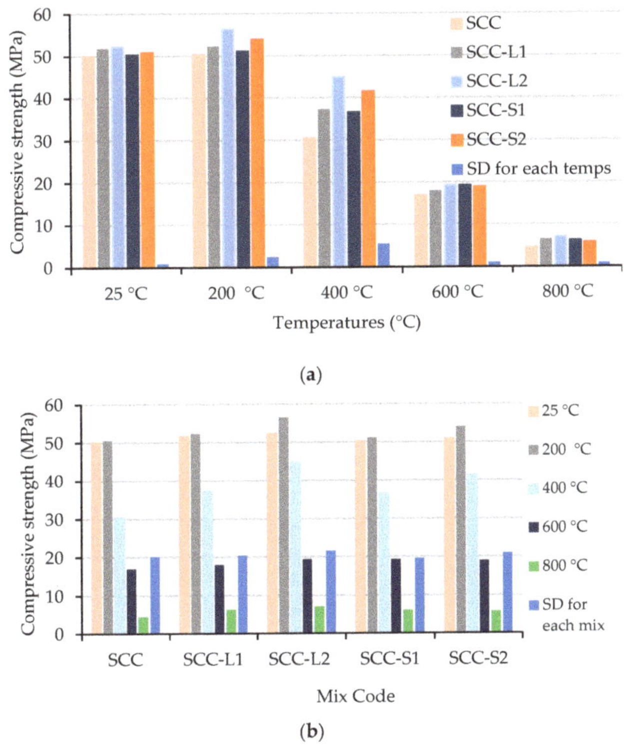

Figure 14. SCC mixtures' compressive strength at high temperature and SD (**a**) for each temperature and (**b**) for each mix.

Figure 15. Failure patterns of SCC samples. (**a**) Specimens without fibers. (**b**) Specimens with fibers.

The compressive strength of all the SCC samples increased when they were heated up to 200 °C, reaching about 0.6% for samples without glass fibers, 1 and 7.8% for samples containing 1% glass fibers with lengths of 13 and 6 mm, and 1.4 and 5.9% for samples containing 0.5% glass fibers with lengths of 13 and 6 mm, respectively. Significant changes in the compressive strength of SCC specimens were prevented at up to 200 °C by the subsequent hydration of unhydrated cement particles as a result of an internal autoclaving effect brought on by high temperature and water evaporation [66,67]. All SCC samples showed an initial rise in compressive strength at about 200 °C followed by a fall at 400 °C. A diminishing trend was seen after increasing the temperature from 200 to 600 °C, especially for the control SCC sample. The physical and mechanical properties of the SCC samples with and without fibers significantly degraded at 300 °C. In samples heated to 600 °C, mechanical characteristics were linked to physical characteristics (appearance of cracking). Comparing samples that were left at room temperature to those that were heated to 400 and 600 °C, the compressive strength values of plain SCC, SCC-$_{L1}$, SCC-$_{L2}$, SCC-$_{S1}$, and SCC-$_{S2}$ decreased by approximately 39 and 66%, 28 and 65%, 27.4 and 61.7%, 14 and 63%, and 18 and 62.7%, respectively. When the temperature reached 800 °C, the compressive strength of each specimen rapidly decreased, falling to 4.6, 6.4, 7.2, 6.2, and 5.8 MPa for the plain SCC, SCC-$_{L1}$, SCC-$_{L2}$, SCC-$_{S1}$, and SCC-$_{S2}$ samples, respectively. This means that at 800 °C, the samples only retained 9.2, 12.4, 13.7, 12.3, and 11.4%, respectively, of their original strength. The loss of SCC compressive strength is related to the collapse of the interface link induced by incompatible volume changes among aggregate and cement paste throughout heating and cooling. Calcium hydroxide dehydrates as the temperature exceeds (400 °C), leading the cement paste to expand. Most quartz-like aggregates achieve a crystalline transition at about 600 °C, generating substantial concrete expansion and fracturing [36]. Additionally, at 600 and 800 °C, the mechanical characteristics of all samples continued to deteriorate due to the heat decomposition of cement components. Above 350 °C, the compressive strength of lime dramatically decreased due to a substantial expansion that occurred along with the rehydration process. Additionally, the disintegration of C-S-H gel began at about 560 °C, which led to a minor decline in compressive strength [67–69].

Comparing plain samples without fibers at 200, 400, 600, and 800 °C, compressive strength is lower in SCC-$_{L1}$ samples by about 3.6, 22, 5.9, and 39%; in SCC-$_{L2}$ samples by about 1.2, 19.6, 13.5, and 34.8% percent; in SCC-$_{S1}$ samples by about 11.9, 47, 14, and 56.5%; and in SCC-$_{S2}$ samples by about 6.9, 35.6, 11.8, 26%, respectively. This could be owing to the elevated temp altering the structure of concrete, as mentioned in the previous statement. The findings of this study are consistent with those of previous studies [30,70].

The fibers are effective until the concrete reaches 700 °C, at which point concrete becomes brittle and crumbly and loses its adhesion strength with the fibers. The degradation of cement-hydrating elements, as well as concrete expansion during heat and vapor pressure from water gel and pores, all help in the creation of cracks [71].

3.3. Spalling

Spalling is the process of removing concrete from a structure's surface [72,73]. The samples are subjected to a complete visual evaluation after being subjected to extreme temps to evaluate visible cracking and spalling on their surfaces. Figure 16 shows the surface characteristics of SCC samples with and without glass fibers at high temps. With temperatures at 200 °C, there was no visible cracking or breakage for SCC samples with and without fibers. Around 400 °C, some SCC samples exhibited some spalling on the corners and edges. Microcracks began to appear in massive amounts above 400 °C. Rapid spalling occurred in SCC samples and they were fully damaged from 600 °C to 800 °C (see Figure 17); the findings of this study confirm Kanema's findings [74]. Low permeability in concrete is related to a dense microstructure that prevents water vapors from dispersing owing to heat, leading to higher pore pressure.

The glass fiber-reinforced self-compacting concrete (GFR-SCC) specimens spalled from 300 °C to 400 °C (Figure 17), became extensively spalled from 400 °C to 600 °C, and were fully damaged from 600 °C to 800 °C. Glass fibers cannot prevent RPC from spalling throughout a fire, according to the findings.

Intense spalling was seen on GFR-SCC samples, most probably due to decreased microfractures and glass fiber permeability. Low permeability combined with a thick SCC microstructure prevents vapor from dispersing during extreme temps, leading to pore water pressure build-up. Because of its poor permeability and thick microstructures, the mortar's bond strength deteriorates faster than other moderate and high-strength mortars. As a result, the matrix's physicochemical bonding characteristics suffer a severe loss [75]. The spalling effect caused by a rise in temperature resulted in a decrease in SCC's compressive strength [76].

Figure 16. Surface properties of SCC samples with and without glass fibers at 200, 400, 600, and 800 °C.

Figure 17. When SCC with and without glass fibers samples are heated to 450 °C, explosive spalling ensues.

Glass fibers can be used to modify pore refining in the microstructure of cement pastes at the initial phases of combustion (200 °C) [77]. The microstructure photo in Figure 18a illustrates these findings. Despite the fact that cracks appeared at 200 °C, the fibers were capable of binding to the concrete structure and offering bridging effects. Around 200 °C, the edge form and pointed ends of fibers could still be seen; however, at 400 °C, they vanished. When glass fiber is steadily heated from the ambient temperature, it becomes softer as the temperature rises. Glass fibers lose their shape as the temperature rises, changing from a stable solid to a pliable phase, as shown in Figure 18b at 400 °C

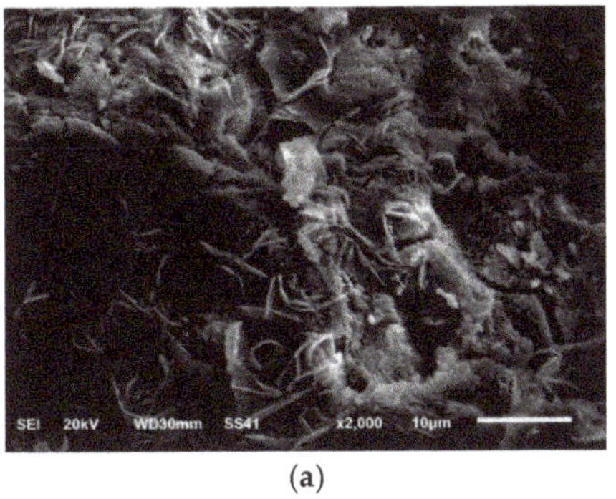

(a)

Figure 18. *Cont.*

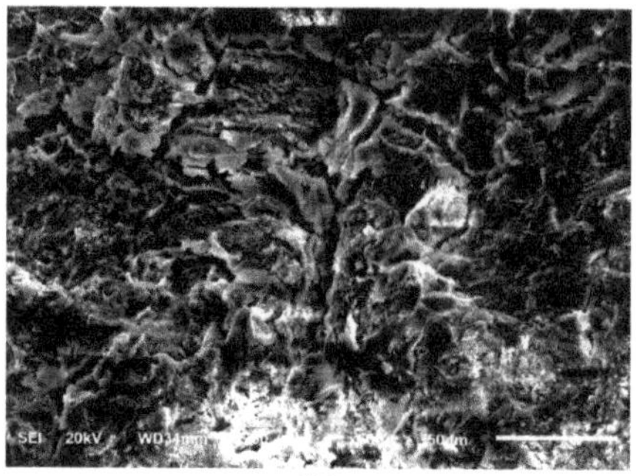

(b)

Figure 18. SEM photos of GFR-SCC specimens after they have been exposed to a temperature of (**a**) 200 °C and (**b**) 400 °C (to the right).

3.4. Splitting Tensile Strength

Table 4 displays the average results of the splitting tensile strength tests performed on three 100 mm by 200 mm cylinders. The tensile strength was tested according to ASTM C496 [57]. This test of each mixture at each stage of heating was performed using three cylinders. Since concrete is often not designed to sustain direct stress, understanding the tensile strength can aid in estimating the load which will induce fracture. Tensile splitting strength is a crucial component in crack development and, as a result, in predicting concrete durability. Crack-free concrete is necessary to sustain structural integrity and, in many circumstances, avoiding corrosion [78].

Table 4. SCC mix tensile strength at 28 days of curing.

Mix	Tensile Strength at Elevated Temp (MPa)				SD	COV
	25 °C	200 °C	400 °C	600 °C		
SCC	4.1	4.20	2.90	0.53	1.707	2.914
SCC-$_{L1}$	4.5	4.70	3.1	0.80	1.797	3.229
SCC-$_{L2}$	4.9	5.00	4.20	1.7	1.542	2.377
SCC-$_{S1}$	4.60	4.70	3.7	1.2	1.630	2.657
SCC-$_{S2}$	4.7	4.8	3.9	1.6	1.489	2.217
SD	0.297	0.295	0.546	0.503		
COV	0.088	0.087	0.298	0.253		

Table 4 and Figures 19 and 20 display the findings for the 28-day tensile strength of SCC with and without glass fibers. Figure 20 displays the data for the splitting tensile strength for SCC specimens with and without glass fibers at room temperature. After 28 days, SCC at room temperature without fibers reached a splitting tensile strength value of 4.1 MPa. The splitting tensile strength of SCC samples containing glass fibers increased at room temperature. SCC containing 1% glass fibers with lengths of 13 and 6 mm increased in splitting tensile strength by about 9.8 and 19.5%, respectively, in comparison to control mixes, while SCC containing 0.5% glass fibers with lengths of 13 and 6 mm increased by about 12.2 and 14.6%, respectively. Increased splitting tensile strength may also be due to improved uniformity from vibration-free manufacture. The findings of this study are consistent with those of previous studies [65,79]. In specimens with fibers, fibers help to hold the cylinder pieces together when they fail, preserving the integrity of the entire component and connection. Figure 15 depicts the failure pattern.

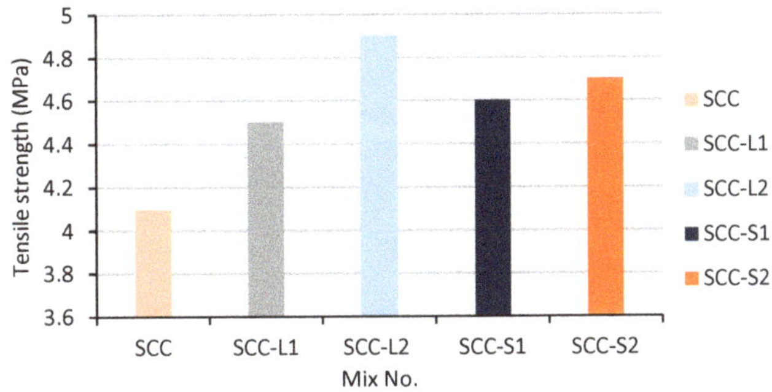

Figure 19. Fiber dosage affects the tensile strength of SCC mixes at room temperatures.

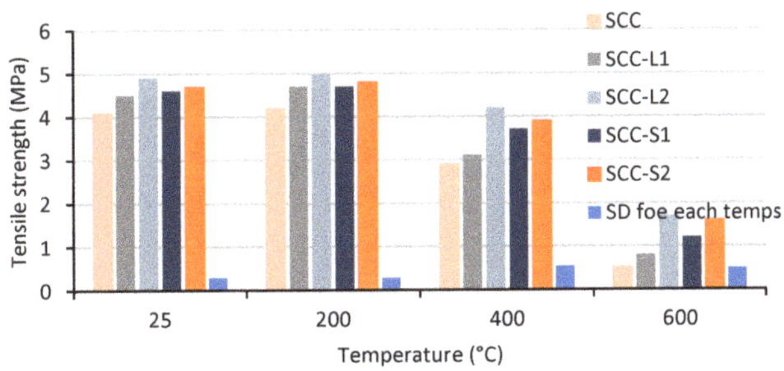

(**a**)

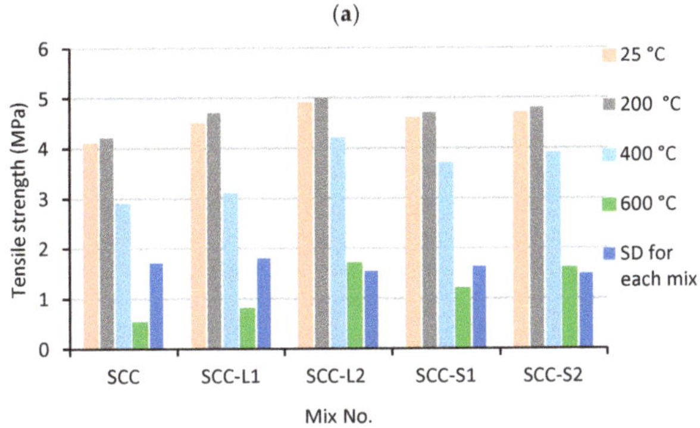

(**b**)

Figure 20. Tensile splitting strengths of SCC mixes at temperatures ranging from 25 to 800 °C and SD (**a**) for each temperature and (**b**) for each mix.

The tensile strength of each SCC sample initially increased at about 200 °C and then decreased at 400 °C. Particularly for the control SCC sample, a decreasing trend was observed after raising the temperature from 200 to 600 °C. When heated to 200 °C, the tensile strength of each SCC sample increased, reaching 2.4% for SCC samples without glass fibers, 4.4% and 2.04% for SCC samples with 1% glass fibers with lengths of 13 and 6 mm, and 2.2% and 2.1% for SCC samples with 0.5% glass fibers with lengths of 13 and 6 mm, respectively. Increased hydration of cement could be the main source of improved tensile splitting strength whenever the temperature goes up to 200 °C.

Tensile strength values of plain SCC, SCC-$_{L1}$, SCC-$_{L2}$, SCC-$_{S1}$, and SCC-$_{S2}$ decreased by roughly 29.3 and 87.1%, 31.1 and 82.2%, 19.6 and 74%, 14.3 and 65.3%, and 17 and 66%, respectively, when samples heated to 400 and 600 °C were compared to samples left at room temperature. All SCC specimens lost all of their original tensile strength as soon as the temperature reached 800 °C. The differential of thermal expansion between both the aggregates and the cement paste and between the dry cement paste and aggregate degradation could all play a role in the loss of strength [80].

The results show a larger improvement in tensile strength in SCC samples with glass fibers than in samples without fibers. The increase is around 11.9, 6.9, and 51 % for SCC-$_{L1}$ samples; 11.9, 27.6, and 126% for SCC-$_{L2}$ samples; 19, 44.8, and 220% for SCC-$_{S1}$ samples; and 14.3, 34.5, and 202% for SCC-$_{S2}$ samples, respectively, at 200, 400, and 600 °C. This might be as a result of the fibers' impact on the structure of the concrete, as demonstrated by the strength qualities [71].

3.5. Flexural Strength

Table 5 shows the average test results of three 100 mm × 100 mm × 300 mm prisms at 28 days of age. ASTM C293 [58] was used to conduct the flexural strength test. In this experiment, flexural testing was carried out on a simple concrete prism. Three prisms were used to conduct this test on each mixture at each heating step.

Table 5. SCC mix flexural strength at 28 days of curing (MPa).

Mix	At Higher Temps, Relative Flexural Strength (MPa)				SD	COV
	25 °C	200 °C	400 °C	600 °C		
SCC	6.375	6.4	2.75	0.10	3.063	9.379
SCC-$_{L1}$	8.60	8.8	4.5	0.40	3.979	15.829
SCC-$_{L2}$	9.8	10.3	7.1	1.40	4.083	16.670
SCC-$_{S1}$	8.6	9.75	6.9	0.64	4.061	16.490
SCC-$_{S2}$	9.0	9.4	7.5	1.8	3.513	12.343
SD	1.272	1.516	2.048	0.709		
COV	1.618	2.297	4.193	0.503		

The results for the 28-day flexural strength of SCC with and without glass fibers are shown in Table 5 and Figures 21 and 22. The data for the flexural strength of SCC specimens with and without glass fibers at room temperature are shown in Figure 22. The flexural strength of SCC at room temperature without fibers was 6.375 MPa after 28 days. At room temperature, the flexural strength of SCC samples containing glass fibers increased. In comparison to control mixes without fibers, the flexural strength of SCC containing 1% glass fibers with lengths of 13 and 6 mm increased by about 34.9 and 53.7%, respectively, while SCC containing 0.5% glass fibers with lengths of 13 and 6 mm increased by about 34.9 and 41.2%, respectively. The higher flexural strength of SCC may result from crack restriction by fibers. The results of this study are in agreement with those of previous studies [64,79]. When prism components in samples with fibers collapse, the fibers aid to keep them together, maintaining the integrity of the entire member and connection. The failure pattern is shown in Figure 15.

At roughly 200 °C, each SCC sample's flexural strength began to rise, and at 400 °C, it began to fall. Increasing the temperature from 200 to 600 °C resulted in a declining trend, especially for the control SCC sample. The flexural strength of each SCC sample increased when heated to 200 °C, reaching 0.4% for SCC samples without glass fibers, 2.3% and 5.1% for SCC samples with 1% glass fibers with lengths of 13 and 6 mm, and 13.3 % and 4.4% for SCC samples with 0.5% glass fibers with lengths of 13 and 6 mm, respectively. The key factor contributing to enhanced flexural strength when the temperature reaches 200 °C may be an increased hydration process. SCC samples lose their flexural strength at high temperatures. When samples heated to 400 and 600 °C were compared to samples left at room temperature, the flexural strength values of plain SCC, SCC-$_{L1}$, SCC-$_{L2}$, SCC-$_{S1}$, and SCC-$_{S2}$ decreased by approximately 56.9 and 98.4%, 47.7 and 95.3%, 27.6 and 85.7%,

19.7 and 92.6%, and 16.7 and 80%, respectively. All SCC specimens lost all of their original flexural strength as soon as the temperature reached 800 °C. This can be a result of the concrete structure being affected by the high temperature. Cement paste expands as a result of calcium hydroxide dehydrating at temperatures over (400 °C). Numerous micro- and macrocracks formed in the samples as a consequence of the heat incompatibility among cement paste and aggregates, lowering flexural strength [80].

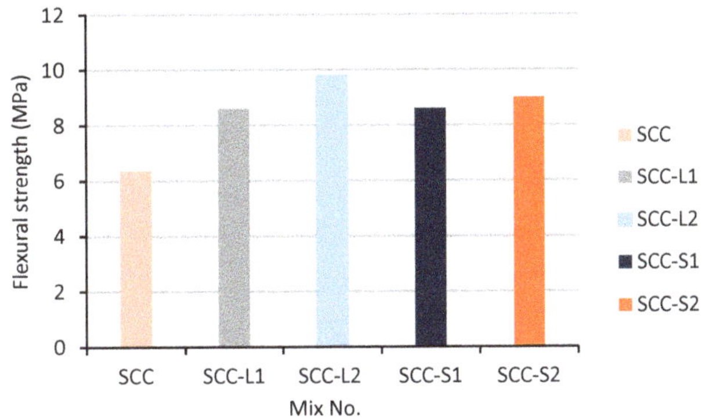

Figure 21. Fiber dosage affects the flexural strength of SCC mixes at room temperature.

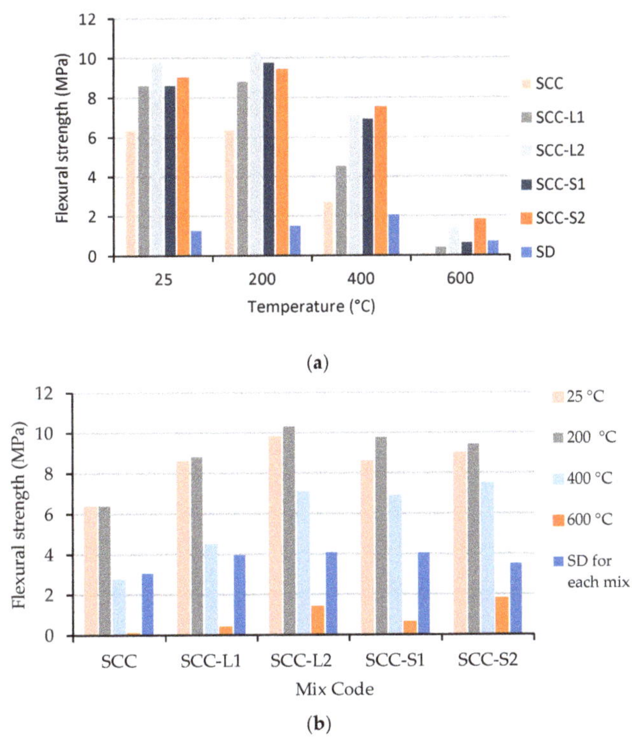

Figure 22. SCC mixes' flexural strengths at various temperatures and SD (**a**) for each temperature and (**b**) for each mix.

The findings reveal that SCC samples with glass fibers improved in flexural strength more than samples without fibers. At 200 °C, the increase is around 37.5% for SCC-L1 samples, 60.95% for SCC-$_{L2}$ samples, 52.35% for SCC-$_{S1}$ samples, and 46.95% for SCC-$_{S2}$ samples, respectively. This might be as a result of the fibers' impact on concrete's structure, as observed by its strength properties [71].

3.6. The Mass Loss Ratio

The difference between the weights before and after heating was used to calculate mass loss. Dehydration, thermal degradation of the cement's components, and spalling from the top layer were the main causes of mass loss at high temperatures [68,81,82]. High temperatures produce cracks that spread and might possibly explode. The mass loss of the examined SCC samples at high temperatures is shown in Figure 23.

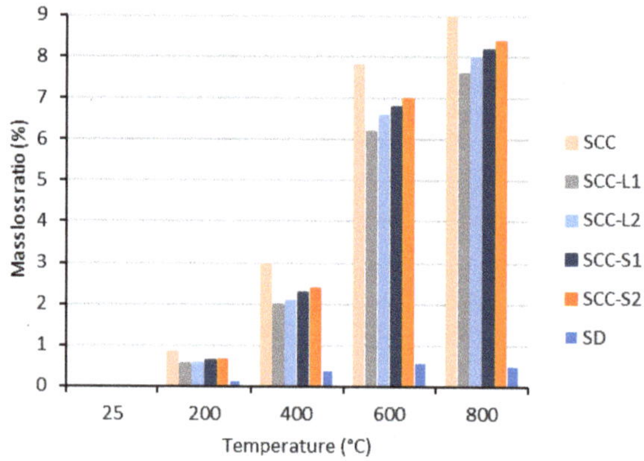

(**a**)

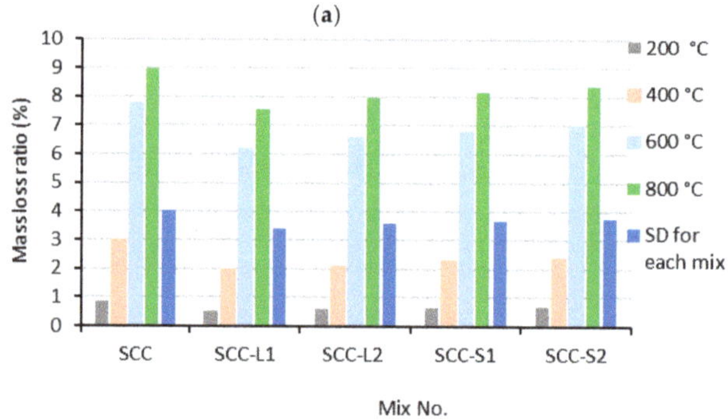

(**b**)

Figure 23. Temperature-to-mass-loss-ratio relationship and SD (**a**) for each temperature and (**b**) for each mix.

No substantial mass loss occurred for the investigated SCC specimens between 200 and 400 °C; however, the evaporation of the bound water is the primary cause of the documented low mass loss % in this temperature range [83]. However, for plain SCC, SCC-$_{L1}$, SCC-$_{L2}$, SCC-$_{S1}$, and SCC-$_{S2}$, there is a sharp increase in mass loss at 600 °C. Compared to 200 °C, their mass loss increased by more than ten times, reaching 7.8%, 6.2%, 6.6%, 6.8%, and 7.0%, respectively. By raising the temperature to 800 °C, the mass loss of plain SCC, SCC-$_{L1}$, SCC-$_{L2}$, SCC-$_{S1}$, and SCC-$_{S2}$ reached 9.0%, 7.6%, 8.0%, 8.2%, and 8.2%, respectively.

3.7. Ultrasonic Pulse Velocity (UPV)

The ultrasonic pulse velocity (UPV) was determined using ASTM C597 [59]. A UPV test was used to determine how much the SCC sample degraded when exposed to high temperatures. In the UPV test, an ultrasonic wave is transmitted through the cube specimens, and the length of time it takes for the wave to pass is measured. The uniformity and quality of the concrete improve with increasing speed, and there are less cracks and voids. The UPV was determined by dividing the width of a structure by the pulse's transit time. The quality range of the concrete sample is shown in Table 6.

Table 6. A concrete velocity-based quality evaluation criterion [84].

Pulse Velocity (km/s)	Concrete Quality Grading
Above 4.5	Excellent
3.5–4.5	Good
3.0–3.5	Medium
Below 3.0	Doubtful

The UPV results of all SCC samples exposed to various high temperatures are shown in Figure 24. Every data point is provided by an average of three SCC cube sampling tests. The SCC's UPV at room temperature without fibers was 4.42 km/s. The UPV of SCC samples with glass fibers increased at room temperature, and SCC-$_{LGF2}$ had a slightly higher UPV than other SCC mixtures. The UPV of SCC containing 1% glass fibers with lengths of 13 and 6 mm increased by approximately 8.4 and 15%, respectively, when compared to control mixes without fibers, while SCC containing 0.5% glass fibers with lengths of 13 and 6 mm increased by approximately 5.5 and 6.5%, respectively. Increased UPV readings could potentially be a result of better homogeneity brought on by vibration-free manufacturing.

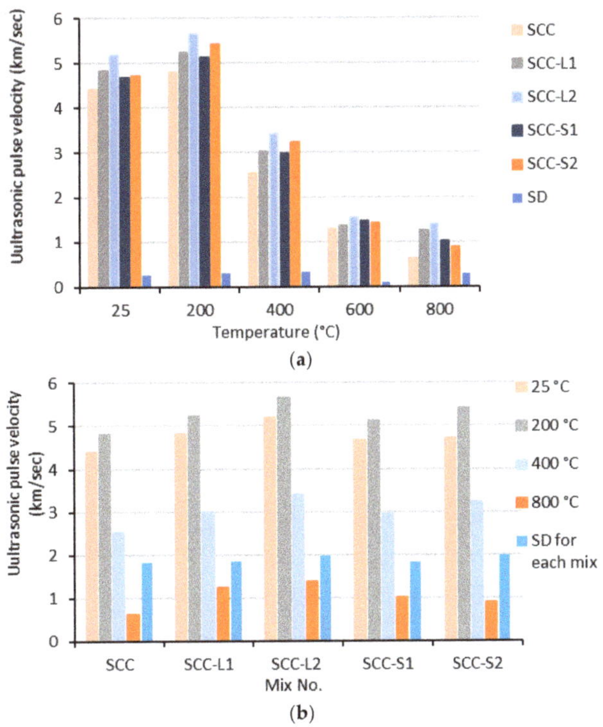

Figure 24. Relationship between UPV and temperature and SD (**a**) for each temperature and (**b**) for each mix.

The UPV of each SCC sample started to increase at around 200 °C and started to decrease at about 400 °C. A diminishing trend was seen when the temperature was raised from 200 to 800 °C, especially for the control SCC sample. When heated to 200 °C, the UPV of each SCC sample rose, reaching 8.1% for plain SCC samples, 7.7% for SCC-$_{L1}$, 8.1% for SCC-$_{L2}$, 8.6% for SCC-$_{S1}$, and 12.8% for SCC-$_{S2}$, respectively. When the temperature hits 200 °C, an accelerated hydration process may be the main factor enhancing UPV. High temperatures cause SCC samples to lose their UPV. The UPV values of plain SCC, SCC-$_{L1}$, SCC-$_{L2}$, SCC-$_{S1}$, and SCC-$_{S2}$ decreased by approximately 42.5 and 70.7%, 37.4 and 71.6%, 34.2 and 70.3%, 36.6 and 68.6%, and 31.4 and 69.9%, respectively, when samples heated to 400 and 600 °C were compared to samples left at room temperature. The rising temperature, which causes total physical degradation of SCC specimens, may be responsible for the low UPV readings for samples. The rate of microcracks increased and the SCC quality decreased at 600 °C, causing a higher drop in UPV values than at 200 and 400 °C. Microcracking has a considerable impact on the way pulses are transmitted through concrete. As a result, as the temperature rises, a crucial signal of material fracture is a decreasing pulse speed. Due to the high temperatures, thermal expansion and drying of the concrete may result in fissures forming. The pulse velocity of the SCC samples increases as a result of the fractures or micro-pathways that lead to more cracks. Microcracks, consequently, result in a slowing of the pulse and low UPV levels [84].

As compared to SCC samples without fibers, the results indicate a greater improvement in UPV in SCC samples with glass fibers. At 200, 400, 600, and 800 °C, the increase is approximately 8, 15, 6, and 11.3% for SCC-$_{L1}$ samples; 15.8, 26.7, 14.6, and 21.5% for SCC-$_{L2}$ samples; 5.1, 16.3, 12, and 9.2% for SCC-$_{S1}$ samples; and 48, 53.6, 36.7, and 27.9% for SCC-$_{S2}$ samples. The strength characteristics of the concrete, which are supported by the fibers' influence on its structural makeup [71], may be the cause of this.

4. Effects or Spin-Offs of the Study in Practice

In practical applications, according to the results of the study, self-compacting glass fiber concrete should be used in construction projects when subjected to elevated temperatures of 200 °C. The addition of silica fume and glass fibers enhanced the concrete's performance for both SCC samples with and without glass fibers and remained intact at 200 °C.

5. Conclusions

Several experiments were carried out throughout this work to analyze variations in the mechanical properties of SCC samples exposed to high temperatures up to 800 °C and to investigate the effects of adding glass fibers on the SCC's fresh and hardening properties. The primary conclusions that can be drawn from the results of this study are as follows:

i. In terms of filling and passing ability, all SCC combinations are regarded as having acceptable consistency and workability. The addition of glass fiber to SCC diminishes workability but it does so within the permissible limit of EFNARC. There is neither bleeding nor segregation while the flow is running or stopped.

ii. When glass fibers were added to SCC samples, their mechanical strength rose. This improvement in mechanical properties can be attributed to fiber influence, which successfully inhibits the development of cracks by forming a strong relationship between the fibers and matrix. The SCC's mechanical strength was increased as a result of the SCC's improved energy absorption capacity. Additionally, fibers can join micro- and macrofractures, delaying the spread of significant fractures and switching the failure mode from brittle to flexible.

iii. Glass fibers were added to SCC samples to improve their mechanical strength, and as a result, the improvements in SCC samples with glass fibers are greater than in those samples without fibers. The influence of fibers, which successfully prevent crack development by forming a strong bond between the fibers and matrix, can be attributed to this improvement in mechanical qualities. As a result, the

SCC's capacity to absorb energy was improved, increasing its mechanical strength. Additionally, fibers have the ability to join micro- and macrofractures, delaying the spread of significant fractures and changing the failure mode from brittle to flexible.

iv. The mechanical properties of specimens containing glass fibers increased at up to 200 °C but then declined at 400 °C, whereas the SCC-L2 mixture exhibited better mechanical properties.

v. SCC samples with and without fibers did not break or show any signs of visual cracking at 200 °C. Some SCC samples showed some corner and edge spalling at temperatures of about 400 °C. Above 400 °C, a large number of microcracks started to develop. Between 600 and 800 °C, SCC samples experienced rapid spalling and complete destruction.

vi. The GFR-SCC specimens cracked between 300 °C and 400 °C, spalled significantly between 400 °C and 600 °C, and were completely destroyed between 600° C and 800°C. According to the results, glass fibers cannot stop RPC from spalling over the course of a fire.

vii. For the examined SCC specimens, there was no significant mass loss between 200 and 400 °C. At 600 °C, however, there is a rapid rise in mass loss, and this mass loss increased by more than ten times when compared to 200 °C.

viii. At room temperature, the UPV of SCC samples with glass fibers grew, and SCC-$_{L2}$ exhibited a somewhat higher UPV than other SCC mixes. The results show that at 200, 400, 600, and 800 °C, SCC samples containing glass fibers exhibit a larger improvement in UPV than SCC samples without fibers. At about 200 °C, the UPV of each SCC sample began to rise, then at roughly 400 °C, it began to fall. When the temperature was increased from 200 to 800 °C, UPV decreased, especially for the control SCC sample.

6. Recommendation for Future Work

Regarding the effect of elevated temperature on GFR-SCC, the following points should be researched.

i. The effects of glass fibers on the mechanical and thermal properties of SCC exposed to varying periods of high temperatures.

ii. The effect of adding carbon fibers to SCC that has been exposed to high temperatures on mechanical and thermal properties.

Author Contributions: Conceptualization, H.K.S. and A.A.A.N.; methodology, H.K.S.; software, H.K.S.; validation, A.A.A.N., H.K.S. and G.F.H.; formal analysis, H.K.S.; investigation, A.A.A.N.; resources, H.K.S.; data curation, H.K.S.; writing—original draft preparation, H.K.S.; writing—review and editing, G.F.H.; visualization, A.A.A.N.; supervision, G.F.H.; project administration, A.A.A.N.; funding acquisition, G.F.H. All authors have read and agreed to the published version of the manuscript.

Funding: This research received no external funding.

Institutional Review Board Statement: Not applicable.

Informed Consent Statement: Not applicable.

Data Availability Statement: The original contributions presented in the study are included in the article, further inquiries can be directed to the corresponding author.

Conflicts of Interest: The authors declare no conflicts of interest.

Abbreviations

SCC	self-compacting concrete
GFR-SCC	glass fiber-reinforced self-compacting concrete
SCC_{-L1}	mixture contains 1% glass fibers of a length of 13 mm
SCC_{-L2}	mixture contains 0.5% glass fibers of a length of 13 mm
SCC_{-S1}	mixture contains 1% glass fibers of a length of 6 mm
SCC_{-S2}	mixture contains 0.5% glass fibers of a length of 6 mm

References

1. Ahmed, G.H.; Ahmed, H.; Ali, B. Alyousef R Assessment of High Performance Self-Consolidating Concrete through an Experimental and Analytical Multi-Parameter Approach. *Materials* **2021**, *14*, 985. [CrossRef] [PubMed]
2. Geiker, M.; Jacobsen, S. Self-compacting concrete (SCC). In *Developments in the Formulation and Reinforcement of Concrete*, 2nd ed.; Mindess, S., Ed.; Woodhead: Vancouver, BC, Canada, 2020; Volume 10, pp. 229–256.
3. Anıl, N. Mechanical Properties of Steel Fiber Reinforced Self-Compacting Concrete. *Int. J. Eng. Technol.-IJET* **2018**, *4*, 2018.
4. Megid, W.A.; Khayat, K.H. Effect of concrete rheological properties on quality of formed surfaces cast with self-consolidating concrete and superworkable concrete. *Cem. Concr. Compos.* **2018**, *93*, 75–84. [CrossRef]
5. El-Chabib, H.; Syed, A. Properties of self-consolidating concrete made with high volumes of supplementary cementitious materials. *J. Mater. Civ. Eng.* **2013**, *25*, 1579–1586. [CrossRef]
6. Sideris, K.K.; Tassos, C.; Chatzopoulos, A.; Manita, P. Mechanical characteristics and durability of self-compacting concretes produced with ladle furnace slag. *Constr. Build. Mater.* **2018**, *170*, 660–667. [CrossRef]
7. Zinkaah, O.H.; Sultan, H.K.; Al-Rifaie, A.; Alridha, Z. Influence of Strut Geometry on the Size Effect of FRP Reinforced Simply Supported Deep Beams: A Theoretical Analysis. *Math. Model. Eng. Probl.* **2022**, *9*, 411–417. [CrossRef]
8. Aziz, H.Y.; Sultan, H.K.; Abbas, B.J. Simulation and Style Design of Bridge Stability Supported on Large Diameter Piles. *Math. Model. Eng. Probl.* **2021**, *8*, 961–966. [CrossRef]
9. Łázniewska-Piekarczyk, B. Investigations on the relationship between porosity and strength of admixtures modified high performance self-compacting concrete. *J. Civ. Eng. Manag.* **2016**, *22*, 520–528. [CrossRef]
10. Huseien, G.F.; Sam, A.R.M.; Alyousef, R. Texture, morphology and strength performance of self-compacting alkali-activated concrete: Role of fly ash as GBFS replacement. *Constr. Build. Mater.* **2020**, *270*, 121368. [CrossRef]
11. Sultan, H.K. Designing reinforced HSC rectangular beams using optimization techniques. *Pollack Period. J.* **2023**, *18*, 20–25. [CrossRef]
12. Sultan, H.K.; Abbas, B.J.; Al Khuzaie, H.M.A.; Alsheakayree, T.K.Q. Designing High Strength Concrete Grade T-Beams at the Lowest Possible Cost. *Math. Model. Eng. Probl.* **2023**, *10*, 1369–1376. [CrossRef]
13. Hameed, A.A.; Al-Sherrawi, M.H. Influence of Steel Fiber on the Shear Strength of a Concrete Beam. *Civ. Eng. J.* **2018**, *4*, 1501–1509. [CrossRef]
14. Faridmehr, I.; Shariq, M.; Plevris, V.; Aalimahmoody, N. Novel hybrid informational model for predicting the creep and shrinkage deflection of reinforced concrete beams containing GGBFS. *Neural Comput. Appl.* **2022**, *34*, 13107–13123. [CrossRef]
15. Sultan, H.K.; Mohammed, A.T. Assessing of the common strengthening methods for existing RC buildings. *Pollack Period. J.* **2023**, *18*, 6–11. [CrossRef]
16. Kumar, P.; Kumar, R.; Gupta, K. Study on Normally Vibrated Concrete to Self-Compacting Concrete. *J. Ceram. Concr. Technol.* **2016**, *1*, 1–18.
17. Sultan, H.K.; Mohammed, A.T.; Qasim, O.D.; Maula, B.H.; Aziz, H.Y. Ductility Factor Evaluation of Concrete Moment Frame Retrofitted by FRP Subjected to Seismic Loads. *Int. Rev. Civ. Eng. (I.RE.C.E.)* **2020**, *11*, 275–282. [CrossRef]
18. Sultan, H.K.; Huseien, G.F. Minimum Shear Reinforcement for Reactive Powder Concrete Beams. *ENG* **2024**, *5*, 801–818. [CrossRef]
19. Alyousef, R.; Benjeddou, O.; Khadimallah, M.A.; Mohamed, A.M.; Soussi, C. Study of the Effects of Marble Powder Amount on the Self-Compacting Concretes Properties by Microstructure Analysis on Cement-Marble Powder Pastes. *Adv. Civ. Eng.* **2018**, *2018*, 6018613. [CrossRef]
20. Domone, P.L. A review of the hardened mechanical properties of self-compacting concrete. *Cem. Concr. Compos.* **2007**, *29*, 1–12. [CrossRef]
21. Foroughi-Asl, A.; Dilmaghani, S.; Famili, H. Bond strength of reinforcement steel in self-compacting concrete. *Int. J. Civ. Eng.* **2008**, *6*, 24–33.
22. Kang, S.-H.; Hong, S.-G.; Moon, J. Shrinkage characteristics of heat-treated ultra-high performance concrete and its mitigation using superabsorbent polymer based internal curing method. *Cem. Concr. Compos.* **2018**, *89*, 130–138. [CrossRef]
23. Zhutovsky, S.; Kovler, K. Influence of water to cement ratio on the efficiency of internal curing of high-performance concrete. *Constr. Build. Mater.* **2017**, *144*, 311–316. [CrossRef]
24. Alyousef, R.; Khadimallah, M.A.; Soussi, C.; Benjeddou, O.; Jedidi, M. Experimental and Theoretical Study of a New Technique for Mixing Self-Compacting Concrete with Marble Sludge Grout. *Adv. Civ. Eng.* **2018**, *2018*, 3283451. [CrossRef]
25. Alabduljabbar, H.; Alyousef, R.; Alrshoudi, F.; Alaskar, A.; Fathi, A.; Mohamed, A.M. Mechanical effect of steel fiber on the cement replacement materials of self-compacting concrete. *Fibers* **2019**, *7*, 36. [CrossRef]

26. Kannan, V.; Jerin, C.; Murali, D.K. A Review on Self Compacting Concrete. *Int. J. Adv. Res. Eng. Manag.* **2015**, *1*, 64–68.
27. Nadeem, A.; Memon, S.A.; Lo, T.Y. The performance of fly ash and metakaolin concrete at elevated temperatures. *Constr. Build. Mater.* **2014**, *62*, 67–76. [CrossRef]
28. Memon, S.A.; Shah, S.F.A.; Khushnood, R.A.; Baloch, W.L. Durability of sustainable concrete subjected to elevated temperature–A review. *Constr. Build. Mater.* **2019**, *199*, 435–455. [CrossRef]
29. Alyousef, R.; Benjeddou, O.; Soussi, C.; Khadimallah, M.A.; Mohamed, A.M. Effects of Incorporation of Marble Powder Obtained by Recycling Waste Sludge and Limestone Powder on Rheology, Compressive Strength, and Durability of Self-Compacting Concrete. *Adv. Mater. Sci. Eng.* **2019**, *2019*, 4609353. [CrossRef]
30. Ahmad, S.; Umar, A.; Masood, A. Properties of normal concrete, self-compacting concrete and glass fibre-reinforced self-compacting concrete: An experimental study. *Procedia Eng.* **2017**, *173*, 807–813. [CrossRef]
31. Hake, S.L.; Shinde, S.S.; Bhandari, P.K.; Awasarmal, P.R.; Kanawade, B.D. Effect of Glass Fibers on Self-Compacting Concrete. *E3S Web Conf.* **2020**, *170*, 6018. [CrossRef]
32. Bogas, J.; Hawreen, A. Capillary Absorption and Oxygen Permeability of Concrete Reinforced with Carbon Nanotubes. *Adv. Civ. Eng. Mater.* **2019**, *8*, 307–326.
33. Hawreen, A.; Bogas, J.A.; Kurda, R. Mechanical Characterization of Concrete Reinforced with Different Types of Carbon Nanotubes. *Arab. J. Sci. Eng.* **2019**, *44*, 8361–8376. [CrossRef]
34. Abu Maraq, M.A.; Tayeh, B.A.; Ziara, M.M.; Alyousef, R. Flexural behavior of RC beams strengthened with steel wire mesh and self-compacting concrete jacketing—Experimental investigation and test results. *J. Mater. Res. Technol.* **2020**, *10*, 1002–1019. [CrossRef]
35. Li, Y.; Pimienta, P.; Pinoteau, N.; Tan, K.-H. Effect of aggregate size and inclusion of polypropylene and steel fibers on explosive spalling and pore pressure in ultra-high-performance concrete (UHPC) at elevated temperature. *Cem. Concr. Compos.* **2019**, *99*, 62–71. [CrossRef]
36. Ali, S.; Saeed, H.; Behzad, T. Residual Strength and Microstructure of Fiber Reinforced Self-Compacting Concrete Exposed to High Temperatures. *Constr. Build. Mater.* **2020**, *230*, 116969. [CrossRef]
37. Cifuentes, H.C.; Leiva, M.F.; Fernández-Pereira, C. Effects of fibers and rice husk ash on properties of heated high-strength concrete. *Mag. Concr. Res.* **2012**, *64*, 457–470. [CrossRef]
38. Zuhair, M.; Deshmukh, S.K. Effect of different fibers on compressive strength of self-compacting concrete at elevated temperature. *IOP Conf. Ser. Mater. Sci. Eng.* **2018**, *410*, 012007. [CrossRef]
39. Qasim, O.A.; Sultan, H.K. Experimental investigation of effect of steel fiber on concrete construction joints of prism. *IOP Conf. Ser. Mater. Sci. Eng.* **2020**, *745*, 012170. [CrossRef]
40. Sultan, H.K.; Alyaseri, I. Effects of elevated temperatures on mechanical properties of reactive powder concrete elements. *Constr. Build. Mater.* **2020**, *261*, 120555. [CrossRef]
41. Ombres, L.; Mazzuca, P.; Verre, S. Effects of Thermal Conditioning at High Temperatures on the Response of Concrete Elements Confined with a PBO-FRCM Composite System. *J. Mater. Civ. Eng.* **2021**, *34*, 04021413. [CrossRef]
42. Bakhtiyari, S.; Allahverdi, A.; Rais-Ghasemi, M.; Zarrabi, B.A.; Parhizkar, T. Self-compacting concrete containing different powders at elevated temperatures–Mechanical properties and changes in the phase composition of the paste. *Thermochim. Acta* **2011**, *514*, 74–81. [CrossRef]
43. Vanwalleghem, H.; Blontrock, H.; Taerwe, L. Spalling tests on self-compacting concrete. In Proceedings of the International RILEM Symposium on Self-Compacting Concrete, Proceedings PRO, Reykjavik, Iceland, 17–20 August 2003.
44. Sultan, H.K.; Zinkaah, O.H.; Rasheed, A.A.; Alridha, Z.; Alhawat, M. Producing Sustainable Modified Reactive Powder Concrete Using Locally Available Materials. *Innov. Infrastruct. Solut.* **2022**, *7*, 342. [CrossRef]
45. Noumowé, A.; Carré, H.; Daoud, A.; Toutanji, H. High-strength self-compacting concrete exposed to fire test. *J. Mater. Civ. Eng.* **2006**, *18*, 754–758. [CrossRef]
46. Haddad, R.H.; Odeh, R.A.; Amawi, H.A.; Ababneh, A.N. Thermal performance of self-compacting concrete: Destructive and nondestructive evaluation. *Can. J. Civ. Eng.* **2013**, *40*, 1205–1214. [CrossRef]
47. Persson, B. Fire resistance of self-compacting concrete, SCC. *Mater. Struct.* **2004**, *37*, 575–584. [CrossRef]
48. Subhan, A.; Arshad, U. Fibre-reinforced Self Compacting Concrete: A Review. *IOP Conf. Ser. Mater. Sci. Eng.* **2018**, *377*, 012117. [CrossRef]
49. Farhad, A.; Jack, K. Assessment and development of high-performance fibre reinforced lightweight self-compacting concrete including recycled crumb rubber aggregates exposed to elevated temperatures F. *Aslani J. Kelin/J. Clean. Prod.* **2018**, *200*, 1009–1025.
50. Asce, M.; Junbo, S.; Guanqi, H. Mechanical Behavior of Fiber Reinforced Self-Compacting Rubberized Concrete Exposed to Elevated Temperatures. *J. Mater. Civ. Eng.* **2019**, *31*, 04019302.
51. Mahapatra, C.K.; Barai, S.V. Temperature impact on residual properties of self-compacting based hybrid fiber reinforced concrete with fly ash and colloidal nano silica. *Constr. Build. Mater.* **2019**, *198*, 120–132. [CrossRef]
52. Yahy AL-Radi, H.H.; Dejian, S.; Sultan, H.K. Performance of Fiber Self-Compacting Concrete at High Temperatures. *Civ. Eng. J.* **2021**, *7*, 2083–2098. [CrossRef]
53. Venkadachalam, G.; Kumar, M.D. Behaviour of Self Compacted Concrete Produced with Steel Fiber, Glass fiber and Polypropylene Fiber Additives Subjected To High Temperature. *IOP Conf. Ser. Mater. Sci. Eng.* **2020**, *981*, 042088. [CrossRef]

54. Pathak, N.; Siddique, R. Properties of self-compacting-concrete containing fly ash subjected to elevated temperatures. *Constr. Build. Mater.* **2012**, *30*, 274–280. [CrossRef]
55. *ASTM C494*; Standard Specification for Chemical Admixtures for Concrete. ASTM: West Conshohocken, PA, USA, 2017.
56. *ASTM C109/C109M-20*; Standard Test Method for Compressive Strength of Hydraulic Cement Mortars. ASTM International: West Conshohocken, PA, USA, 2020.
57. *ASTM C496*; Standard Test Method for Splitting Tensile Strength of Cylindrical Concrete Specimens; American Society for Testing and Materials Standard Practice C496. American Society for Testing and Materials: Philadelphia, PA, USA, 2017.
58. *ASTM C293*; Standard Test Method for Flexural Strength of Concrete (Using Simple Beam with Center Point Loading). ASTM International: West Conshohocken, PA, USA, 2016.
59. *ASTM C597-16*; Standard Test Method for Pulse Velocity through Concrete. ASTM International: West Conshohocken, PA, USA, 2016.
60. EFNARC. *Specification and Guidelines for Self-CompactingConcrete*; European Federation for Specialist Construction Chemicals and Concrete: Brussels, Belgium, 2005.
61. Mezzal, S.K.; Khalid, B.; Najim, K.N.; Al-Azzawi, Z. Residual Mechanical Properties of High Strength Self-Compacting Concrete with Reused Steel Fibers at High Temperatures. *J. Green Eng. (JGE)* **2020**, *10*, 14.
62. Qasim, O.A. Ahmed AS High temperature effect on shear transfer strength of steel fiber reinforced self-compacting concrete. *J. Eng. Appl. Sci.* **2019**, *14*, 3158–3174. [CrossRef]
63. Kumar, S.L.; Manasa, V.; Harish, M. Evaluation of workability characteristics of self-compacting concrete creep behavior of self-compacting concrete. *Int. Res. J. Eng. Technol. (IRJET)* **2018**, *5*, 1193–1198.
64. Umar, A.; Al-Tamimi, A. Acritical study of the effect of viscosity modifying admixture and glass fibers on the properties of self-compacting concrete (SCC). *J. Struct. Eng.* **2011**, *38*, 153–1621.
65. Salami, F.; Shami, N. Self-compacting concrete incorporating steel and polypropylene fibers: Compressive and tensile strengths, moduli of elasticity and rupture, compressive stress—Strain curve, and energy dissipated under compressive. *Compos. Part B Eng.* **2013**, *53*, 121–133.
66. Lim, S.; Mondal, P. Effects of Nanosilica Addition on Increased Thermal Stability of Cement-Based Composite. *ACI Mater. J.* **2015**, *112*, 305–315. [CrossRef]
67. Ma, Q.; Guo, R.; Zhao, Z.; Lin, Z.; He, K. Mechanical properties of concrete at high temperature—A review. *Constr. Build. Mater.* **2015**, *93*, 371–383. [CrossRef]
68. Wang, G.; Zhang, C.; Zhang, B.; Li, Q.; Shui, Z. Study on the high-temperature behavior and rehydration characteristics of hardened cement paste. *Fire Mater.* **2015**, *39*, 741–750. [CrossRef]
69. Arioz, O. Effects of elevated temperatures on properties of concrete. *Fire Saf. J.* **2007**, *42*, 516–522. [CrossRef]
70. Ezziane, M.; Kadri, T.; Molez, L.; Jauberthie, R.; Belhacen, A. High temperature behaviour of polypropylene fibres reinforced mortars. *Fire Saf. J.* **2015**, *71*, 324–331. [CrossRef]
71. Poon, C.; Shui, Z.; Lam, L.J. Compressive behavior of fiber reinforced high-performance concrete subjected to elevated temperatures. *J. Cem. Concr. Res.* **2004**, *34*, 2215–2222. [CrossRef]
72. Ali, M.H.; Dinkha, Y.Z.; Haido, J.H. Mechanical properties and spalling at elevated temperature of high performance concrete made with reactive and waste inert powders. *Eng. Sci. Technol. Int. J.* **2017**, *20*, 536–541. [CrossRef]
73. So, H.-S. Spalling Prevention of High Performance Concrete at High Temperatures. In *High Performance Concrete Technology and Applications*; InTech: London, UK, 2016.
74. Kanema, M. Influence des Paramètres de Formulation Etmicrostructurauxsur le Comportement à Haute Température des Bétons. Ph.D. Thesis, Université de Cergy-Pontoise, Cergy, France, 2007.
75. Parka, J.-J.; Yoob, D.-Y.; Kimb, S.; Kima, S.-W. Benefits of synthetic fibers on the residual mechanical performance of UHPFRC after exposure to ISO standard fire. *Cem. Concr. Compos.* **2019**, *104*, 103401. [CrossRef]
76. Horszczaruk, E.; Sikora, P.; Cendrowski, K.; Mijowska, E. The effect of elevated temperature on the properties of cement mortars containing nanosilica and heavyweight aggregates. *Constr. Build. Mater.* **2017**, *137*, 420–431. [CrossRef]
77. Kamali, M.; Ghahremaninezhad, A. An investigation into the hydration and microstructure of cement pastes modified with glass powders. *Constr. Build. Mater.* **2016**, *112*, 915–924. [CrossRef]
78. Behnood, A.; Ghandehari, M. Comparison of compressive and splitting tensile strength of high-strength concrete with and without polypropylene fibers heated to high temperatures. *Fire Saf. J.* **2009**, *44*, 1015–1022. [CrossRef]
79. Rathish, K.; Kumar Srikuth, K. Mechanical Characteristics of Polypropylene Fiber Reinforced Self Compacting Concrete. *Asian J. Civ. Eng. (Build. Hous.)* **2008**, *9*, 647–657.
80. Corinaldesi, V.; Moriconi, G. Characterization of self-compacting concretes prepared with different fibers and mineral additions. *Cem. Concr. Compos.* **2011**, *33*, 596–601. [CrossRef]
81. Düğenci, O.; Haktanir, T.; Altun, F. Experimental research for the effect of high temperature on the mechanical properties of steel fiber-reinforced concrete. *Constr. Build. Mater.* **2015**, *75*, 82–88. [CrossRef]
82. Yüksel, I.; Siddique, R.; Özkan, Ö. Influence of high temperature on the properties of concretes made with industrial by-products as fine aggregate replacement. *Constr. Build. Mater.* **2011**, *25*, 967–972. [CrossRef]

83. Gencel, O. Effect of elevated temperatures on mechanical properties of high-strength concrete containing varying proportions of hematite. *Fire Mater.* **2012**, *36*, 217–230. [CrossRef]
84. *IS 13311-1*; Method of Non-Destructive Testing of Concrete, Part 1: Ultrasonic Pulse Velocity [CED 2: Cement and Concrete]. Bureau of Indian Standards: New Delhi, India, 1992.

eng

MDPI

Article

Minimum Shear Reinforcement for Reactive Powder Concrete Beams

Hussein Kareem Sultan [1] and Ghasan Fahim Huseien [2,3,*]

[1] Civil Engineering Department, College of Engineering, Al-Muthanna University, Al-Muthanna, Samawah 66001, Iraq; hussein.ksz@mu.edu.iq
[2] Guangzhou Institute of Energy Conversion, Chinese Academy of Sciences, Guangzhou 510640, China
[3] Department of the Built Environment, School of Design and Environment, National University of Singapore, Singapore 117566, Singapore
* Correspondence: eng.gassan@yahoo.com; Tel.: +65-8-305-7143

Abstract: The aim of this research was to determine the minimal requirements for shear reinforcement for reactive powder concrete (RPC) rectangular cross-sectional beams with a compressive strength of 157 MPa and a steel fiber volume content of 2.0% that remained constant for all the tested beams. Additionally, the recommendations of KCI-2012 and AFGC-2013 for the design of RPC beams as well as the shear design requirements of ACI 314-2014 when applied to RPC beams were studied. Utilizing a three-dimensional finite element program, a computational model was designed for forecasting the deformations and shear strength of the examined RPC beams. Both the shear-span-to-depth relationship (a/d) and the minimal reinforcement web ratio, represented by the distance between stirrups and the diameter of the stirrup bars, are the key study parameters in this regard. According to this study's experimental findings, increasing the given reinforcement of the web ratio has little influence on both the ultimate shear strength as well as the diagonal cracking strength of the beams. Additionally, the findings demonstrated that the ACI 318-2014 maximum stirrup spacing requirement of 0.5 d can safely be extended to 0.75 d for beams that are relatively short. Compared to what ACI 318-2014 mandates, the suggestions of AFGC-2013 and KCI-2012 are more cautious and safe. According to the AFGC-2013 criteria, the mean proportion of V_{fb} to projected $V_{u,AFGC}$ is roughly 58.3%, whereas the mean proportion of vs. and V_c is just 41.7%. The deformation response and ultimate shear strength of the examined RPC beams were well predicted by the designed model using finite elements when metal fibers were taken into account.

Keywords: ACI 318-14 code; finite element model (EEM); reactive powder concrete (RPC); shear reinforcement; RPC beam

Citation: Sultan, H.K.; Huseien, G.F. Minimum Shear Reinforcement for Reactive Powder Concrete Beams. *Eng* **2024**, *5*, 801–818. https://doi.org/10.3390/eng5020043

Academic Editor: Alessio Cascardi

Received: 3 April 2024
Revised: 1 May 2024
Accepted: 6 May 2024
Published: 8 May 2024

1. Introduction

Reactive powder concrete (RPC) is a kind of concrete that was introduced as a result of the building industry's ongoing growth over the past few decades to serve a variety of structural applications, offering a considerably greater level of strength and more durability [1]. The outstanding performance of RPC is due to the microstructure improvement processes applied to cementitious materials, including excluding coarse aggregates, using super-plasticizing agents to lower the water/cement ratio, and incorporating fibers and cementitious components like silica fume [2–4]. RPC is being employed more and more in applications in civil engineering owing to its excellent structural performance to date, particularly for high-rise, long-span structures; bridges; highways; municipal transit; etc. [5–8]. Several studies have looked into the mechanical features [9–14], fiber dispersion features [15–20], bond efficiency [21–27], and shrinkage behaviors of RPC [28,29]. But there is limited data in the literature regarding the shear properties of RPC deep and narrow beams both without and with reinforcement of the web [30–36]. The design of RPC beams is not addressed in any of the current international codes, including Eurocode 2 (EC-2) [37] and the ACI 318-2014 Building Code [38]. As a result, there have been several inquiries into the minimal reinforcement for shear required for RPC beams. The first recommendations for design for RPC construction projects were introduced in 2002 by the France Association of Civil Engineers (AFGC)-2002 [39], and they were modified in 2013 by the AFGC-2013 [40] so as to account for the role of steel fibers in the shearing design. In 2008, the Japan Society of Civil Engineers (JSCE) [41] and, in 2012, the Korea Concrete Institute (KCI-2012) [42] also made suggestions for the design of RPC beams.

Based on that, this study focuses on reporting the results of shear tests of RPC beams provided with minimum shear reinforcement. The RPC beams' shear design was examined following the KCI-2012 and AFGC-2013 recommendations. Additionally, anumerical model for predicting the RPC beams' shear strength and deformations was created utilizing a 3-dimensional finite element program.

2. Design Guidelines for the Shear of RC Beams

2.1. Shear Design of RC Beams under ACI318-2014 Code

According to the commonly used approach for the investigated codes, the total design shear strength, V_{ud}, is calculated as the sum of the shear sustained by the shear reinforcement, V_s, and the concrete, V_c, i.e.,

$$V_{ud} = V_c + V_s \tag{1}$$

The following are the specifications for concrete's design shear strength, V_c:

$$V_c = 0.17\sqrt{f'c}\, b_w d \tag{2}$$

where b_w and d are the beam's width and depth, respectively, and $f'c$ is the concrete's cylinder compressive strength in MPa. It is possible to calculate the shear strength offered by shear reinforcement V_s by using the following formula:

$$V_s = \left(\frac{A_v}{s_v}\right) f_{yv} \cdot d \tag{3}$$

where A_v and f_{yv} represent the area and required yield strength, respectively, of vertical reinforcement on the web located within a distance between stirrups (s_v). The specified vertical web strengthening ratio ($\rho_v = A_v/b_w\, s_v$) should be greater than the minimal value ($\rho_{v,min}$) demanded by this code, which is equal to the higher of the two values $0.35/f_{yv}$ or $(0.062 f'c/f_{yv})$.

2.2. The ACI318-2014 Code for RC Beams Specifies the Minimal Amount of Shear Reinforcement

According to this code, beams are structural members whose clear spans cannot be more than four times the total component depth (h), or whose concentrated loads are located at a distance of two times the distance ($2h$) from the face of the support, and that satisfy these requirements.

The following are the minimal areas for the horizontal and vertical reinforcement of the web, ($A_{h,min}$) and ($A_{v,min}$), respectively:

$$A_{h,min} = 0.0025 b_{sh} \tag{4a}$$

$$A_{v,min} = 0.0025 b_{sv} \tag{4b}$$

where b is the beam's breadth, sh is the distance between longitudinal reinforcements that are dispersed horizontally, and sv is the distance between vertical web reinforcements. In order to prevent sh and sv from exceeding the minimum value of 300 mm or $d/5$, wherein d is the effective depths, the reinforcement of the web needs to be designed in this way.

2.3. Limits of Spacing for Shear Reinforcement

The shear reinforcement spacing limits are outlined in Section 9.7.6.2.2 of ACI 318-14 (2014) [38]. The highest permitted distance between stirrups ($s_{v,max}$) is equal to a minimum of 600 or 0.5 d mm; however, if the shear reinforcement's contribution to the shear strength is above (0.33 $\sqrt{f'c}\ b_w\ d$), $s_{v,max}$ should be decreased by half. The spacing limitations recommened by EC2 [33] are 0.75 d or 600 mm.

But the present design recommendations for RPC members do not specify the shear reinforcement spacing limitations.

2.4. Design Recommendations for RPC Beams in Regard to Shear

The following equation shows how the design shear strength V_{ud} of RPC is determined:

$$V_{ud} = V_c + V_{fb} + V_s \tag{5}$$

where V_s, V_c, and V_{fb} are the shear strengths contributed by the shear reinforcement, cement matrix, and steel fiber, respectively.

The France Association of Civil Engineers presented the initial design recommendations for RPC in 2002 (AFGC-2002) [39] and updated them in 2013 (AFGC-2013) [40]. The Korea Concrete Institute (KCI-2012) [36] has made a further design suggestion for RPC.

2.5. The France Association of Civil Engineers (AFGC-2013)

The cement matrix's design shear strength V_c is specified as follows:

$$V_c = \left(\frac{0.21}{\gamma_{cf}\gamma_E}\right) k\sqrt{f'c}b_w d \tag{6}$$

where γ_{cf} is the assumed value of 1.30 for the partial safety factor regarding fibers, k is the factor pertaining to the situation in which pre-stressing has been applied, and γ_E is a safety factor, with $\gamma_{cf}\,\gamma_E$ being equal to 1.5. The following formula can be used to compute the contribution of steel fibers, V_{fb}:

$$V_{fb} = \frac{A_{fv}\,\sigma_{Rd,f}}{tan\theta} \tag{7}$$

where θ is the angle between the primary compression stress and the beam axis, which can be taken to be at least 30°; A_{fv} is the area of the a fiber effects, which can be taken as $bw\ z$ for rectangular sections, where z = 0.9 d; and $\sigma_{Rd,f}$ is the remaining tensile strength, which can be calculated as follows:

$$\sigma_{Rd,f} = \left(\frac{1}{k\,\sigma_{cf}}\right)\left(\frac{1}{w_{lim}}\right)\int_0^{w_{lim}} \sigma_f\ (w)\ d_w \tag{8}$$

where $\sigma_f(w)$ is a function of tensile stress and fracture breadth, where w_{lim} = max (w_u, w_{max}); K is the orientation of the fiber factor, which can be considered to be equal to 1.25; and w_{max} is the widest fracture possible.

The vertical shear reinforcement's shear strength is calculated as follows:

$$V_s = \left(\frac{A_v}{s_v}\right) z\ f_{yv}{\cdot}cot\theta \tag{9}$$

2.6. Korea Concrete Institute (KCI-2012)

The cement matrix's shear strength is given as follows:

$$V_c = 0.18\ \phi_b\sqrt{f'c}\ b_w d \tag{10}$$

where b_w is the beam's breath, d is the beam's effective depth, $f'c$ is the cylinder's compressive strength, and ϕ_b is a member reduction factor of 0.77. According to the equation below, the steel fibers' shear strength (V_{fb}) is

$$V_{fb} = \phi_b \left(\frac{f_{vd}}{\tan \beta u} \right) b_w \, z \tag{11}$$

where βu is the angle between the diagonal tensile fracture plane and the axial direction of the beam and must be greater than 30°, f_{vd} is the mean splitting tensile strength design in a location perpendicular to the diagonal tensile break, and z is the span from the location affected by compressive stresses to the centroid of tension steel, which is typically equal to $d/1.15$. This formula is used to determine f_{vd}'s value:

$$f_{vd} = \left(\frac{1}{w\ v} \right) \int_0^{w_v} \phi_c \sigma_k\ (w)\ d_w = \left(\frac{1}{w\ v} \right) \int_0^{w_v} \sigma_{kd}\ (w)\ d_w \tag{12}$$

where w_v is the ultimate fracture width at the area at which outer-fiber peak stress is applied; ϕ_c is the material's reducing factor, considered to be 0.8; $\sigma_k\ (w)$ is the tensile softening curve; w_v = max (w_v, 0.3 mm); and $\sigma_d\ (w)$ is equivalent to $\phi_c\ \sigma_k\ (w)$.

The vertical shear reinforcement's shear strength is calculated as follows:

$$V_s = \phi_b \left[\frac{A_v\ f_{yv}(\ sin\alpha s + cos\alpha s\ )}{s_v} \right] d \tag{13}$$

where α_s is the angle between a beam's longitudinal direction and the shearing reinforcement, A_v is the cross-sectional area of the reinforcement, f_{yv} is its design yield strength, and s_v is its spacing.

3. Materials and Methods

3.1. Materials Used

Materials utilized for this investigation include quartz sand, silica fume, cement, water, fibers, steel reinforcing bars, and superplasticizer. In the RPC mix, Portland cement type CEM I 52.5 N with a mean compressive strength of 53.8 MPa at 28 days, a mean specific surface area of 3300 cm^2/g, and a medium diameter of particles of 15 μm was utilized. It was made from natural sand that was 0.5 mm in size and had a specific density of 2.65. In addition to silica fume powder, which has a specific gravity of 2.20, a specific surface area of 170,000 cm^2/g, and an average diameter of around 0.15 mm, crushed quartz powder with an average diameter of 1 um to 100 um, a Blaine fineness of 3100 cm^2/gm, and a specific gravity of 2.85 was also utilized. All mixes contained a superplasticizer of a new generation of polycarboxlic ether. The steel fibers employed in this investigation were a form of locally accessible hook-ended straight fiber. The steel fibers had an equivalent diameter and length of 13.0 mm and 2.0 mm, respectively. By performing a direct tension test, the fibers' yield strength and tensile strength were calculated, and the results were 550 MPa and 820 MPa, respectively. The volume fraction of steel fibers was maintained at a consistent level of 2.0% for each tested beam. The regular tap water that was used complied with the standards for concrete mixing water.

3.2. Concrete Mix Proportions and Specimen Casting

Table 1 provides a summary of the mixture ratios for 1.0 m^3 of RPC. A high-speed mixer was used to combine the mixture's contents for 10 min. Then, 75% of the water was added to the mixer after all powders and natural sand had been dry-mixed for 2 min at low speed. The mixer was stopped for 1 min after 2 min of mixing at low speed (140 ± 5 rpm). The mixture was then blended for three minutes while the remaining water and superplasticizer were added. The final step involved mixing the mixture for two minutes at high speed (285 ± 10 rpm). For the first 24 h, concrete specimens were allowed to cure at ambient temperature (21 ± 2 °C). The samples were damp-cured up until the testing day after demolding. This RPC mix's cube concrete compressive strength (f_{cu}) is 157 MPa depending on a typical of three cube specimens, and its cylinder concrete compressive strength ($f'c$) and splitting cylinder tensile strength (f_{sp}) are 144.3 MPa and 11.9 MPa, respectively. Its flexural strength (f_r) is 39.7 MPa based on 100 mm × 100 mm × 500 mm prisms.

Table 1. Ratio of RPC mix for a cubic meter (kg/m^3).

C [a], kg	FS [b], kg	QP [c], kg	SF [d], kg	W [e], kg	SP [f], kg
760	1026	228	190	144.4	30.4

a, b, c, d, e, and f represent cement, fine sand, quartz powder, silica fume, water, and superplasticizer, respectively.

The beams were put to the test in a 200-ton loading frame. Each specimen was equipped with enough instrumentation to determine the strains in the stirrups and longitudinal bars as well as the mid-span deflection. The induction and spread of cracks were noted, and measurements of the crack widths inside the shear span zone were made.

According to ASTM C109 [43], ASTM C496 [44], and ASTM C78 [45], the compressive strength of 100 mm cubic molds, splitting tensile strength of the cylinder (diameter of 100 mm × depth of 200 mm), and flexural strength of 100 mm × 100 mm × 500 mm prisms were calculated, respectively, and the average values of the three tested specimens were adopted.

3.3. Experimental Program for the RPC Beams

Information about the test specimens can be found in Table 2. The test specimens, which consist of five RPC beams composed of an identical concrete mixture, were subjected to two point loads with simple supports.

Table 2. Specifications of the RPC beams that were examined in regard to shear.

Beam	f_{cu}, MPa	$b \times h$ (mm)	a/d Ratio	Main Longitudinal Bars		Provided Stirrups (Vertical Web Reinforcement)			Requirements of ACI 318-14 for $s_{v,max}$ and $\rho_{v,min}$	
				Lower	Upper	s_v, (mm)	d_v, (mm)	ρ_v, (%)	$s_{v,max}$, (mm)	$\rho_{v,min}$ (%)
B-1	157	140 × 220	3.0	6Φ18	2Φ10	100	6	0.20	90	0.235
B-2	157	140 × 220	3.0	6Φ18	2Φ10	200	6	0.10	90	0.235
B-3	157	140 × 220	3.0	6Φ18	2Φ10	100	8	0.36	90	0.25
B-4	157	140 × 220	3.0	6Φ18	2Φ10	200	8	0.18	90	0.25
B-5	157	140 × 220	3.2	6Φ18	2Φ10	200	8	0.18	90	0.25

As presented in Figure 1, the beam specimen's cross-section had the following dimensions: a width of 140, a depth of 220 mm, and a length of 2000 mm. The appropriate a/d ratio was achieved by varying the distance (c) between the two loads while maintaining the beams' effective spans, which were 1750 mm each. Six distorted 18 mm diameter bars arranged in two layers were used as the principal longitudinal tension reinforcement of the examined beams in order to guarantee shear failure. These bars have a yield strength of 496 MPa. Two deformed bars with a yield strength (f_{yl}) of 410 MPa served as the compression reinforcement for the beams. Each bar was 10 mm in diameter. The yield strengths (f_{yv}) of stirrup bars with diameters of 6 mm and 8 mm are 330 MPa and 310 MPa, respectively.

The existing vertical reinforcement of the web was altered in the tested beams by using different stirrup spacings (s_v = 200 and 100 mm) and different bar stirrup diameters (d_v = 6 and 8 mm) so as to test the suitability of the minimal vertical reinforcement of the web required according to ACI 318-14 when applied to RPC beams. In accordance with ACI 318-14, $s_{v,max}$ is equal to the lesser of 600 mm or 0.5d, and $\rho_{v,min}$ is equivalent to the larger of $0.35/f_y$ or ($0.062\sqrt{fc'}/f_{yv}$), while $s_{v,max}$ must be decreased by half in cases where the shear strength given by shear reinforcement (V_s) is above (0.33 $\sqrt{fc'}$ bd). Table 2 compares the provided vertical web reinforcement ratio ($\rho_v = A_v/(b_w \cdot s_v)$) of the examined RPC beams with the minimum requirements ($\rho_{v,min}$) of the code, as well as the distance between stirrups (s_v) of the examined RPC beams with the highest permitted distance between stirrups ($s_{v,max}$) required by the ACI 318-2014 code. Clearly, it can be observed that the provided area of vertical web reinforcement (A_v) for all the tested beams satisfies the minimum requirements ($A_{v,min}$) of the ACI 318-2014 code. The specified stirrup spacing (s_v) of each of the examined beams, however, is significantly greater than the highest stirrup distance ($s_{v,max}$) allowed by the aforementioned code. It should be noted that the stated stirrup spacing (s_v) of all examined beams is more than double the highest stirrup spacing ($s_{v,max}$) permitted by ACI 318-2014.

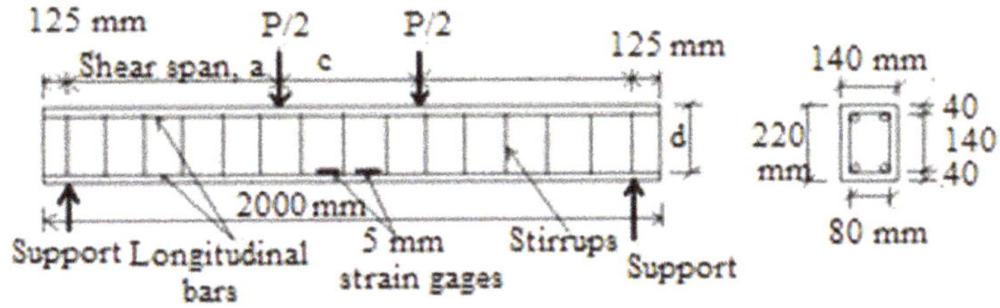

Figure 1. Details of the reinforcement employed in the RPC beam samples (depending on a/d ratio 3 or 3.2 the a: 465 or 496 mm; c: 758 or 820 mm; d: 155 mm).

4. Analysis Software and Model Calibration

Nonlinear finite element software was utilized to perform the analysis. This R/C-focused software features a wide variety of constitutive models for concrete, reinforcement, metals, and robust nonlinear solvers. Hence, it meets the requirements of the analysis. Several criteria were set to evaluate the numerical results. Primarily, it was determined that the numerical and experimental failure modes for the R/C beam should be identical. Furthermore, numerical analysis was conducted to predict parameters such as the maximum load (P_{max}) and the corresponding displacement (δ_{max}) but also the ultimate load (P_u) and the corresponding displacement (δ_u). The final parameter was the energy absorption capacity of the beam, which is equal to the area below the force–deflection response curve [44,45].

4.1. Concrete and Reinforcement Constitutive Models

Utilizing the software program ABAQUS [46], a 3-dimensional nonlinear finite element model (FEM) was utilized to forecast the overall response of the reinforced concrete beams, including displacement, stress and strain distributions, ultimate shear stresses and modes of failure, and crack patterns. SOLID C3D8R, a three-dimensional RC element that can collapse under compression and fracture under tension, was used to mimic concrete. This component is composed of eight nodes, each of which has three translational degrees of freedom (x, y, and z). Utilizing a bar element (T2D3) inside the concrete solid 65 component, the main and web reinforcement were modeled. It was presumed that the bar element was embedded into the concrete solid element.

4.2. Finite Element Modelling and Analysis Procedure

Concrete was modelled using eight-node isoparametric solid elements, with each side equal to 2 cm. Cables and reinforcement were modeled using 2-node truss elements, which were embedded in concrete elements. Load was applied as prescribed displacement, which was applied in small steps, in order to simulate the experimental procedure and avoid local failures. To evaluate the force–deflection response curves of the specimens, two monitoring points were utilized. The first was placed in the loading plate and monitored the applied load, which was essentially the reaction of the beam to the applied displacement. The second was located on the bottom surface of the R/C beam, right below the first one, and monitored the corresponding displacement. The Modified Newton–Raphson iterative scheme was applied with appropriate convergence criteria and the maximum number of iterations. Table 3 summarizes the concrete and reinforcement model calibrations.

Table 3. Summary of model calibrations [47].

		Details
Concrete	Finite element type	8-node isoparametric solid elements
	Failure mode	Fracture under tension, plasticity under compression
	Crack formulation	Smeared
	Numerical modification	Shear factor reduction
Reinforcement	Finite element type	2-node truss elements
	Constitutive model	Uniaxial multilinear law
	Modeling type	Embedded reinforcement
	Bond type	Full reinforcement–concrete bond

5. Results and Discussion

5.1. Damage and Crack Patterns

Figure 2 provides images of the examined beams that demonstrate typically observable fracture patterns and modes of failure. Except for the tested beam, B-3, which failed in a compression region in the middle of the spans prior to a shear failure, all of the examined RPC beams failed under shear.

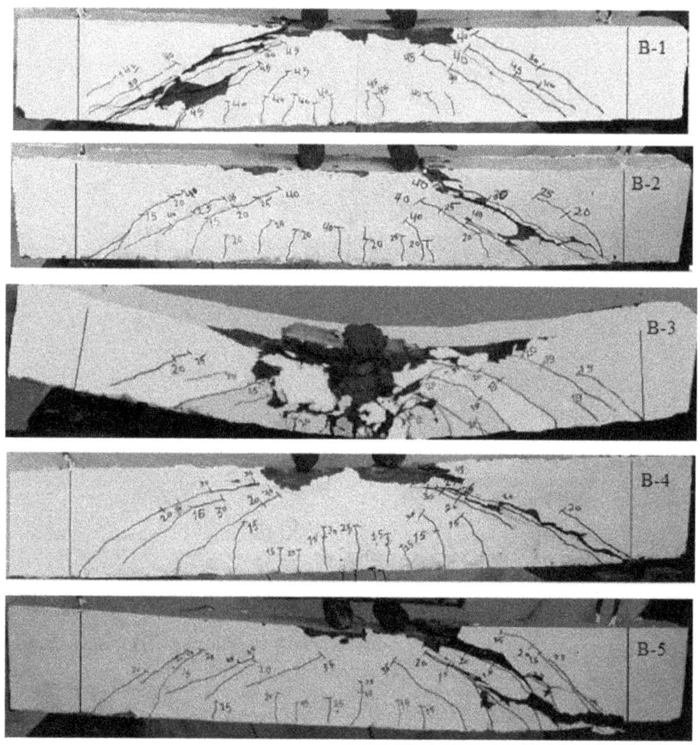

Figure 2. Images of the tested beams' shear cracks.

The ultimate strength ($P_{u,exp}$) and diagonal cracking strength (P_{cr}) of the examined beams in this study are shown in Table 4. Following the development of flexural fractures in the middle of the span, diagonal cracks often developed in the beam's two shear spans. The breadth of the flexural crack narrowed significantly after the formation of diagonal shear cracks. Typically, a diagonal shear crack began abruptly in the center of the span of a shear crack and spread toward the supports and load points as a result of an increase in the load being applied. The already-present diagonal shear cracks only slightly spread further as the load applied was increased, but a few new inclined cracks were also created. Finally, the concrete fractured abruptly across the inclined crack due to shear failure. The images demonstrate that the shear spans of the examined beams B-1, B-2, B-4, and B-5 failed due to significant concrete degradation, whereas beam B-3 fell due to crushing of the compression zone in the middle of the span with an *a/d* of 3.0. According to the test results, flexural failure may occur before shear failure when minimal reinforcement for shear is provided within the distance of 0.5*d* recommended in ACI code [34]. The yielding of shear reinforcement, however, may be seen before the yielding of flexural reinforcement and the compression failure of beams with a distance that is higher than the minimal levels required by code.

Table 4. Summary of the test findings for the shear-tested RPC beams.

Beam	*a/d*	$2V_{cr}$ (kN)	$2V_{u,exp}$ (kN)	V_{cr} (kN)	$V_{u,exp}$ (kN)	$\frac{Vcr}{Vu,exp}$	$\frac{Vu,exp}{bd\sqrt{fc}}$
B-1	3.0	130	403	65	201.5	0.323	0.638
B-2	3.0	110	369	55	184.5	0.298	0.584
B-3	3.0	170	450	85	225	0.378	0.713
B-4	3.0	120	369	60	184.5	0.325	0.584
B-5	3.2	110	310	55	155	0.355	0.491

5.2. Load–Displacement Relationships

In Figure 3, the mid-span load (*P*)–deflection curves for the examined RPC beams with various *a/d* ratios and varying vertical web reinforcement ratios (ρ_v) are displayed.

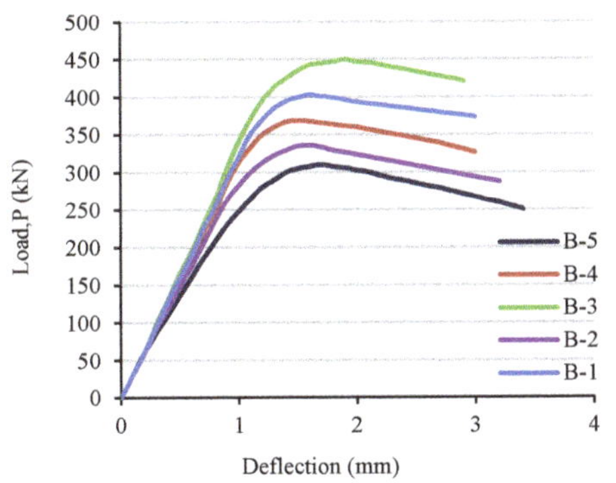

Figure 3. Mid-span load–deflection relationships for the shear-tested beams.

The beams exhibited fully elastic behavior when they were initially loaded. Overall, the upward portion of the load–deflection curves was not significantly affected when altering the tested beams' ρv% when the a/d ratio was kept constant. The deflection values of the examined beams with an equal *a/d* ratio and various ρ_v% values showed only slight variations after being subjected to the ultimate load. Figure 3 shows that beam B-2 with an *a/d* ratio of 3.0 and a supplied ρv of 0.10, the latter of which is lower than the minimum specified by the code, exhibits a stiffness like that of beam B-1, with the same a/d ratio but providing a ρ_v of 0.20. The rigidity of the examined beams significantly decreases as the a/d ratio rises. In comparison to beam B-4, which has an a/d ratio of 3.0, beam B-5, which has an a/d ratio of 3.2, is less stiff.

5.3. Strain Response

The maximum bending moment in the middle of the span of the examined beams' longitudinal steel strain revealed that the forms of tensile strain there are nearly uniform at all load levels. All of the tested beams failed before the longitudinal bars gave way. The tensile steel strain rose at an approximately constant rate. The longitudinal bars' strain values were unaffected by the development of inclined diagonal shear cracks. The longitudinal bars of the tested RPC beams' strain measurements are roughly identical for the same a/d ratio. Figure 4 compares the recorded strain in the stirrups' vertical legs in the shear span of the examined beams. As can be seen, the stirrup legs' tensile steel strain starts out very low and increases roughly at a consistent rate. Upon initiation of the diagonal shear crack, the strain rate accelerates and eventually reaches the yield level just prior to the imposed ultimate load. When the stirrup leg yields, the measured strain grows very quickly while the load being applied decreases. At the same time, the shear crack breadth increases rapidly till the concrete in the zone of shear is crushed. The measured strains of the examined beams at a similar load decreased for a similar a/d ratio after enlarging the stirrup's diameter and decreasing the distance between stirrups.

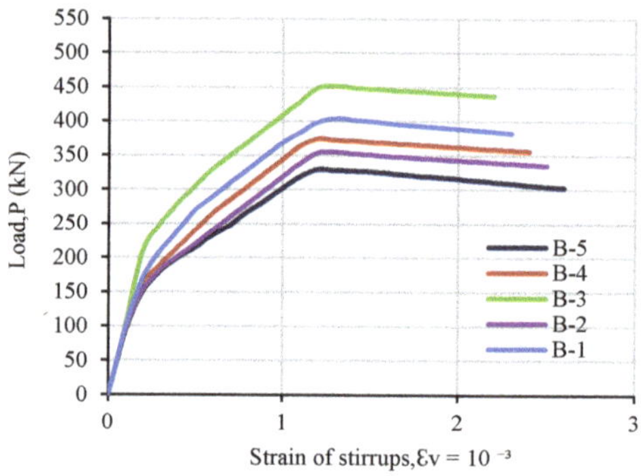

Figure 4. Relationship between the examined beams' total load and strain in the stirrups' vertical legs.

5.4. Effect of Web Reinforcement Ratio

Table 4 and Figure 5 show the impact of the given vertical reinforcements in terms of the web ratio (ρ_v%) on the examined beams' ultimate shear strength and the strength of diagonal cracks in terms of the distance between stirrups and the diameter of the stirrups' bars. It is evident that while diagonal cracks formed more slowly with a smaller distance between stirrups (s_v), they formed more quickly with a smaller diameter of the stirrup bars (d_v). The growth of diagonal cracks was significantly slower for beam B-1, with a 100 mm stirrup spacing (s_v), than it was for the comparable B-2 beam, with a 200 mm stirrup spacing (s_v). The emergence of diagonal cracking occurred more slowly in beam B-4, with an 8 mm stirrup bar diameter and a 200 mm stirrup spacing (s_v), than in beam B-2, with a 6 mm stirrup bar diameter and a 200 mm stirrup spacing (s_v). As can be observed, decreasing the distance between stirrups had a greater inhibitory effect on the growth of diagonal cracks than increasing stirrup diameter. The given stirrup spacings (s_v) for all the examined RPC beams are noticeably greater than the maximum stirrup spacing ($s_{v,max}$) stipulated in the ACI code [34]. The RPC beams strengthened using the widest possible stirrup spacing, however, exhibited good overall performance.

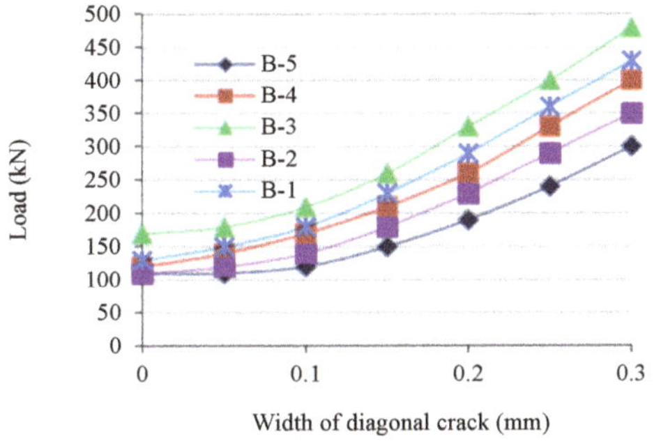

Figure 5. Development of diagonal cracks in RPC beams with various a/d ratios.

For RPC beams with relatively low heights, the $s_{v,max}$ requirements of ACI 318-2014 are not practically appropriate and can be safely adjusted to 0.75d instead of 0.50d. Table 3 shows that for the beams with an identical a/d ratio, the ultimate shear strength improves marginally as the given vertical reinforcement of the web ratio (ρ_v%) rises. It should be noticed that, in accordance with ACI 318-2014, the given reinforcement of the web ratio (ρ_v%) for beams B-1, B-2, B-4, and B-5 is lower than the minimal web reinforcement ratio ($\rho_{v,min}$). Nevertheless, excellent general performance was seen for all of the examined RPC beams that were strengthened via reinforcement of the web ratios (ρ_v%) smaller than the minimal web reinforcement ratios ($\rho_{v,min}$) stipulated by the ACI code. This shows that when used with RPC beams with 2.0% steel fibers, the minimal vertical reinforcement of the web ratio specified by the ACI code can be properly lowered.

Stirrups play a critical role in reinforced concrete (RC) beams, including those made using reactive powder concrete (RPC). The configuration of stirrups significantly affects the performance of RPC beams in terms of shear strength, ductility, crack control, flexural capacity, construction considerations, and material optimization. Proper detailing and optimization of stirrups are essential for ensuring the structural integrity and longevity of RPC beams in various applications.

5.5. Analyzing the RPC Beam Test Results in Relation to the ACI 318-2014 Code's Shear Requirements

The ACI 318-14 code's shear design formulas were employed to determine the ultimate shear strength of the examined RPC beams. It must be mentioned that Table 2 compares the supplied web reinforcements (ρ_v) of the examined beams with the minimal demands of the ACI 318-14 code, while Table 4 gives a ratio between the estimated ultimate shear strength ($V_{u,cal}$) utilizing the ACI 318-14 code and the estimated experimental ultimate shear strength ($V_{u,exp}$). As can be observed, the average value for the evaluated beams' $V_{u,cal}$ to $V_{u,exp}$ ratio according to the ACI 318-14 code is 0.403. This shows that despite every one of the examined beams having stirrups with a spacing (s_v) significantly greater than the maximum stirrup spacing ($s_{v,max}$) required by ACI 318-14, the measured readings of $V_{u,exp}$ for all of the examined beams were significantly higher than those specified by the code. This demonstrates that the shear strength estimation formulae established by ACI 318-14 are not suitable for RPC beams as they do not account for the significant role that steel fibers play in providing resistance to shear stresses. It should be noted that, in accordance with ACI 318-14, steel fibers may be utilized as the beam's reinforcement for shear if their normalized shear strength is more than 0.29 (for $f'c \leq$ MPa, $d \leq 600$ mm). For all of the examined RPC beams with a fiber percent of 2.0%, the normalized shear strength values in Table 3 are significantly higher than 0.29, having an average value of 0.60.

5.6. Comparison of Test Results for RPC Beams with AFGC-2013 and KCI-2012 Design Recommendations

Two design suggestions have been put forth, with the first stemming from AFGC-2013 and the second from KCI-2012, to account for the role that steel fibers play in the construction of RPC structures. The ultimate shear strength of this study's examined RPC beams was estimated using the shearing design techniques specified in KCI-2012 and AFGC-2013, and the results were contrasted with the experimental ultimate shearing strength shown in Tables 5 and 6.

Table 5. Comparison of the experimental findings with the maximum shear strength specified by the ACI 318-14 code.

Beam	$V_{u,exp}$ (kN)	ACI 318-14 Code			
		V_c (kN)	V_s (kN)	$V_{u,cal}$ (kN)	$\frac{Vu,cal}{Vu,exp}$
B-1	201.5	45.6	32.3	77.9	0.387
B-2	184.5	45.6	16.2	61.8	0.335
B-3	225	45.6	52.7	98.3	0.437
B-4	184.5	45.6	26.4	72	0.390
B-5	155	45.6	26.4	72	0.465

Table 6. Comparison of testing results and maximum shear strength computed using the RPC design guidelines.

Beam	$V_{u,exp}$ (kN)	KCI-2012					AFGC-2013				
		V_c (kN)	V_{fb} (kN)	V_s (kN)	$V_{u,KCI}$ (kN)	$\frac{Vu,exp}{Vu,KCI}$	V_c (kN)	V_{fb} (kN)	V_s (kN)	$V_{u,AFGC}$ (kN)	$\frac{Vu,exp}{Vu,AFGC}$
B-1	201.5	37.2	87.4	24.9	149.5	1.348	37.6	90.4	29.1	157.1	1.283
B-2	184.5	37.2	87.4	12.5	137.1	1.346	37.6	90.4	14.5	142.5	1.295
B-3	225	37.2	87.4	40.6	165.2	1.362	37.6	90.4	47.4	175.4	1.283
B-4	184.5	37.2	87.4	20.3	144.9	1.273	37.6	90.4	23.7	151.7	1.216
B-5	155	37.2	87.4	20.3	144.9	1.07	37.6	90.4	23.7	151.7	1.022

As can be observed, the average value for the examined beams' $V_{u,exp}$ and $V_{u,AFGC}$ ratios is 1.280, while the average value for their $V_{u,exp}$ and $V_{u,KCI}$ ratios is 1.220. This shows that, when used for RPC beams equipped with shear reinforcement below the minimum requirement allowed by ACI 318-14, the KCI-2012 and AFGC-2013 forecasts for the ultimate shear strength are secure and cautious. When AFGC-2013 and KCI-2012 recommendations were compared for all of the examined beams, it was found that there were only very slight variations among the projections of the ultimate shear strength. In actuality, the tiny variation in the safety parameters taken into account by each methodology is what caused the small variance in the forecasts of the two recommendations. In accordance with the KCI-2012 recommendations, Table 5 demonstrates that the mean percent of the forecast contributions of the fibers of steel (V_{fb}) in comparison to the forecast ultimate shear strength of the examined beams ($V_{u,\mathrm{KCI}}$) is roughly 59.2%, while the mean percent of the forecast contributions of the shear reinforcement (V_s) and concrete (V_c) is only 40.8%. According to the AFGC-2013 criteria, the mean proportion of V_{fb} to projected $V_{u,AFGC}$ is roughly 58.3%, whereas the mean proportion of V_s to V_c is just 41.7%.

5.7. RPC Beam Analytical Modeling Utilizing Finite Element Software

The RPC was represented using a mesh of eight-node linear 3D-brick solid elements with reduced integration (C3D8R) based on their capability to simulate the physical behavior of RPC beams [48,49] with possible nonlinearities emanating from the plastic behavior of RPC under shear loading. Longitudinal and shear reinforcement bars were schematized with two-node linear 3D truss elements (T3D2). To avoid stress concentrations and simulate the experimental setup of shear tests, supports and loading points were provided with 50 mm × 100 mm × 150 mm steel plates, which were also schematized with C3D8R element types.

To simulate the four-point bending experimental setup [50], the model was constrained by roller and pin boundary conditions at the left and right supports, respectively (see Figure 6). Displacement boundary conditions were applied at the loading plates with respect to the required a/d ratio.

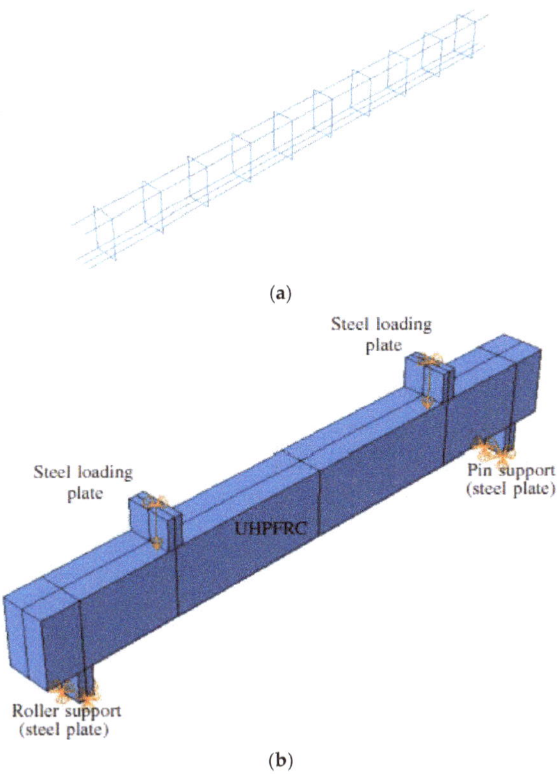

Figure 6. (**a**) A wire-frame model reinforcement cage profile showing longitudinal and transverse reinforcements and (**b**) a shaded model showing the loading and boundary condition profile with the load applied taken to be P/2.

The tie constraint was used to define the contact interaction of the RPC beam with the four steel plates, while the embedded region interaction property was used for the steel reinforcing bar–RPC interaction, for both transverse and longitudinal reinforcements (see Figure 6). After a careful mesh convergence study and taking into account node compatibility, a mesh size of 25 mm was found to be adequate for the RPC matrix, steel plates, and longitudinal and transverse steel reinforcement bars.

Each example was created in accordance with the size and reinforcement details. The ABAQUS program's documentation contains comprehensive information on FEM for steel reinforcement and concrete. As seen in Figure 7 for B-1, the tested beams' FEM is visible in the corresponding figure. Nonlinear RPC and reinforcement constitutive models were introduced in this model. Poisson's ratio was set to 0.30, and the steel's elastic modulus (E_s)

was assumed to be 200 GPa. It was believed that the concrete and reinforcement would always stick together perfectly.

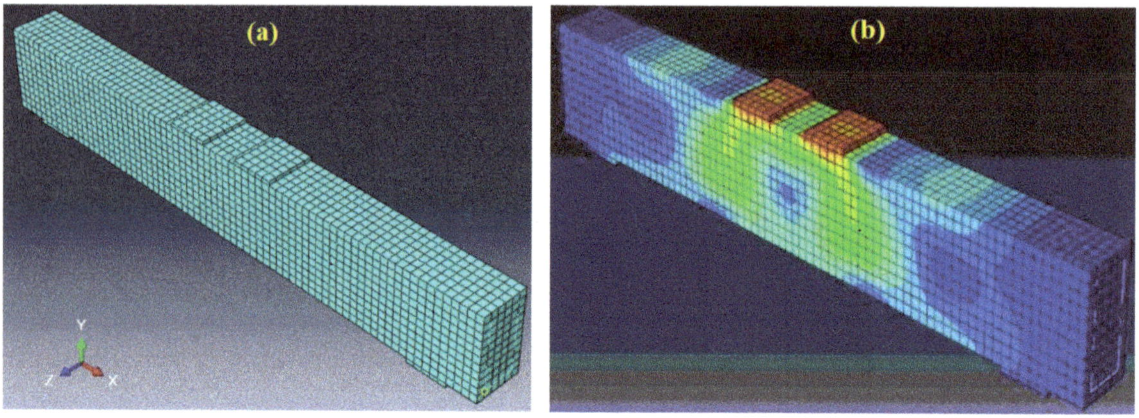

Figure 7. RPC beam B-1 meshing and stress distribution determined using a finite element model (**a**) designed been (**b**) loaded beam.

According to the results of this study's experimental tests, the following proposed simple equation was used to determine the elasticity modulus (E_c) of RPC:

$$E_c = 3737\sqrt{f'c} \tag{14}$$

For the tested beams with a compressive strength (f_{cu}) equal to 157 MPa, the normative value of E_c is equal to 48,929.5 MPa. The Poisson's ratio is assumed to be equal to 0.20, and the tensile strength of RPC is considered to be 11 MPa.

The acquired results were contrasted with the outcomes for the RPC beams examined in the current study in order to assess the precision of the nonlinear finite element model. Table 7 compares the ultimate load, the experimental crack load, and the projected results for the examined beams derived from the FEM. For RPC beams, the average proportion of $V_{cr,exp}$ to $V_{cr,NUM}$ is equivalent to 1.034, whereas the average proportion of $V_{u,exp}$ to $V_{u,NUM}$ is equivalent to 1.007. Additionally, the proportion of $\Delta_{cr,exp}$ to $\Delta_{cr,NUM}$ has a mean value of 0.92, while the proportion of $\Delta_{u,exp}$ to $\Delta_{u,NUM}$ has a mean value of 0.979. This demonstrates that for the examined RPC beams, the FEM model can accurately predict the ultimate shear load and the diagonal cracking load. Figure 8 displays a comparison of the empirical and numerical load–displacement curves for the examined beams. It can be observed that the load–displacement response of the examined RPC beams can be well predicted by the suggested nonlinear FEM.

Table 7. Comparison of the experimental data with the numerical model using a finite element model.

Beam	Cracking Load and Ultimate Load						Cracking Displacement and Maximum Displacement					
	$V_{cr,NUM}$ (kN)	$V_{cr,exp}$ (kN)	$\frac{V_{cr,exp}}{V_{cr,NUM}}$	$V_{u,exp}$ (kN)	$V_{u,NUM}$ (kN)	$\frac{V_{u,exp}}{V_{u,NUM}}$	$\Delta_{cr,NUM}$ (mm)	$\Delta_{cr,exp}$ (mm)	$\frac{\Delta_{cr,exp}}{\Delta_{cr,NUM}}$	$\Delta_{u,exp}$ (mm)	$\Delta_{u,NUM}$ (mm)	$\frac{\Delta_{u,exp}}{\Delta_{u,NUM}}$
B-1	72	76.5	1.06	201.5	185	1.089	0.5	0.34	0.68	1.75	1.95	0.897
B-2	51	57	1.12	184.5	178	1.037	0.48	0.4	0.83	1.85	2.1	0.881
B-3	88	83.5	0.95	225	255	0.882	0.31	0.33	1.07	2.2	1.98	1.11
B-4	58	65	1.12	184.5	174	1.06	0.24	0.26	1.08	1.78	1.88	0.947
B-5	60	55	0.92	155	160.5	0.966	0.47	0.44	0.94	2.1	1.98	1.06

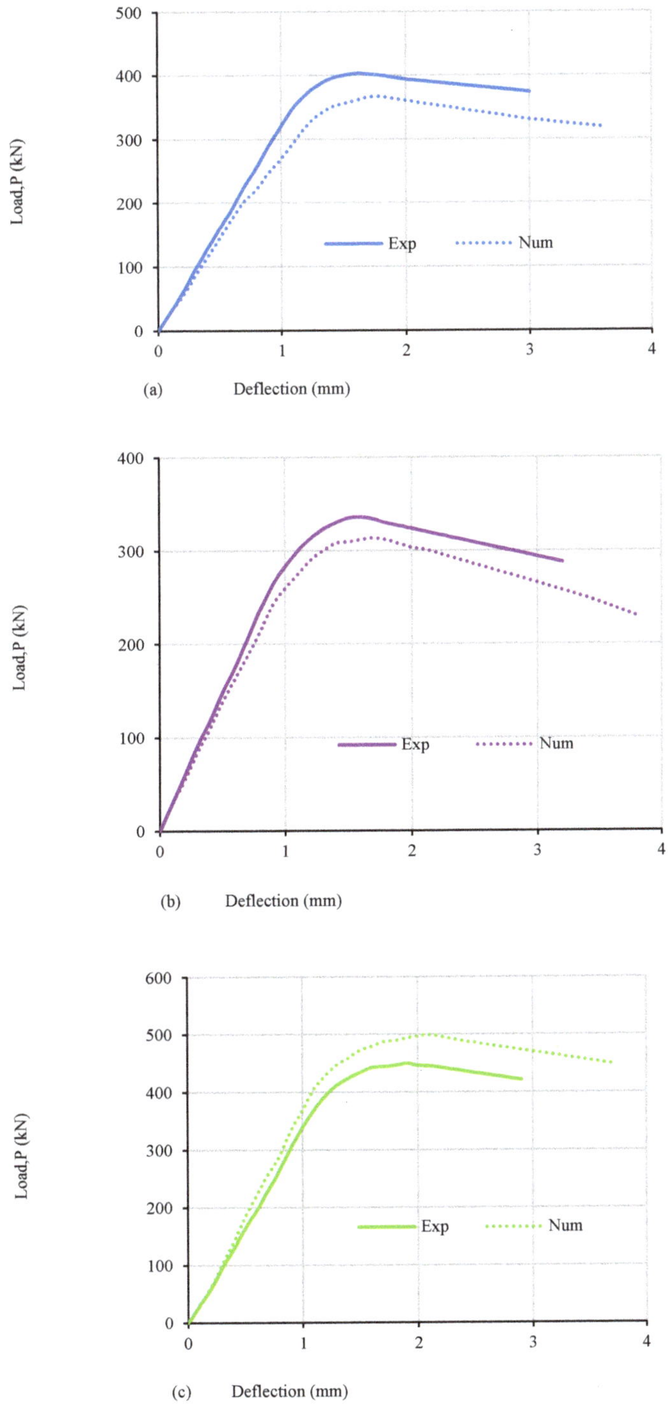

Figure 8. *Cont.*

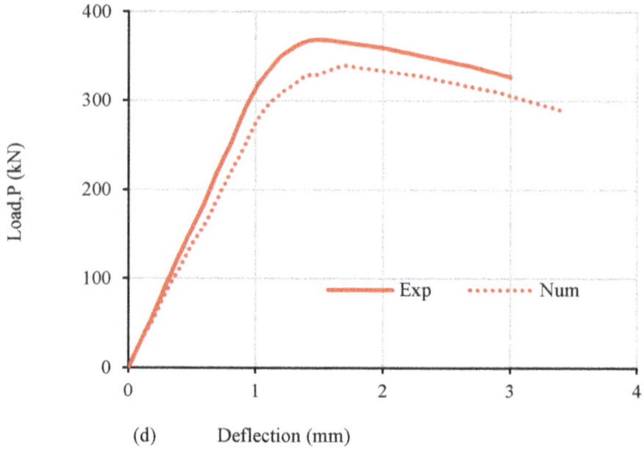

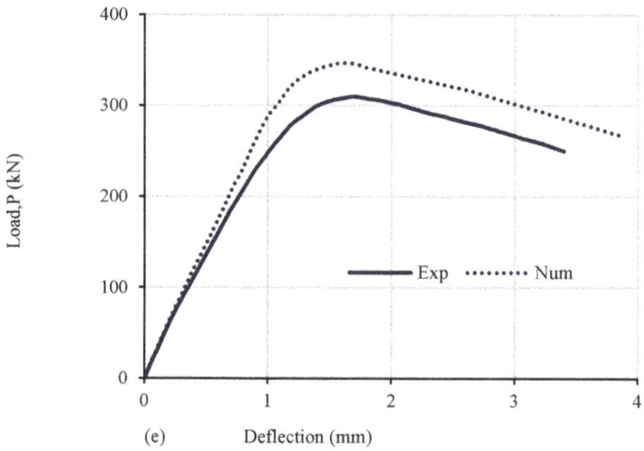

Figure 8. Comparison of the load–displacement curves for the examined beams using experimental and numerical data for (**a**) B-1, (**b**) B-2, (**c**) B-3, (**d**) B-4, and (**e**) B-5.

6. Conclusions

The following conclusions can be drawn from this study's findings regarding RPC slender beams under shear:

i The offered shear reinforcement barely affects the maximum shear strength of the tested RPC beams with a volume content of 2.0% steel fibers. These steel fibers are crucial in helping RPC beams endure shear loads. Despite the fact that the examined RPC shallow beams' vertical web reinforcement ratio was far below the smallest proportion specified by ACI 318-14, all of the examined RPC shallow beams displayed excellent performance in general.

ii ACI 318-14's shear strength calculation formulas significantly understate the shear strength of the examined RPC beams that have a minimal vertical web reinforcement ratio. As a result, these formulas are not suitable for RPC beams as they do not account for the significant role that steel fibers play in resisting shear stresses. For RPC beams with relatively low heights, the highest distance between stirrups ($s_{v,max}$) specified according to the ACI 318-14 guidelines can safely be extended from 0.50 d to 0.75 d.

iii In light of the predicted ultimate shear strengths of the examined beams, the design suggestions for RPC specified by KCI-2012 and AFGC-2013 are secure and restrained. The ultimate shearing strength forecasts made by KCI-2012 and AFGC-2013 are roughly equivalent (the mean proportion of the experimental ultimate shearing strength and the predicted ultimate shearing strength utilizing KCI-2012 and AFGC-2013 are approximately 1.462 and 1.446, respectively).

iv According to the AFGC-2013 criteria, the mean proportion of V_{fb} to projected $V_{u,AFGC}$ is roughly 58.3%, whereas the mean proportion of V_s to V_c is just 41.7%.

v The deformation response and the ultimate shear strength of the examined RPC beams with vertical reinforcement of the web ratio much below the lowest value permitted by the code were reliably predicted by the suggested FEM when steel fibers were taken into account.

Author Contributions: Conceptualization, H.K.S. and G.F.H.; methodology, H.K.S.; software, H.K.S.; validation, H.K.S. and G.F.H.; formal analysis, H.K.S.; investigation, G.F.H.; resources, H.K.S.; data curation, H.K.S.; writing—original draft preparation, H.K.S.; writing—review and editing, G.F.H.; visualization, H.K.S.; supervision, G.F.H.; project administration, H.K.S.; funding acquisition, G.F.H. All authors have read and agreed to the published version of the manuscript.

Funding: This research received no external funding.

Institutional Review Board Statement: Not applicable.

Informed Consent Statement: Not applicable.

Data Availability Statement: The original contributions presented in the study are included in the article, further inquiries can be directed to the corresponding author.

Acknowledgments: The authors gratefully acknowledge the financial support for this research by Guangzhou Institute of Energy Conversion, Chinese Academy of Sciences.

Conflicts of Interest: The authors declare no conflicts of interest.

References

1. Hou, X.M.; Cao, S.J.; Rong, Q.; Zheng, W.Z.; Li, G. Effects of steel fiber and strain rate on the dynamic compressive stress-strain relationship in reactive powder concrete. *Constr. Build. Mater. J.* **2018**, *170*, 570–581. [CrossRef]
2. Shen, L.; Ding, M.; Chen, K.; Feng, C.; Yang, B.; Elchalakani, M. The mechanical behavior of RPC under combined shear and compressive loads. *Cem. Concr. Compos.* **2021**, *121*, 104071. [CrossRef]
3. Hiremath, P.N.; Yaragal, S.C. Performance evaluation of reactive powder concrete with polypropylene fibers at elevated temperatures. *Constr. Build. Mater. J.* **2018**, *169*, 499–512. [CrossRef]
4. Abid, M.; Hou, X.; Zheng, W.; Hussain, R.R. High temperature and residual properties of reactive powder concrete—A review. *Constr. Build. Mater. J.* **2017**, *147*, 339–351. [CrossRef]
5. Wang, Z.; Wang, J.Q.; Tang, Y.C.; Liu, T.X.; Gao, Y.F.; Zhang, J. Seismic behavior of precast segmental UHPC bridge columns with replaceable external cover plates and internal dissipaters. *Eng. Struct.* **2018**, *177*, 540–555. [CrossRef]
6. Paschalis, S.A.; Lampropoulos, A.P. Ultra-high-performance fiber reinforced concrete under cyclic loading. *ACI Mater. J.* **2016**, *113*, 419–427. [CrossRef]
7. Xu, S.C.; Wu, C.Q.; Liu, Z.X.; Han, K.P.; Su, Y.; Zhao, J.; Li, J.C. Experimental investigation of seismic behavior of ultra-high performance steel fiber reinforced concrete columns. *Eng. Struct.* **2017**, *152*, 129–148. [CrossRef]
8. Yu, R.; Spiesz, P.; Brouwers, H.J.H. Energy absorption capacity of a sustainable Ultra-High Performance Fibre Reinforced Concrete (UHPFRC) in quasi-static mode and under high velocity projectile impact. *Cement Concr. Compos.* **2016**, *68*, 109–122. [CrossRef]
9. *ACI Committee 363*; Ultra-High Performance Concrete, A State-of-the-Art Report for the Bridge Community, Publication No. FHWA-HRT-13-060. Federal Highway Administration: Washington, DC, USA, 2013; pp. 13–44.
10. Ji, W.Y.; Li, W.W.; An, M.Z.; Zhu, L. Shear capacity of T-section girders made of reactive powder concrete. *J. Bridge Eng.* **2018**, *23*, 04018041. [CrossRef]
11. Zinkaah, O.H.; Sultan, H.K.; Al-Rifaie, A.; Alridha, Z. Influence of Strut Geometry on the Size Effect of FRP Reinforced Simply Supported Deep Beams: A Theoretical Analysis. *Math. Model. Eng. Probl.* **2022**, *9*, 411–417. [CrossRef]
12. Zheng, H.; Fang, Z.; Chen, B. Experimental study on shear behavior of prestressed reactive powder concrete I-girders. *Front. Struct. Civ. Eng.* **2019**, *13*, 618–627. [CrossRef]
13. Kaufmann, W.; Amin, A.; Beck, A.; Lee, M. Shear transfer across cracks in steel fibre reinforced concrete. *Eng. Struct.* **2019**, *186*, 508–524. [CrossRef]
14. Sultan, H.K.; Alyaseri, I. Effects of elevated temperatures on mechanical properties of reactive powder concrete elements. *Constr. Build. Mater.* **2020**, *261*, 120555. [CrossRef]
15. Sultan, H.K.; Zinkaah, O.H.; Rasheed, A.A.; Alridha, Z.; Alhawat, M. Producing Sustainable Modified Reactive Powder Concrete Using Locally Available Materials. *Innov. Infrastruct. Solut.* **2022**, *7*, 342. [CrossRef]

16. Lantsoght, E.O.L. How do steel fibers improve the shear capacity of reinforced concrete beams without stirrups? *Compos. B Eng.* **2019**, *175*, 107079. [CrossRef]
17. Biswas, R.K.; Bin Ahmed, F.; Haque, M.E.; Provasha, A.A.; Hasan, Z.; Hayat, F.; Sen, D. Effects of steel fiber percentage and aspect ratios on fresh and harden properties of ultra-high performance fiber reinforced concrete. *Appl. Mech.* **2021**, *2*, 501–515. [CrossRef]
18. He, J.; Chen, W.; Zhang, B.; YU, J.; Liu, H. The mechanical properties and damage evolution of UHPC Reinforced with glass fibers and high-performance polypropylene fibers. *Materials* **2021**, *14*, 2455. [CrossRef] [PubMed]
19. Yoo, D.Y.; Yang, J.M. Effects of stirrup, steel fiber, and beam size on shear behavior of high-strength concrete beams. *Cement Concr. Compos.* **2018**, *87*, 137–148. [CrossRef]
20. Głodkowska, W.; Ziarkiewicz, M. Cracking behavior of steel fiber reinforced waste sand concrete beams in flexure-experimental investigation and theoretical analysis. *Eng. Struct.* **2018**, *176*, 1–10. [CrossRef]
21. Wille, K.; Naaman, A.E.; Parra-Montesions, G.J. Ultra-High-Performance Concrete with Compressive Strength Exceeding 150 MPa (22 ksi), A simpler Way. *ACI Mater. J.* **2011**, *108*, 46–54.
22. Sultan, H.K. Designing reinforced HSC rectangular beams using optimization techniques. *Pollack Period. J.* **2023**, *18*, 20–25. [CrossRef]
23. Wang, C.; Xiao, J.; Liu, W.; Ma, Z. Unloading and reloading stress-strain relationship of recycled aggregate concrete reinforced with steel/polypropylene fibers under uniaxial low-cycle loadings. *Cem. Concr. Compos.* **2022**, *131*, 104597. [CrossRef]
24. Wang, C.; Wu, H.; Li, C. Hysteresis and damping properties of steel and polypropylene fiber reinforced recycled aggregate concrete under uniaxial low-cycle loadings. *Constr. Build. Mater.* **2022**, *319*, 126191. [CrossRef]
25. Samadi, M.; Baghban, M.H.; Kubba, Z.; Faridmehr, I.; Abdul Shukor Lim, N.H.; Benjeddou, O.; Huseien, G.F. Flexural behavior of reinforced concrete beams under instantaneous loading: Effects of recycled ceramic as cement and aggregates replacement. *Buildings* **2022**, *12*, 439. [CrossRef]
26. Wang, C.; Wu, H.; Li, C. Hysteretic deteriorating behaviors of fiber-reinforced recycled aggregate concrete composites subjected to cyclic compressive loadings. *J. Build. Eng.* **2022**, *49*, 104087. [CrossRef]
27. Sultan, H.K. Hysteresis Damping Ratio for Moment Steel Frame Evaluation Using DDBD. *Pollack Period. J.* **2024**, *19*, 33–40. [CrossRef]
28. Sultan, H.K.; Mohammed, A.T. Assessing of the common strengthening methods for existing RC buildings. *Pollack Period. J.* **2023**, *18*, 6–11. [CrossRef]
29. Yahy Al-Radi, H.H.; Dejian, S.; Sultan, H.K. Performance of Fiber Self-Compacting Concrete at High Temperatures. *Civ. Eng. J.* **2021**, *7*, 2083–2098. [CrossRef]
30. Sultan, H.K.; Mohammed, A.T.; Qasim, O.D.; Maula, B.H.; Aziz, H.Y. Ductility Factor Evaluation of Concrete Moment Frame Retrofitted by FRP Subjected to Seismic Loads. *Int. Rev. Civ. Eng. (IRECE)* **2020**, *11*, 275–282. [CrossRef]
31. Qasim, O.A.; Sultan, H.K. Experimental Investigation of Effect of Steel Fiber on Concrete Construction Joints of Prism. *IOP Conf. Ser. Mater. Sci. Eng.* **2020**, *745*, 012170. [CrossRef]
32. Voo, Y.; Poon, W.K.; Foster, S.J. Shear strength of steel fiber-reinforced ultrahigh-performance concrete beams without stirrups. *J. Struct. Eng.* **2010**, *136*, 1393–1400. [CrossRef]
33. Aziz, O.Q.; Hasan, M.; Ali, H. Shear Strength and Behavior of Ultra-High Performance Fiber Reinforced Concrete (UHPC) Deep Beams without Web Reinforcement. *Int. J. Civ. Eng. (IJCE)* **2013**, *2*, 85–96.
34. Baby, F.; Marchand, P.; Toutlemonde, F. Shear Behavior of Ultra-High Performance Fiber-Reinforced Concrete Beams, I: Experimental investigation. *J. Struct. Eng.* **2014**, *140*, 04013111. [CrossRef]
35. Lim, W.-Y.; Hong, S.-G. Shear Test for Ultra-High Performance Fiber Reinforced concrete (UHPFRC) Beams with shear reinforcement. *Int. J. Concr. Struct. Mater.* **2016**, *10*, 177–188. [CrossRef]
36. Qi, J.; Ma, Z.J.; Wang, J. Shear Strength of UHPFRC Beams: Mesoscale Fiber Matrix Discrete Model. *J. Struct. Eng.* **2016**, *143*, 1–10. [CrossRef]
37. European Committee for Standardization. *Eurocode 2: Design of Concrete Structures—Part 1*; European Committee for Standardization: Brussels, Belgium, 2004.
38. *ACI Committee 318*; Building Code Requirements for Structural Concrete (ACI 318-14) and Commentary (ACI 318R-14). American Concrete Institute: Farmington Hills, MI, USA, 2014.
39. Association Francaise du Genil Civil (AFGC). *Ultra-High Performance Fiber Reinforced Concrete*; AFGC: Paris, France, 2002.
40. Association Francaise du Genil Civil (AFGC). *Ultra-High Performance Fiber Reinforced Concrete*; AFGC: Paris, France, 2013.
41. Japan Society of Civil Engineers (JSCE). *Recommendations for Design and Construction of Ultra-High-Strength Fiber Reinforced Concrete Structures*; Concrete Engineering Series; Japan Society of Civil Engineers: Tokyo, Japan, 2008.
42. Korea Concrete Institute. *Design Recommendations for Ultra High Performance Concrete (K-UHPC), KCI-M-12-003*; Korea Concrete Institute: Seoul, Republic of Korea, 2012.
43. *ASTM C109/C109M-20b*; Standard Test Method for Compressive Strength of Hydraulic Cement Mortars (Using 2-in. or [50 mm] Cube Specimens). ASTM International: West Conshohocken, PA, USA, 2020.
44. *ASTM C496*; Standard Test Method for Splitting Tensile Strength of Cylindrical Concrete Specimens. American Society for Testing and Materials Standard Practice C496. ASTM International: West Conshohocken, PA, USA, 2017.
45. *ASTM C78*; Standard Test Method for Flexural Strength of Concrete (Using Simple Beam with Center Point Loading). ASTM International: West Conshohocken, PA, USA, 2010.

46. *ABAQUS: Abaqus Analysis User's Manual*, Version 6.9; Dassault Systems Corp.: Providence, RI, USA, 2009.
47. Katsamakas, A.A.; Papanikolaou, V.K.; Thermou, G.E. Tree-dimensional numerical modeling of RC beam strengthened in shear with steel reinforced grout (SRG). In Proceedings of the 1st Fib Italy YMG Symposium on Concrete and Concrete Structures, Krakow, Poland, 27–29 May 2019; pp. 64–71.
48. Ombres, L.; Verre, S. Shear strengthening of reinforced concrete beams with SRG (Steel Reinforced Grout) composites: Experimental investigation and modeling. *J. Build. Eng.* **2021**, *42*, 103047. [CrossRef]
49. Bahij, S.; Adekunle, S.K.; Al-Osta, M.; Ahmad, S.; Al-Dulaijan, S.U.; Rahman, M.K. Numerical investigation of the shear behavior of reinforced ultra-high-performance concrete beams. *Struct Concr.* **2018**, *19*, 305–317. [CrossRef]
50. Ahmad, S.; Bahij, S.; Al-Osta, M.; Adekunle, S.K.; Al-Dulaijan, S.U. Shear behavior of ultra-high-performance concrete beams reinforced with high-strength steel bars. *ACI Struct. J.* **2019**, *116*, 3–14. [CrossRef]

eng

MDPI

Article

Numerical Investigation of the Seismic Performance of an Innovative Type of Buckling-Restrained Brace (BRB)

Ali Naghshineh [1,*], Oliver Fischer [1], Nasreen B. Pathan [2], Logan Couch [2] and Fariborz M. Tehrani [2]

1 School of Engineering and Design, Concrete and Masonry Structures, Technical University of Munich (TUM), 80290 Munich, Germany
2 Department of Civil & Geomatics Engineering, California State University, Fresno, CA 93740, USA
* Correspondence: a.naghshineh@tum.de

Abstract: Previous studies have demonstrated that the inclusion of tire-derived aggregate (TDA) enhances the damping, ductility, and toughness of concrete mixtures. The effectiveness of tire-derived aggregate as a ductile material with a higher damping ratio and lower density in buckling-restrained braces has been examined at California State University's Structures Laboratory (CSU). Through experimental and theoretical investigations, this study compares the structural application of buckling-restrained braces with TDA and with conventional concrete infill subjected to various ground motions as well as artificial excitations. The evaluations include modeling a full-scale experimental setup equipped with a single-leg BRB utilizing ETABS 2016 and OpenSees 2000 software. The effectiveness of the application is demonstrated through a comparison of accelerations, displacements, stiffness, and damping ratios between TDA and concrete filling. Additionally, a design guideline for TDA-filled buckling-restrained braced frames is provided.

Keywords: tire-derived aggregate; seismic performance; experimental; steel frame; buckling-restrained braced frame; ductility; toughness; passive control; dynamic response

Citation: Naghshineh, A.; Fischer, O.; Pathan, N.B.; Couch, L.; Tehrani, F.M. Numerical Investigation of the Seismic Performance of an Innovative Type of Buckling-Restrained Brace (BRB). *Eng* **2023**, *4*, 2978–2990. https://doi.org/10.3390/eng4040167

Academic Editor: Alessio Cascardi

Received: 14 September 2023
Revised: 25 October 2023
Accepted: 22 November 2023
Published: 1 December 2023

1. Introduction

Buckling-restrained braced frames (BRBFs) offer an alternative to conventional braced frames, which have displayed poor performance during past earthquakes, including buckling failure, limited ductility, fraction in connections, and asymmetric hysteresis behavior. A typical BRBF consists of a ductile steel core embedded in concrete and encased concrete in a steel tube to prevent brittle failure modes.

This study explores the potential of modified concrete using waste tire chips that has a lower strength and stiffness yet a high toughness. However, when its strength falls too low, it becomes unsuitable for construction. The potential for increased strength and stiffness at a reduced cost by substituting waste tires in the form of fibers instead of chips was investigated [1]. A total of 42 cylinders underwent testing for compressive strength, split tensile strength, and modulus of elasticity. Twelve cylinders were prepared with waste tire chips, while thirty specimens were constructed using waste tire fibers, varying in terms of length and stiffness. The results indicated that using waste tire fibers in modified concrete as opposed to waste tire chips leads to increased stiffness and strength. Moreover, analytical studies confirmed that waste tire fibers could also reduce stress concentrations. Further studies, such as beam fatigue tests, are necessary to evaluate the practical use of rubberized concrete in constructions. Rubberized concrete showed an improved heat transfer, but reduced sound absorption compared to standard concrete [2]. Testing involved 64 specimens with varying levels of crumb rubber as a fine aggregate replacement. The results showed a decreased compressive and tensile strength, improved sound absorption, and higher electrical resistivity in crumb rubber-modified blocks.

Aslani [3] investigated the effects of rubber types and content on various properties including compressive strength, tensile strength, flexural strength, modulus of elasticity, strain, and stress–strain curves. The investigation of large-size TDA material showed that the unit weight of TDA depends on placement and compaction [4]. A whole yield and the secant friction angles of 24 and 60 were set as the upper bound and lower bound, respectively. The low unit weight of these materials makes them suitable for landslide repair and weak foundation soils. The shear behavior tests of TDA with large-size particles were performed, and the related properties and interface shear strength against concrete were measured [5]. The internal failure was nonlinear, with a decreasing peak secant friction angle from 39.6 to 30.2. In contrast, the TDA–concrete interface failure exhibited linear failure, with a higher initial shear stiffness than the TDA's internal shear test. Rubberized concrete exhibits a superior impact resistance and enhanced damping properties for cost-effective and flexible concrete barriers, in contrast to conventional concrete. However, increased tire chips decrease the compressive strength while improving energy absorption [6–11]. The impact of crumb rubber concrete on the ductility of profiled steel sheeting in composite slabs was also examined to achieve the desired performance [12].

The ductility and energy absorption capacity of RC columns by incorporating concrete waste tire rubber with varying compressive strengths, tire rubber size, and content were investigated [13]. Twelve column specimens were tested using two types of concrete with a compressive strength of 24 and 28 MPa and two different crumb rubber particle sizes (0.6 and 1 mm). The results indicated a reduction in the modulus of elasticity, compressive strength, and compressive load capacity when using waste tire rubber in concrete. Moreover, the lateral deformations and the energy dissipation capacity of column specimens utilizing waste tire rubber concrete were double those of the conventional concrete specimens. In addition, the curvature ductility increased to 45–90% with the inclusion of 0.5–1% waste tire particles in concrete mixtures. Other published research on rubberized concrete reported similar characteristics when incorporating rubber content in concrete mixtures. Increasing the rubber content resulted in decreased flexural and compression strengths but increased ductility, damping, toughness, sound isolation, and reduced density [14,15]. Therefore, the application of ductile materials with improved damping properties, such as tire-derived lightweight aggregate concrete, can further enhance the overall performance of the system.

In contrast to concrete, rubber is significantly softer and can tolerate far more deformation before failing. Crumb rubber consists of smaller particles, while tire chips are larger rubber particles used to replace fine and coarse aggregate in concrete [16]. The stress–strain curve of rubber is nonlinear, initially exhibiting a lower stiffness before increasing, while cement paste/mortals demonstrate brittle behavior with a higher compressive strength compared to tensile/flexural strength, following a linear stress–strain curve. Additionally, the inclusion of rubber-based aggregates results in decreased mechanical parameters, like compressive strength, tensile strength, flexural strength, and modulus of elasticity. Consequently, concrete containing a substantial proportion of rubber aggregate exhibited enhanced ductility compared to conventional concrete [17–22].

In the present study, the application of an innovative single-leg BRB with TDA and concrete infill is verified against the experimental tests conducted at the Structures Laboratory of California State University (CSU), Fresno. Based on the experimental tests, four models for a BRB with TDA and concrete filling are developed using ETABS and OpenSees software [23,24]. The models are subjected to various artificial loadings, including harmonic, periodic, and impulse loadings, as well as different ground motion loadings. The results allow for a comparison between the experimental and analytical studies, covering the acceleration, displacement, stiffness, and damping ratios of BRBs with TDA and concrete infill.

2. Tire-Derived Aggregate (TDA)

Tire-derived aggregate is derived from scrap tires accessible in a standard range of practical sizes. TDA is classified into two types, Type A (75–100 mm) and Type B

(150–300 mm) [25]. The mechanical properties from two experimental investigations are employed in the numerical simulations.

In the experimental study on tire-derived lightweight aggregate (TDLWA), the mechanical properties, including the compressive, splitting tensile, flexural strength, and flexural toughness, and an impact flexure test of rubberized lightweight aggregate, were investigated using 38 cylindrical and 36 beam specimens [26]. A target strength of 21 MPa was specified, and the cylinder and beam specimens were composed of shale lightweight coarse aggregate, natural sand fine aggregate, cement, and water. The TDA was then incorporated into the cylinder and beam specimens via a replacement with replacement ratios ranging from 0% to 100%. It was concluded that as the rubber content increased, the static mechanical strength decreased, while the ductility and toughness increased. Hence, these materials were found to be functional, particularly when energy absorption was a critical aspect. Nevertheless, dynamic testing was limited, and further comprehensive testing, both full-scale and small-scale, is warranted to gather more practical data.

The mechanical properties for TDA and concrete were conducted using six cylindrical specimens (0.1 m × 0.2 m) in accordance with ASTM C39 and C78, and ASTM C496 standards [27]. The test included compression and split tensile tests, with loads applied at rates of 0.24 MPa per second for compression and 48.9 kN per second for tensile testing. The test results for the concrete and TDA cylinders are presented in Table 1.

Table 1. Cylinder test results for concrete and TDA [27].

Cylinder Testing	Compression Strength	Cylinder Testing
Conventional concrete	34.6	11.2
TDA concrete	10.6	0.36

3. Buckling-Restrained Braces

Conventional steel-braced frames are used for lateral load resistance but have shown poor performance in past earthquakes due to factors like limited ductility and connection fractures. An alternative, buckling-restrained braces (BRBs), consist of a ductile steel core encased in concrete, providing the necessary yielding mechanism and the potential for enhanced seismic resistance. Buckling-restrained braced frames (BRBFs) can be categorized as one of the new types of seismic force-resisting systems, with the two primary bracing configurations being single-leg and concentratedly braced frames. While a BRBF shares some geometric similarities with a conventional CBF, it exhibits noticeable distinctions in terms of connections, members, hysteresis behavior, and ductility. Figure 1 presents the BRBF details [28–30].

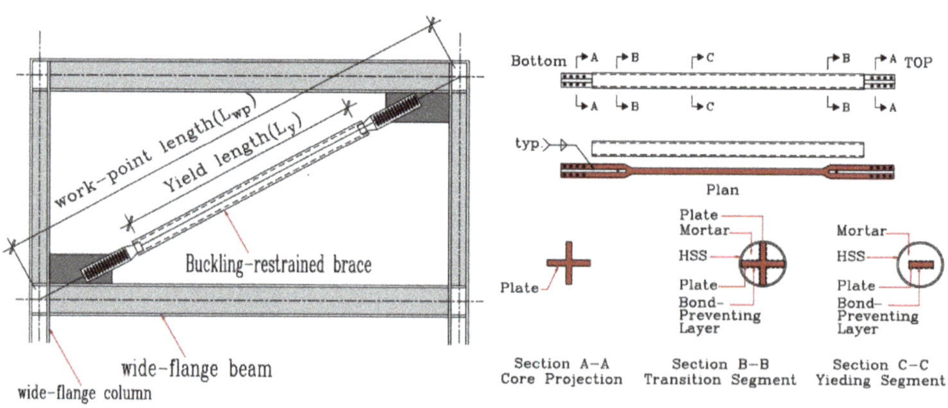

Figure 1. Single-leg buckling-restrained bracing and details [28–31].

The primary aim of this study is to demonstrate TDA's effectiveness as a filling material in comparison to conventional concrete filling. Figure 2 provides the specifics of the experimental buckling-restrained brace, with an effective length of 2.31 m (7′–7.0″) and a total length of 2.51 m (8′–3.0″). The steel core cross-section, consisting of an A36 plate measuring 6.23 cm in × 0.32 cm (2.45 in × 0.125 in), is designed to withstand a lateral load of 27.7 kN.

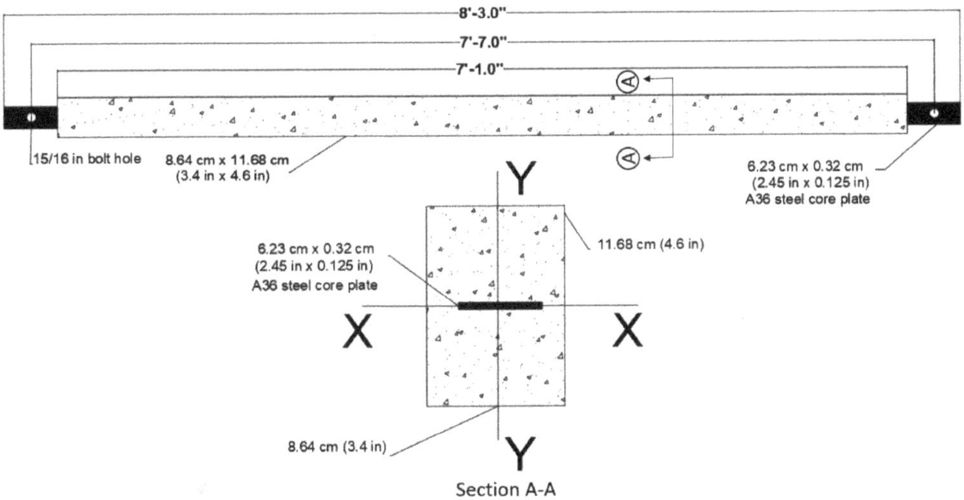

Figure 2. The experimental model of buckling-restrained brace [30].

The initial stiffness (K_{in}) of the buckling-restrained brace for both the TDA and concrete fillings is determined using Equation (1), utilizing the experimental results. In this equation, A_{sc} represents the steel core, E stands for the modulus of elasticity, and L_{wp} is the work point length, which can be predicated assuming minor angle changes. The axial deformation, Δ_{bx}, is calculated as the product of the drift angle, q_x, and the work point length, L_{wp}, multiplied by sin(2α), where α represents the BRB angle. Subsequently, the yield length ratio is computed by dividing the length of the yielding region, L_y, by the work point length, L_{wp}. The initial stiffness is established based on the modulus of elasticity of the steel core (A36) [32], resulting in a value of 71.255 kip/in for both the BRB with TDA and concrete fillings.

$$k_{in} = \frac{A_{sc}E}{L_{wp}} \tag{1}$$

4. Experimental Setup

The setup consists of one bay in the X direction with a 2.02 m span length and two bays in the Y direction, each spanning 0.762 m. The frame has a height of 2.44 m and utilizes W6 × 9 and W6 × 15 (A992) for the columns and beams. Two concrete blocks, each weighing 1035 Kg, are placed atop the frame. Lateral translation is controlled by two wires with a 4.4 kN capacity, and the shake table measures 2.44 m × 2.06 m in the X and Y directions. Three accelerometers were employed to measure the frame vibrations, located at the northeast, southwest, and base of the frame (Figure 3).

Figure 3. A 3-D and elevation view of the frame [30].

5. Loadings

The selected excitations were acquired from the Pacific Earthquake Engineering Research (PEER) database [33]. The scaling of ground motions was restricted to the base displacement of 12.17 cm (5″) due to the limitations of the shake table. The frame was subjected to increasing amplitude FEMA loadings [34] with varying time steps (0.00625 s, 0.003125 s, 0.0125 s) and with maximum displacements of 20.32 mm (0.80″) and 40.46 mm (1.60″) in FEMA-2D. The analysis also included a 5 mm (0.2″) frequency-increasing sweep loading and impulse loading, as presented in Figures 4 and 5 [30].

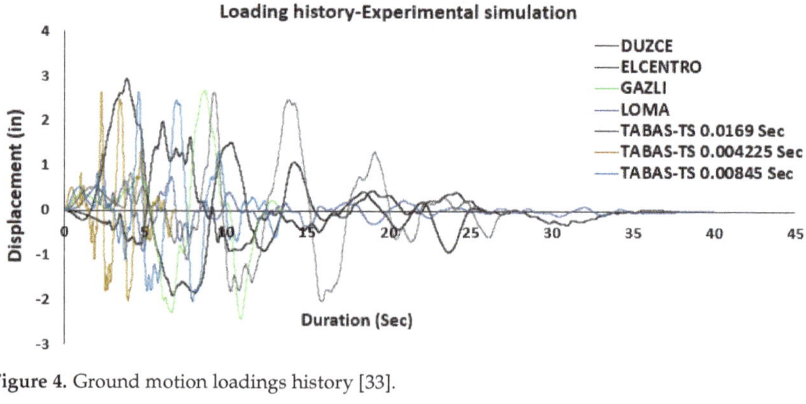

Figure 4. Ground motion loadings history [33].

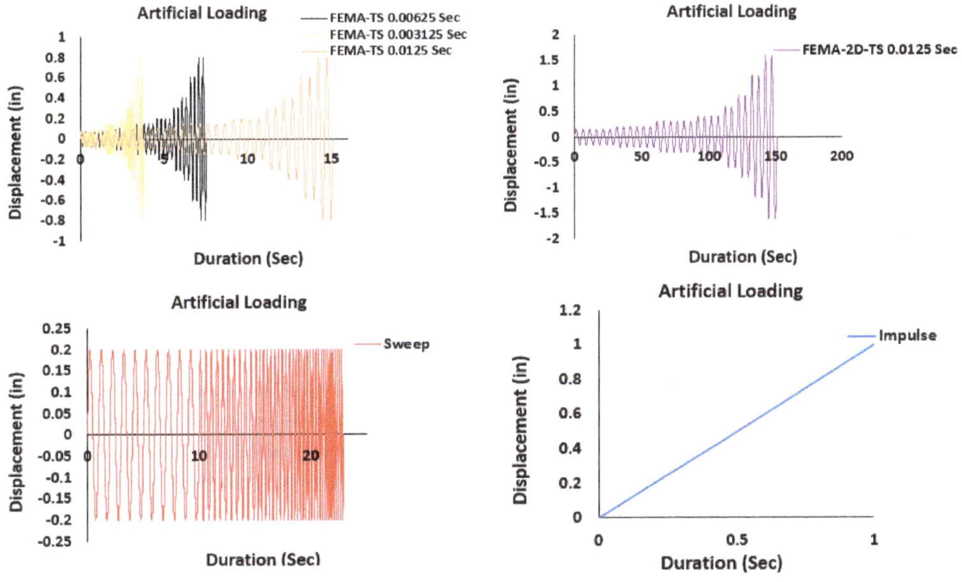

Figure 5. Artificial loadings history [30].

The time steps for each loading are calculated according to ASCE 41-17 [35] and FEMA P-2208 revisions [36]; these values are tabulated in Table 2.

Table 2. Selected time steps for acceleration and displacement.

Loadings	T_S (s)	T/100 (s)	T_{90} (s)	Selected T_S-Acceleration (s) (Min: [1] T_S, T/100, T_{90}, and 0.01_s)	Selected T_S-Displacement (s)
DUZCE	0.0050	0.0023	0.145	0.0023	0.0050
ELCENTRO	0.0050	0.0023	0.145	0.0023	0.0050
GAZLI	0.0066	0.0023	0.145	0.0023	0.0066
LOMA	0.0050	0.0023	0.145	0.0023	0.0050
TABAS 1	0.0169	0.0023	0.145	0.0023	0.0169
TABAS 2	0.004225	0.0023	0.145	0.0023	0.0042
TABAS 3	0.00845	0.0023	0.145	0.0023	0.0084
FEMA 1	0.00625	0.0023	0.145	0.0023	0.0062
FEMA 2	0.003125	0.0023	0.145	0.0023	0.0031
FEMA 3	0.0125	0.0023	0.145	0.0023	0.0125
FEMA-2D-4	0.0125	0.0023	0.145	0.0023	0.0125
Sweep	0.0125	0.0023	0.145	0.0023	0.0125

[1] T is the fundamental period of the structure, and T_{90} is the highest mode when T reaches 90% of modal mass participation.

6. Simulation of Experimental Model

Four numerical models were developed based on the experimental tests using ETABS and OpenSees software [23,24]. The weight of the existing blocks is applied as two-point loads, each with a magnitude of 2.3 kip (10.21 kN) on the middle beam (2-L1&L2) and four-point loads, each with a magnitude of 1.15 kip (5.1 kN) on the outer beams (1&3-L1&L2).

OpenSees is open-source, object-oriented software designed for earthquake engineering. It can be used to simulate structural responses through finite element computer applications [37–40]. The simulated experimental BRB frame with both TDA and concrete is depicted in Figure 6.

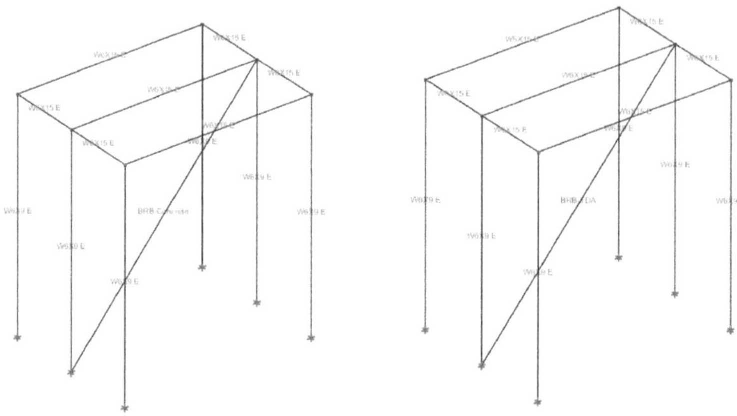

Figure 6. Simulation of experimental BRB frame using OpenSees software.

7. Results and Discussions

Selected roof acceleration–time diagrams for the BRB with concrete and TDA fillings, based on the experimental findings in the tension, compression, and an average of the tension and compression along with the analytical results from ETABS and OpenSees software [23,24], are presented in Figure 7 (concrete filling) and Figure 8 (TDA). All diagrams demonstrate a strong agreement between the experimental and analytical results. Figure 9 displays the maximum roof acceleration outcomes for the BRB with TDA and concrete fillings. The highest roof acceleration for the BRB with concrete infill was observed during the Loma Prieta ground motion, with a maximum of 0.49 g in the tension, 0.44 g in the compression, and an average acceleration of 0.46 g (experimental) and 0.42 g (simulation). For the frame subjected to the El Centro ground motion for the BRB with concrete infill, the minimum acceleration recorded was 0.24 g in the tension and 0.25 g in the compression, with an average acceleration of 0.245 g in the experimental study and a similar value of 0.23 g for the analytical work. In the case of the BRB with TDA filling, Tabas (TS-0.004225) exhibited the highest acceleration, with 0.83 g in the tension and 0.64 g in the compression, and an average of 0.735 g in the experimental study, compared to 0.89 g in the analytical work. Duzce displayed the lowest acceleration for the BRB with TDA filling, with an average of 0.075 g in the tension and compression in the experimental study and 0.08 g in the analytical study.

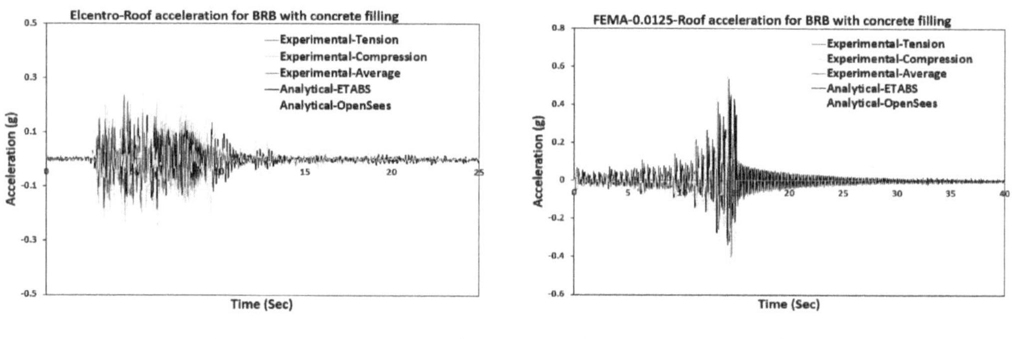

Figure 7. Comparison of roof acceleration for BRB with concrete filling subjected to different ground motions.

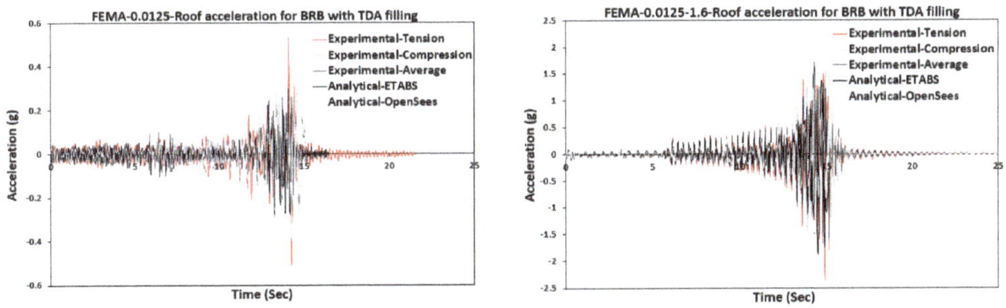

Figure 8. Comparison of roof acceleration for BRB with TDA filling subjected to different ground motions.

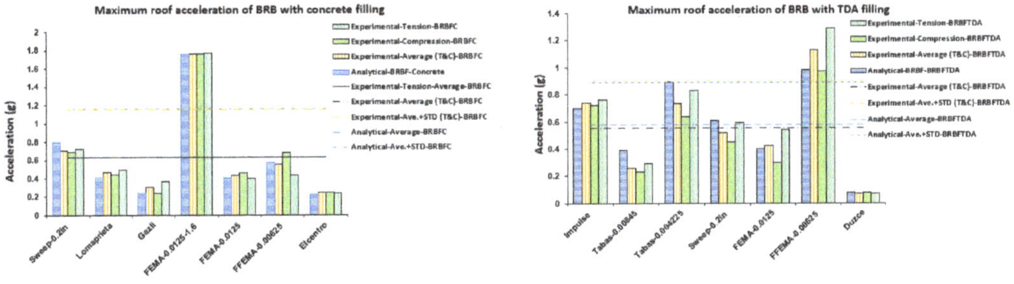

Figure 9. Selected maximum roof acceleration for BRB with concrete (**left**) and TDA filling (**right**).

Hysteresis loops of the BRB with TDA and concrete infills are presented in Figure 10. It is evident that the BRB with TDA infill dissipates less energy compared to the BRB with concrete infill. This observation aligns with the existing literature, indicating that the addition of rubber content in concrete mixtures leads to reduced flexural and compressive properties. A similar conclusion is drawn from the experimental work, which can be attributed to the unbinding of the rubber content within concrete mixtures.

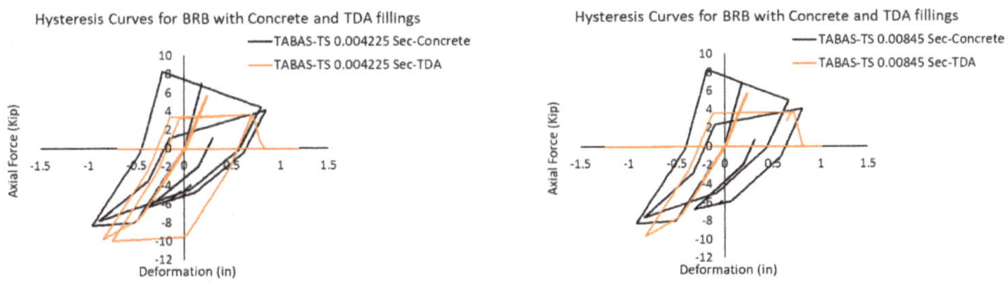

Figure 10. Results of hysteresis curves for BRB with TDA and concrete infills subjected to Tabas motion: TS-0.004225 (**left**), TS-0.00845 (**right**).

Figure 11 compares the backbone curves of the analytical and the experimental test for both BRB models with TDA and concrete fillings under FEMA 4 loading. The analytical and experimental results exhibit a consistent trend, revealing that the BRB with TDA filling is less ductile compared to the conventional concrete.

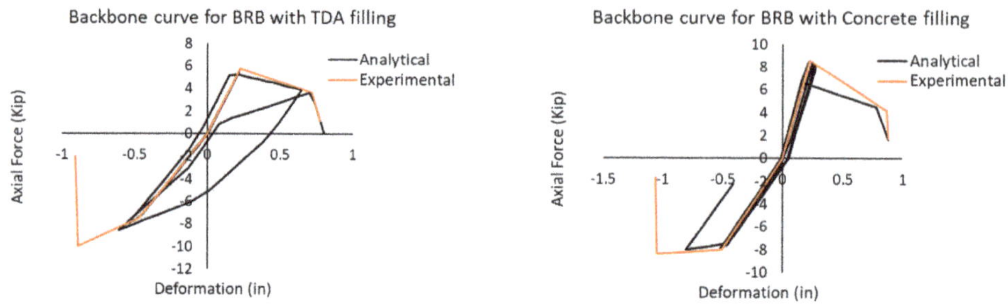

Figure 11. Comparison of hysteresis curves of analytical and experimental works for BRB with TDA (**left**) and with concrete filling (**right**).

The maximum roof displacements for the BRB with TDA and concrete infills are presented in Figure 12. The BRB frame with concrete infill showed the maximum roof displacements of 3.08 (in) in the tension and 2.97 (in) in the compression during the Gazli ground motion, with minimum roof displacements of 0.38 (in) and 0.28 (in) subjected to sweep loading. Analytically, the highest and lowest displacements were 2.75 (in) for the Gazali loading and 0.49 (in) for sweep loading. The BRB frame with TDA under the Tabas (TS-0.00845) ground motion had the maximum displacement of 2.67 (in) in the tension and 2.3 (in) in the compression for the experimental study, and 2.7 in the analytical work. The minimum roof displacement was 0.63 (in) in the tension and 0.43 (in) in the compression for the experimental frame, and 0.7 (in) for the simulated work under FEMA-0.0125 loading.

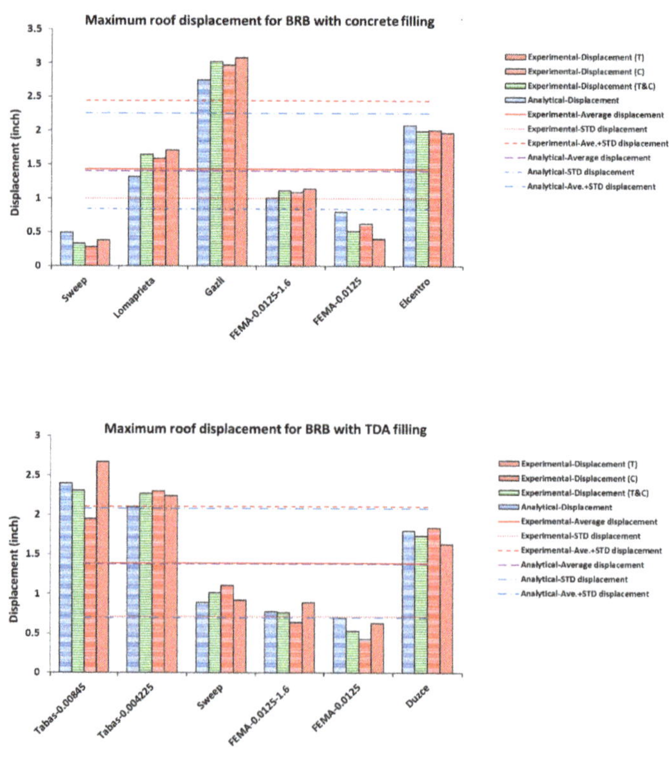

Figure 12. Maximum displacements for BRB with TDA and concrete fillings.

The effective stiffness is determined using a 1-inch impulse load, while system damping is derived from FEMA 4 loading based on the deformation response data. In both the experimental and analytical setting, the average effective stiffness of the BRB with concrete filling is 26.5 Kip/inch and 28 Kip/inch, respectively. For the BRB with TDA filling, the effective stiffness decreases to 20 Kip/inch in the experimental work and 21 Kip/inch in the analytical work. The damping ratios for the BRB with TDA filling exhibited a significant increase compared to the BRB with concrete infill, as shown in Table 3. The BRB with TDA demonstrated a notable 51% increase in damping compared to the conventional system.

Table 3. Buckling-restrained braced frame (BRBF)—damping ratio and stiffness (comparison of experimental and analytical results).

Description	Damping Ratio (%)			
	Experimental		Analytical	
	Tension	Compression	Tension	Compression
BRBF with TDA filling	16%	26%	25%	25%
BRBF with concrete filling	13%	17%	14%	14%
	Stiffness (Kip/in)			
	Experimental		Analytical	
	Tension	Compression	Tension	Compression
BRBF with TDA filling	19	21	21	21
BRBF with concrete filling	26	27	28	28

8. Design Guideline

ASCE 7 [41] establishes essential system design parameters, system-independent criteria, seismic hazard levels, redundancy, limitations, and conditions related to structural irregularities. AISC 341 [29] offers comprehensive design and detailing guidelines for individual structural elements, connections, and requirements to ensure the desired ductile behavior. In the context of a buckling-restrained braced frame (BRBF) system, the structure is expected to withstand significant inelastic deformation demands. Therefore, ASCE 7 [41] assigns the most prominent response modification coefficient (R = 8). The following steps and design procedures are recommended for a BRBF system with TDA infill:

The application of ductile tire-derived lightweight aggregate concrete significantly enhances the overall system performance. To achieve this, it is crucial to assess the mechanical properties of both the TDA and concrete including the compressive, flexural, and splitting tensile strength, toughness, and modulus of elasticity, among other important criteria, prior to the design process. This study relies on the experimental tests to determine these mechanical properties [15,16].

The reduction factors for buckling-restrained braced frames (BRBFs) depend on the building height and bracing configurations. To obtain more realistic results, it is recommended to calculate the response modification factor using pushover response curves [35]. However, in accordance with the current American code, ASCE 7 [41], BRBFs shall be designed based on the prescribed reduced reduction factor R. Then, select the appropriate load combinations.

The strength of the steel core can be defined as either the actual yield stress obtained from a coupon test or the specified minimum yield stress of the steel core. The design of the steel core must ensure its ability to resist axial forces within the brace [27].

The analysis and design of the BRB and control inelastic design level of the BRB strain and drift should adhere to the provisions outlined in AISC 341, ASCE 7, and ASCE 41 [29,35,41]. For the expected deformation, the greater value of either 0.02 times the story height (h_{sx}) or 2 times the story drift (Δ_x) should be considered.

Subsequently, the forces on the BRB are transferred to the connections, columns, and beams. Therefore, the design of the beams, columns, and connections should accommodate the adjusted brace strength while remaining within the elastic zone. Additionally, all beam to column joints must guarantee sufficient shear strength to handle the extra forces from adjoining braces.

9. Conclusions

- This study compares experimental and analytical outcomes for a single-story steel frame utilizing buckling-restrained braces filled with TDA and concrete. It demonstrates the effectiveness of using tire-derived aggregates (TDAs) as an alternative material. It examines the innovative buckling-restrained brace with TDA infill, considering its potential for sustainability through reduced CO_2 emissions. A BRB with TDA contributes to sustainability and improves the damping properties compared to a conventional BRB with concrete infill. It provides numerical and experimental insights into structural responses, enhancing the existing knowledge. This study yields the following key conclusions:
- Analytical and experimental investigations on buckling-restrained braces (BRBs) with TDA infill have demonstrated a 25% increase in frame damping compared to a conventional BRB with a damping rate of approximately 14%. When higher damping is required, both the experimental and analytical findings suggest the use of a BRB with TDA infill. However, it is worth noting that a BRB with TDA filling offers a superior damping but reduced ductility compared to regular concrete, owing to the lower strength of TDA concrete, which cannot sufficiently delay steel core plate buckling as conventional concrete does [30].
- In general, a BRB with TDA filling reduced acceleration by approximately 20% compared to conventional concrete filling. The experimental study showed average acceleration values of 1.15 g with concrete and 0.91g with TDA fillings, while the analytical study results were 1.16 g with concrete and 0.89g with TDA fillings.
- The displacement value for the frame with the BRB and concrete infill was 2.44 (in) in the experimental and 2.26 (in) in the analytical study. Consequently, the BRB with TDA filling exhibits a lower ductility compared to the BRB with conventional concrete filling.
- A comparison of the hysteresis curves in both the analytical and experimental works revealed a similar trend in the backbone curves for the BRB with TDA and concrete infills subjected to FFEMA 4 loading. Furthermore, it was determined that the BRB with TDA infill absorbs less energy than the BRB with concrete infill, implying a lower ductility in the BRB with TDA infill compared to conventional concrete.

Additional experimental tests and numerical analyses in the future are necessary to fully understand the impact of a BRB with TDAFRC infill on the system performance. These studies may explore the potential for enhancing the ductility by incorporating fibers.

Author Contributions: Conceptualization, A.N., N.B.P. and L.C.; Methodology, A.N. and N.B.P.; Software, A.N.; Validation, A.N. and N.B.P.; Formal analysis, A.N.; Supervision, O.F. and F.M.T. All authors have read and agreed to the published version of the manuscript.

Funding: This research received no external funding.

Informed Consent Statement: Not applicable.

Data Availability Statement: Data are contained within the article.

Conflicts of Interest: The authors declare no conflict of interest.

References

1. Li, G.; Garrick, G.; Eggers, J.; Abadie, C.; Stubblefield, M.A.; Pang, S.-S. Waste Tire Fiber Modified Concrete. *Compos. Part B Eng.* **2004**, *35*, 305–312. [CrossRef]
2. Sukontasukkul, P. Use of Crumb Rubber to Improve Thermal and Sound Properties of Pre-Cast Concrete Panel. *Constr. Build. Mater.* **2009**, *23*, 1084–1092. [CrossRef]
3. Aslani, F. Mechanical Properties of Waste Tire Rubber Concrete. *J. Mater. Civ. Eng.* **2016**, *28*, 04015152. [CrossRef]

4. Ahn, I.S.; Cheng, L.; Fox, P.J.; Wright, J.; Patenaude, S.; Fujii, B. Material Properties of Large-Size Tire Derived Aggregate for Civil Engineering Applications. *J. Mater. Civ. Eng.* **2015**, *27*, 04014258. [CrossRef]
5. Ghaaowd, I.; McCartney, J.S.; Thielmann, S.S.; Sanders, M.J.; Fox, P.J. Shearing Behavior of Tire-Derived Aggregate with Large Particle Size. I: Internal and Concrete Interface Direct Shear. *J. Geotech. Geoenviron. Eng.* **2017**, *143*, 04017078. [CrossRef]
6. Mohammed, B.S.; Anwar Hossain, K.M.; Eng Swee, J.T.; Wong, G.; Abdullahi, M. Properties of Crumb Rubber Hollow Concrete Block. *J. Clean. Prod.* **2012**, *23*, 57–67. [CrossRef]
7. Topçu, I.B.; Avcular, N. Collision Behaviours of Rubberized Concrete. *Cem. Concr. Res.* **1997**, *27*, 1893–1898. [CrossRef]
8. Zheng, L.; Huo, X.S.; Yuan, Y. Strength, Modulus of Elasticity, and Brittleness Index of Rubberized Concrete. *J. Mater. Civ. Eng.* **2008**, *20*, 692–699. [CrossRef]
9. Atahan, A.O.; Sevim, U.K. Testing and Comparison of Concrete Barriers Containing Shredded Waste Tire Chips. *Mater. Lett.* **2008**, *62*, 3754–3757. [CrossRef]
10. Aiello, M.A.; Leuzzi, F. Waste Tyre Rubberized Concrete: Properties at Fresh and Hardened State. *Waste Manag.* **2010**, *30*, 1696–1704. [CrossRef]
11. Atahan, A.O.; Yücel, A.Ö. Crumb Rubber in Concrete: Static and Dynamic Evaluation. *Constr. Build. Mater.* **2012**, *36*, 617–622. [CrossRef]
12. Mohammed, B.S. Structural Behavior and m–k Value of Composite Slab Utilizing Concrete Containing Crumb Rubber. *Constr. Build. Mater.* **2010**, *24*, 1214–1221. [CrossRef]
13. Son, K.S.; Hajirasouliha, I.; Pilakoutas, K. Strength and Deformability of Waste Tyre Rubber-Filled Reinforced Concrete Columns. *Constr. Build. Mater.* **2011**, *25*, 218–226. [CrossRef]
14. Siddique, R.; Naik, T.R. Properties of Concrete Containing Scrap-Tire Rubber—An Overview. *Waste Manag.* **2004**, *24*, 563–569. [CrossRef] [PubMed]
15. Tehrani, F.M.; Miller, N.M. Tire-Derived Aggregate Cementitious Materials: A Review of Mechanical Properties. In *Cement Based Materials*; Saleh, H.E.-D.M., Rahman, R.O.A., Eds.; InTech: London, UK, 2018; ISBN 978-1-78984-153-4.
16. Tehrani, F.M.; Carreon, J.; Miller, N. An Investigation of Tire-Derived Lightweight Aggregate Concrete. *ACI Symp. Publ.* **2019**, *334*, 68–98. [CrossRef]
17. Treloar, L.R.G. Stress-Strain Data for Vulcanized Rubber under Various Types of Deformation. *Trans. Faraday Soc.* **1944**, *40*, 59–70. [CrossRef]
18. Wood, L.A. Uniaxial Extension and Compression in Stress-Strain Relations of Rubber. *J. Res. Natl. Bur. Stand.* **1977**, *82*, 57–63. [CrossRef] [PubMed]
19. Hertz, D.L., Jr. An analysis of rubber under strain from an engineering perspective. *J. Elastomerics* **1991**. Available online: https://www.sealseastern.com/pdf/rubberunderstrain.pdf (accessed on 2 November 2021).
20. Yang, H.; Jiang, L.; Zhang, Y.; Qi, P. Flexural Strength of Cement Paste Beam under Chemical Degradation: Experiments and Simplified Modeling. *J. Mater. Civ. Eng.* **2013**, *25*, 555–562. [CrossRef]
21. Toutanji, H.A. The Use of Rubber Tire Particles in Concrete to Replace Mineral Aggregates. *Cem. Concr. Compos.* **1996**, *18*, 135–139. [CrossRef]
22. Humphrey, D.; Blumenthal, M. The Use of Tire-Derived Aggregate in Road Construction Applications. In *Green Streets and Highways 2010*; American Society of Civil Engineers: Denver, CO, USA, 2010; pp. 299–313. [CrossRef]
23. CSI. Integrated Building Design Software, ETABS; *CSI Analysis Reference Manual, Computers and Structures*; Walnut Creek, CA, USA, 2016; Available online: https://www.csiamerica.com/products/etabs (accessed on 5 January 2021).
24. The Regents of the University of California. OpenSees User. 2000. Available online: https://opensees.berkeley.edu/wiki/index.php/OpenSees_User (accessed on 15 March 2021).
25. *ASTM D6270*; Standard Practice for Use of Scrap Tires in Civil Engineering Applications. ASTM: West Conshohocken, PA, USA, 2008. Available online: https://www.astm.org/d6270-08.html (accessed on 2 November 2021).
26. Miller, N.M.; Tehrani, F.M. Mechanical Properties of Rubberized Lightweight Aggregate Concrete. *Constr. Build. Mater.* **2017**, *147*, 264–271. [CrossRef]
27. Tehrani, F.M.; Masswadi, N.A.; Miller, N.M.; Sadrinezhad, A. An Experimental Investigation of Dynamic Properties of Fiber-Reinforced Tire-Derived Lightweight-Aggregate Concrete. *Eur. J. Eng. Technol. Res.* **2020**, *5*, 702–707. [CrossRef]
28. Kersting, R.A.; Fahnestock, L.A.; Lopez, W.A. *Seismic Design of Steel Buckling-Restrained Braced Frames: A Guide for Practicing Engineers*; National Institute of Standards and Technology: Gaithersburg, MD, USA, 2016; p. NIST GCR 15-917-34.
29. *ANSI/AISC 341*; Seismic Provisions for Structural Steel Buildings. American Institute of Steel Construction: Reston, VA, USA, 2016. Available online: https://www.aisc.org/2016-Seismic-Provisions (accessed on 5 November 2021).
30. Pathan, N.B.; Couch, L.; Tehrani, F.M.; Naghshineh, A.; Fischer, O. Experimental Seismic Evaluation of Novel Buckling-Restrained Braced Frames Containing Tire-Derived Aggregate Concrete. *Civ. Eng.* **2023**, *4*, 551–566. [CrossRef]
31. Naghshineh, A.; Tehrani, F.M.; Fischer, O. Application of Sustainable Concrete in the Seismic Evaluation of an Innovative Type of Buckling Restrained Brace. *Sustainability* **2022**, *14*, 16344. [CrossRef]
32. *ASTM A36/A36M-19*; Standard Specification for Carbon Structural Steel. ASTM International: West Conshohocken, PA, USA, 2019. [CrossRef]

33. Ancheta, T.D.; Darragh, R.B.; Stewart, J.P.; Seyhan, E.; Silva, W.J.; Chiou, B.S.J.; Wooddell, K.E.; Graves, R.W.; Kottke, A.R.; Boore, D.M.; et al. *PEER 2013/03—PEER NGA-West2 Database, PEER 2013/03*; Pacific Earthquake Engineering Research Center: Berkeley, CA, USA, 2013. Available online: https://ngawest2.berkeley.edu (accessed on 15 March 2021).
34. *FEMA 461*; Interim Testing Protocols for Determining the Seismic Performance Characteristics of Structural and Nonstructural Components. Federal Emergency Management Agency: Washington, CD, USA, 2007. Available online: https://www.atcouncil.org/pdfs/FEMA461.pdf (accessed on 15 March 2021).
35. *ASCE 41-17*; Seismic Evaluation and Retrofit of Existing Buildings (41-17). Available online: https://sp360.asce.org/PersonifyEbusiness/Merchandise/Product-Details/productId/233163464 (accessed on 10 January 2021).
36. *FEMA P-2208*; NEHRP Recommended Revisions to ASCE/SEI 41-17, Seismic Evaluation and Retrofit of Existing Buildings. Available online: https://www.fema.gov/sites/default/files/documents/fema_p-2208.pdf (accessed on 10 January 2021).
37. McKenna, F. Object-Oriented Finite Element Programming: Frameworks for Analysis, Algorithms and Parallel Computing. Ph.D. Thesis, Civil Engineering, University of California, Berkley, CA, USA, 1997.
38. McKenna, F.; Fenves, G.L.; Scott, M.H. *Open System for Earthquake Engineering Simulation*; University of California: Berkeley, CA, USA, 2000. Available online: http://opensees.berkeley.edu (accessed on 15 March 2021).
39. Mazzoni, S.; McKenna, F.; Scott, M.H.; Fenves, G.L. *OpenSees Command Language Manual*; University of California: Berkeley, CA, USA, 2006. Available online: http://opensees.berkeley.edu/manuals/usermanual (accessed on 15 March 2021).
40. McKenna, F.; Scott, M.H.; Fenves, G.L. Nonlinear finite-element analysis software architecture using object composition. *J. Comput. Civ. Eng.* **2010**, *24*, 95–107. [CrossRef]
41. *ASCE/SEI 7-22*; Minimum Design Loads and Associated Criteria for Buildings and Other Structures. American Society of Civil Engineers: Reston, VA, USA, 2022.

eng

Article

Assessment of Groundwater Contamination in the Southeastern Coast of Brazil: A Potential Threat to Human Health in Marica Municipality

Maria Cristina M. Publio [1,2], Jessica F. Delgado [1,2], Bruno S. Pierri [1], Leonardo da S. Lima [1,2], Christine C. Gaylarde [2,3], José Antônio Baptista Neto [1,2,4], Charles V. Neves [2] and Estefan M. Fonseca [1,2,*]

1 Programa de Pós-Graduação em Dinâmica dos Oceanos e da Terra, Universidade Federal Fluminense, Niterói 24210-346, RJ, Brazil; publiomcm@gmail.com (M.C.M.P.); jessicadelgado@id.uff.br (J.F.D.); pierri.bruno@hotmail.com (B.S.P.); leodslima@gmail.com (L.d.S.L.); jabneto@id.uff.br (J.A.B.N.)
2 Aequor-Laboratório de Inteligência Ambiental, R. Joaquim Eugênio dos Santos, 408-Eldorado, Maricá 24901-040, RJ, Brazil; cgaylarde@gmail.com (C.C.G.); contato@charlesneves.com.br (C.V.N.)
3 Department of Microbiology and Plant Biology, Oklahoma University, 770 Van Vleet Oval, Norman, OK 73019, USA
4 LAGEMAR—Laboratório de Geologia Marinha, Department of Geology and Geophysics, Instituto de Geociências, Universidade Federal Fluminense, Avenida Litorânea s/n, Niterói 24210-340, RJ, Brazil
* Correspondence: oceano25@hotmail.com

Abstract: Groundwater pollution is a current issue that may result in considerable negative effects on human health and the ecological balance. In the present study, the authors evaluated pollutants in groundwater in Maricá Municipality, located on the east side of Rio de Janeiro state in Brazil. The evaluated parameters were temperature, pH, electrical conductivity, Eh, dissolved oxygen, chlorides, nitrates, dissolved organic carbon, total inorganic carbon, phosphates, and total and thermotolerant coliforms. Due to the large number of evaluated points, they were divided into zones according to the respective hydrographic basin. The local accelerated urbanization accompanying income from oil production has led to uncontrolled population growth and associated groundwater pollution. The results of the present study suggest that sewage pollution of Maricá groundwater is already a significant issue. The lack of investment in basic sanitation has led to an imbalance in the local groundwater reservoir. In certain locations of the municipality, dissolved organic carbon (DOC), nutrient, and bacteria concentrations increase and spread in the aquifers because of domestic waste disposal. As aquifers are the main source of freshwater for the residents, contamination of them represents a potential threat to local public health.

Keywords: organic matter; pollution indicators; sewage; domestic pollution

Citation: Publio, M.C.M.; Delgado, J.F.; Pierri, B.S.; Lima, L.d.S.; Gaylarde, C.C.; Baptista Neto, J.A.; Neves, C.V.; Fonseca, E.M. Assessment of Groundwater Contamination in the Southeastern Coast of Brazil: A Potential Threat to Human Health in Marica Municipality. *Eng* **2023**, *4*, 2640–2655. https://doi.org/10.3390/eng4040151

Academic Editor: Alessio Cascardi

Received: 4 September 2023
Revised: 10 October 2023
Accepted: 13 October 2023
Published: 17 October 2023

1. Introduction

Groundwater is the largest global source of freshwater, representing a fundamental reservoir for humanity that is exploited for domestic, agricultural, and industrial purposes [1]. About 30% of the Earth's population uses groundwater as drinking water [2]. It is especially essential in arid and semi-arid areas, where rain is scarce and there are few surface water reserves [3]. Preserving an unpolluted and renewable source of groundwater for human demands represents one of the biggest challenges of sustainable development for every country [4]. This is even more important in the present scenario of global warming and the consequent rise in sea levels, which may result in the salinization of several subterranean water reservoirs located in coastal regions [5]. Additionally, uncontrolled human occupation and agglomeration, excessive urbanization, agricultural and industrial activities, and uncontrolled water exploitation all negatively impact groundwater quality and availability [6].

Groundwater contamination due to organic pollutants such as hydrocarbons and pesticides and inorganic substances like heavy metals, microplastics, and endocrine disruptors poses a substantial threat to human health [7,8]. During the last three decades, several authors have considered the chemical pollution of groundwater [6,9–13]. Although the major pollutants are of geologic origin, resulting from dissolution of the aquifer rock within the Earth's crust [14–16], the recent increase in the world's population has led to increases in anthropogenic contaminants. The most impacted areas are those that are experiencing rapid economic development [17–19]. Several researchers have focused on the impact of megacities on groundwater [20,21], but few scientists have evaluated small coastal cities, where urbanization has recently intensified.

The Brazilian coastal area currently comprises 40% of the country's total population. Although some coastal cities in Brazil have moderate groundwater demand, the constant population increase tends to intensify groundwater exploitation [22]. Because of the recent discovery of the huge petroleum reservoir in the oceanic pre-salt layer and oil production expansion, certain Brazilian municipalities located on the southeast coastal zones have benefited from fees charged by the oil extraction industry and are now facing new social and environmental impacts [23]. One of the main results is exponential migration and the resulting increase in population accompanying new job opportunities. This has occurred in the coastal city of Maricá.

The present study provides a diagnosis and spatial groundwater chemical and microbiological characterization of wells from sedimentary and fissured aquifers in this recently urbanized coastal area. It discusses contamination and salinization of aquifers, identifying patterns that may be occurring in many recently urbanized coastal regions in Brazil. One of the main objectives of the present study is, through the assessment of groundwater quality in the coastal municipality of Maricá, to inquire about whether the royalty payments from the oil industry to these municipalities have helped mitigate secondary socio-environmental impacts arising from oil production or whether these same payments have generated new problems, particularly in water quality, due to the local population increase resulting from the pursuit of new income opportunities.

2. Study Site

Maricá is a municipality of the Rio de Janeiro metropolitan area, located on the Atlantic coast. Its geology is marked by the presence of metamorphic, igneous, and sedimentary rocks, with great geodiversity. It includes the Jaconé beachrocks, whose sedimentary formations were described by Darwin and are considered natural heritage of high geological, environmental, landscape, and cultural relevance.

The coastal geomorphology of Maricá presents an ecologically important lagoonal system, composed of four main lagoons and two sandy barriers that isolate the coastal system from the adjacent sea. Its sedimentary history, architecture, and evolution took place in the Quaternary. It is composed of six lithological units forming three sedimentary sequences compartmented by erosive surfaces [24]. As for the geological aspects, the coast located to the west of the Cabo Frio point undergoes a sudden inflection, having an east–west alignment, and is called the Coastal Cords coast [21]. In structural terms, it is part of a geological complex composed predominantly of granite and gneiss, represented in the landscape by coastal massifs.

The aquifers found in the area are of the granular type, comprising heterogeneous alluvial deposits and portions of soil and weathered rock. Alluvial sediments are composed predominantly of quartz sands and, subordinately, silty and clayey soils, which occur in the vicinity of the lagoons. These are due to the formation of the coastal lowland during the last transgression and regression movements in the Holocene, between 5000 and 7000 years ago. The coastal lowland is formed by lateral variations parallel to the coastline, with sandy highs and low vegetation typical of restinga and sandy-clay wetlands. These make up the marshes and mangroves in the region. The underground water horizon level is generally near the soil surface, at depths from a few centimeters to a few meters. In the area of the sands, which corresponds to the shallowest places, the water is generally less brackish than in the swampy lowlands [25]. The municipality has groundwater as its main source of freshwater. With the growing demand for water resources underground, supply and contamination problems are expected, in addition to the phenomenon of seawater intrusion in these aquifers. The groundwater resource in the target area has been suffering degradation in its quality in response to the advance of the marine saline wedge, brought about by excessive uncontrolled pumping and organic pollution [25].

3. Methodology

A total of 124 groundwater samples were collected through pre-installed wells already in use by the population in the plain area of the Maricá Basin (Figure 1) during the winter season. The points were divided into nine zones with different urban occupation degrees and with different hydrographic basins. All of the groundwater samples were kept in a portable refrigerator and transported back to the lab at 4 °C to be analyzed within one week. Chemical examination of the groundwater included determination of the temperature, pH, electrical conductivity, Eh, turbidity, dissolved oxygen, dissolved solids, chlorides, nitrates, dissolved organic carbon, dissolved inorganic carbon, phosphates, and total and thermotolerant coliforms. Proper sampling techniques and handling were used to produce high-quality data. Brown glass sampling bottles, which had been previously cleaned with deionized water, underwent a triple rinse with water collected from each sampling location before being used for sampling. Each rinsing cycle utilized a volume of 2.5 L. Subsequently, these bottles were transported to the laboratory within 24 h for preliminary processing. The water samples were analyzed in accordance with the testing procedures outlined in the Groundwater Quality Standard (GB/T 14848-2017 [26]).

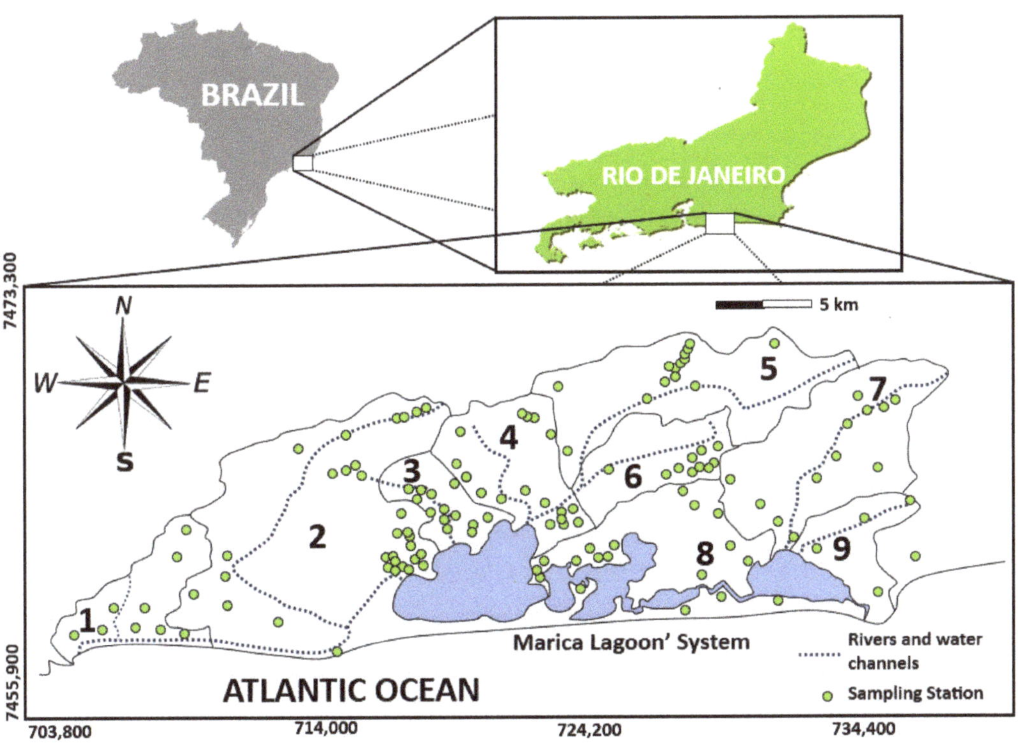

Figure 1. Study site and sampling stations in the 9 pre-determined zones.

The first five parameters (temperature, pH, electrical conductivity, Eh, dissolved oxygen) were measured in situ using a multiparameter Horiba U10 probe. The chloride analysis used back titration with potassium thiocyanate. Dissolved organic carbon (DOC) was determined via the method defined by Van Hall et al. [27]. Interferences from particulate carbon and inorganic carbon were removed before analysis via filtration through glass fiber filters and sparging with CO_2-free gas after acidification of the sample [28]. Phosphate and nitrate were analyzed via colorimetric methods.

To determine bacterial levels, water samples were plated on mFC (membrane fecal coliform agar) and typical colonies identified as *E. coli* with EC-MUG (*Escherichia coli*-methylumbelliferyl-β-D-glucuronide) medium. The 124 samples obtained during the present study were also evaluated directly for *E. coli* through modified mTEC (membrane thermotolerant *Escherichia coli*) agar [29].

Data for all parameters were tested for homoscedasticity and normality before being submitted to one-way ANOVA (variance analyses) to test whether there was a statistical difference between the zones. The Tukey test was used to compare average values when necessary. Pearson's correlation analysis was performed between the total coliform data and the other parameters. All statistical analyses were performed using the GraphPrism 8.0 software (GraphPad Software—v.10.0.3), accepting 5% variation as significant.

4. Results and Discussion

The chemical constitution of groundwater is directly influenced by the composition of the water percolating the soil, as well as the resulting by-products and kinetically controlled reactions within the aquifer basin and overlying subsoil coverage [30]. As a universal solvent, water mobilizes minerals from the soil and bedrock with which it comes into contact. The physicochemical characteristics of drinking water, on the other hand (temperature, pH, dissolved oxygen levels, conductivity, turbidity, organic matter and NH3, potential contaminants and other chemical constituents), eventually impact its portability, in some cases influencing consumers' health, possibly as a result of the survival of harmful microorganisms [31].

Surface water temperature controls the metabolism of the aquatic ecosystem. For instance, relatively high water temperatures may reduce its ability to hold essential dissolved gasses like oxygen, potentially killing fish and other water organisms [32]. Variations in groundwater temperature may trigger changes in biogeochemical mechanisms in the subsurface environment, impacting water quality [33,34]. Laboratory assays allow for the identification of processes like carbonate precipitation, silicate dissolution, mobilization of cations, trace elements, and dissolved organic carbon (DOC) [35–39]. In the present study, the temperature values varied between 21.5 and 29.61 °C, with one anomalous record of 35.1 °C (average of 24.83 °C). Opportunistic pathogens, including *Legionella* spp., can grow in the water systems in buildings, representing a critical public health issue [40,41]. According to World Health Organization, water temperature is important for the control of *Legionella* spp., and water temperatures should be maintained above 50 °C to prevent the growth of this organism [42].

PH represents one of the main physicochemical parameters influencing the behavior of water-quality parameters as well as pollutant concentrations in aquatic ecosystems [43]. This parameter also influences the communities of bacteria and other microorganisms. In general, relatively high or low pH conditions can make water improper for certain purposes. At higher pH levels, metals tend to precipitate, whereas other compounds like ammonia become toxic to aquatic life, releasing bad odors and tastes [44]. At lower pH levels, heavy metals tend to dissolve, becoming bioavailable, and chemicals like cyanide and sulfide become more toxic. The pH values determined in the present study are presented in Table 1.

Of the 124 sampling stations, 47 showed pH values in disagreement with Brazilian legislation, which represents almost 38% of the monitored points. Every zone had at least two sampling stations presenting values above the minimum allowed by the Brazilian law CONAMA 357/05. It is important to emphasize that the minimum pH value in zone 02

(3.11) is alarming. This value is close to the pH for solutions such as vinegar and orange juice. Additionally, five other sampling stations in zone 02 recorded pH values below 5.00. There was no statistically significant difference in pH values among the zones ($p > 0.05$).

Table 1. Summary of the pH values of all areas sampled.

Zone	Mean	Minimum Value	Maximum Value	Brazilian Legislation Limits	N° of Points in Disagreement with Brazilian Legislation	Proportion of Points in Disagreement with Brazilian Legislation, %
01	5.92	5.01	6.88		06	66.66
02	5.83	3.11	6.97		10	34.48
03	6.52	4.92	7.54		02	11.76
04	6.18	4.95	7.91		07	43.75
05	6.17	5.23	7.20	6.00 to 9.00	03	50.00
06	6.23	5.43	7.09		06	50.00
07	6.25	5.69	6.97		03	30.00
08	5.95	3.45	7.25		07	38.88
09	6,04	4,88	6,79		03	42,85

The U.S. Environmental Protection Agency and the equivalent Arabian legislation suggest that the pH of water sources should be maintained at between 6.5 and 8.5. According to Nasseem et al. (2022), it is better to have greater alkalinity, below 7.0, in human drinking water, since it keeps the water safe for drinking. According to the same authors, acidic water with a pH of less than 6.5 suggests potential contamination with pollutants, making it unsafe for drinking purposes. Additionally, water with a pH < 6.5 could be corrosive, leaching metal ions such as Fe, Mn, Cu, Pb, Ni, Cr, and Zi from the aquifer rocks or transporting piping nets [6]. Throughout the present study, water pH values were more acidic than those acceptable by the legislation, varying at around 6.04, and several times being below 6. In fact, 81 sampling stations (approximately 65%) showed unacceptable pH values.

Dissolved oxygen (DO) levels have a direct impact on groundwater quality by regulating the valence state of heavy metals and the microbial catabolism of dissolved organic compounds [45]. A decrease in dissolved oxygen levels can result in anaerobic patterns, which negatively impacts aquatic organisms. The organic matter potentially released from surface sources through a groundwater reserve, on the other hand, can rapidly decrease the dissolved oxygen in the groundwater, turning it into a reducing underground environment more susceptible to the dissolution of the Fe and Mn that compose the aquifer rocks [46]. The Brazilian law CONAMA 357/05 suggests a minimum dissolved oxygen concentration of drinking water of 6.00 $mg \cdot L^{-1}$. In the present study, the recorded oxygen levels (Figure 2) suggest a high oxygen demand in some zones. Zones 05 and 07 showed DO concentrations of 4.21 and 4.93 $mg \cdot L^{-1}$, respectively. On the other hand, the other seven zones did not suggest a high oxygen demand, with DO concentrations varying at around 6 mg/L. When we compared the DO concentration of zones 05 and 07 with that in zone 08, we found a statistically significant difference between them.

The dissolved organic carbon (DOC) sometimes showed high concentrations, especially in the most urbanized zones. The DOC averages varied. All zones had at least two collection points whose values exceeded the CONAMA 357/05 limit of 10 $mg \cdot L^{-1}$. Only one zone (zone 03) had an average value above the maximum allowed by law. Approximately 35% of all sampled stations (43 points of all zones) had values above the limits allowed by CONAMA 357/05. Zones 05, 06, 07, and 03 were those with the highest proportion of points above this limit. The summary of the DOC values is shown in Table 2.

The potability of groundwater is directly linked to DOC concentrations, which influence water chemistry and microbial levels [47–51]. As a result, there have been several studies on the concentrations, sources, and diffusion of natural DOC in aquifers [52–54]; however, to date, few have focused on human health. DOC consists of the carbon part of dissolved organic matter (DOM), typically representing more than 90% of the whole organic carbon content in natural groundwater reserves [55]. Nevertheless, levels above background concentrations may indicate organic pollution [48,54,56].

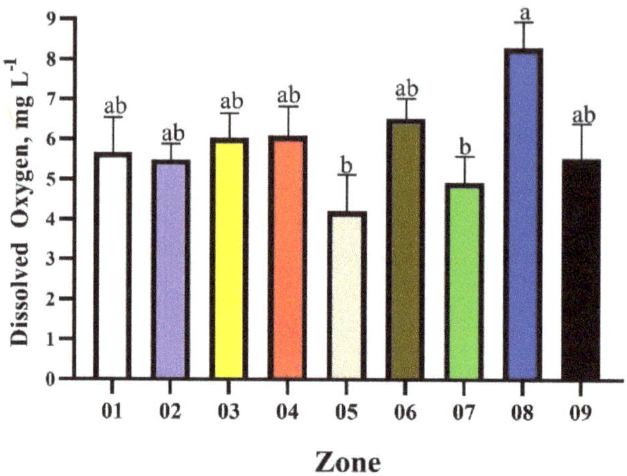

Figure 2. Dissolved oxygen (mg·L^{-1}) of all nine zones. The values are shown as mean ± SD. Different letters show statistically significant differences between the zones ($p < 0.05$).

Table 2. Summary of the DOC values of all areas sampled.

Zone	Mean	Minimum Value	Maximum Value	Brazilian Legislation Limits	N° of Points in Disagreement with Brazilian Legislation	Proportion of Points in Disagreement with Brazilian Legislation, %
01	8.01	1.90	35.30		02	22.22
02	5.40	0.50	29.20		03	44.82
03	11.56	1.20	49.10		08	47.05
04	7.97	0.00	28.00		05	31.25
05	7.77	0.90	18.40	10 mg·L^{-1}	04	66.66
06	9.91	0.60	35.00		06	50.00
07	6.41	0.50	15.10		05	50.00
08	9.36	0.50	37.00		07	38.88
09	6.37	0.50	12.00		03	42.85

Comparing the results of the present study with the Canadian legislation for untreated drinking water (4 mg·L^{-1} of DOC), we observed that the percentage of points above the allowed value increased to 53.22%, which corresponds to 66 sampling stations. This means that more than half of the water wells would be in unfit condition for consumption according to Canadian legislation.

High concentrations of DOC in water may result in aesthetic and odor problems, in addition to the potential stimulation of pathogenic bacteria [57–59]. High DOC levels also result in a significant impact on the geochemical dynamic of other pollutants, such as pesticides, pathogens, and pharmaceuticals; lower oxygen conditions may decrease the degradation of carbon-based compounds [60]. High levels of COD in groundwater may also suggest the development of trihalomethanes in water disinfected with active chlorine. When compared to other environments, the values obtained in the present study were significantly high (Table 3).

Table 3. Groundwater dissolved organic carbon obtained in different studies.

Site	Reference	Max DOC ($mg \cdot L^{-1}$)	Min DOC ($mg \cdot L^{-1}$)	Average DOC ($mg \cdot L^{-1}$)
Maricá, Brazil	Present study	49.10	<0.5	8.06
Puducherry, India	[61]	29	0.5	3.6
Petit Hermitage, France	[62]	-	-	4.4
Sri Lanka	[63]	2.08	1.35	1.69
Bangladesh	[64]	20.39	1.29	4.90
Anna Bay, Australia	[65]	15.1	1.0	8.0
Bangladesh	[66]	5.9	0.20	2.79
Macquarie River, Australia	[67]	-	-	8.26
Bell River, Australia	[67]	-	-	1.87
Elfin Crossing, Australia	[67]	-	-	1.34
Silurian S, Scotland	[68]	3.05	0.45	-

In a way similar to that of other substances, chloride in groundwater originates from both geogenic and anthropogenic sources. In the second case, agricultural, industrial, and/or domestic wastes are considered potential causes [69,70]. Chloride concentrations in uncontaminated waters often stay below 10 $mg \cdot L^{-1}$ and sometimes below 1 $mg \cdot L^{-1}$. On the other hand, high levels of chlorides are suggestive of pollution [71]. Although they may be present in most freshwater ecosystems, large concentrations are potentially toxic to freshwater organisms. The results are shown in Table 4.

Table 4. Summary of the chloride values of all areas sampled.

Zone	Mean	Minimum Value	Maximum Value	Brazilian Legislation Limits	N° of Points in Disagreement with Brazilian Legislation	Proportion of Points in Disagreement with Brazilian Legislation, %
01	170.4	32.0	853.6		01	11.11
02	157.9	12.2	1259.6		04	13.79
03	185.8	32.5	483.2		05	29.41
04	93.0	21.0	317.0		01	6.25
05	327.5	16.4	1472.4	250 $mg \cdot L^{-1}$	02	33.30
06	58.8	11.5	178.3		00	0.00
07	14.0	77.1	1.04		00	0.00
08	148.9	739.8	27.54		01	5.55
09	382.8	29.5	1466.2		03	42.85

The by-products of chlorine disinfection potentially affect consumers' health. Their magnitude is influenced by a number of variables, such as period of action, levels, and frequency of exposure. Compared to other locations, the studied area revealed extremely high values. The results found in this study ranged between 11.5 and 1466.2 $mg \cdot L^{-1}$. The U.S. Environmental Protection Agency (EPA) and the Brazilian legislation recommend a maximum value of 250 $mg \cdot L^{-1}$ for drinking water. In the present study, several sampling stations recorded values significantly higher than that. Chloride compounds have two main natural sources: soil chloride dissolved during water runoff and seawater intrusion during high tides. An addition of 1% seawater can increase the chloride content to 190 $mg \cdot L^{-1}$ [72]. Anthropogenic chloride sources are mainly industrial wastewater discharge in densely occupied areas; industrial wastewater and domestic sewage represent significant sources of chloride in water bodies. In the present study, chloride values varied between 1472.4 and 11.5 $mg \cdot L^{-1}$ (average 154 $mg \cdot L^{-1}$). Aside from the potential health threats related to high blood pressure, these chloride concentration limits have been established to protect water from tasting salty and to prevent a corrosive effect on plumbing.

Nitrogen and phosphorus levels may indicate wastewater contamination and can be a threat to human health. Excessive levels of nitrate may cause disease. High levels of nitrate in drinking water may decrease blood oxygen transport, resulting in health issues such as blue baby syndrome [73,74]. According to the China Bureau of Quality and Technical Supervision, the acceptable limits of nitrate for drinking groundwater is 20.0 mg·L^{-1} [3]. In the present study, all values of nitrate remained below these values. Brazilian CONAMA 357 legislation limits the nitrate concentration in drinking water to 10 mg·L^{-1}.

The nitrate concentration found in the present study ranged from 0.00 to 27.98 mg·L^{-1}. Zones 04, 05, and 07 showed 100% of their collection stations, in accordance with CONAMA 357. All zones yielded mean and median values below the maximum allowed by Brazilian law. However, zones 08 and 09 had the highest proportion of points in disagreement, with approximately 15% being above the Brazilian law limits. A summary of the nitrate values can be found in Table 5.

Table 5. Summary of the nitrate values of all areas sampled.

Zone	Mean	Minimum Value	Maximum Value	Brazilian Legislation Limits	N° of points in Disagreement with Brazilian Legislation	Proportion of Points in Disagreement with Brazilian Legislation, %
01	4.15	0.00	16.89	10 mg·L^{-1}	01	11.11
02	3.37	0.05	15.04		02	6.89
03	1.89	0.05	15.84		01	5.88
04	2.66	0.00	7.00		00	0.00
05	0.16	0.05	0.41		00	0.00
06	2.23	0.05	11.47		01	8.33
07	0.28	0.05	1.04		00	0.00
08	4.86	0.05	27.54		03	16.66
09	4.51	0.08	27.98		01	14.28

Although the findings for nitrate concentrations were relatively positive, the total organic nitrogen concentration (TON) presented a slightly more worrying scenario. As for nitrate, zone 05 presented 100% of its collection points with TON concentrations in accordance with CONAMA 357. However, zone 09 had the highest average TON, as well as the point with the highest overall concentration (36.23 mg·L^{-1}). Zones 01 and 08 also showed average values above the legal maximum. A total of 50% of the points in Zone 08 had TON concentrations above the legal limit. The ANOVA showed a statistical difference between zone 03 and zones 08 and 09, with zones 08 and 09 presenting TON means significantly higher than that of zone 03 ($p < 0.05$). The same results were observed for zones 05, 08, and 09, with zones 08 and 09 presenting TON means significantly higher than that of zone 05 ($p < 0.05$). There was no difference between the other studied areas. Figure 3 presents the statistical analysis across all zones, and Table 6 shows the summary of the TON information for all areas studied.

Table 6. Summary of the TON values of all areas.

Zone	Mean	Minimum Value	Maximum Value	Brazilian Legislation Limits	N° of Points in Disagreement with Brazilian Legislation	Proportion of Points in Disagreement with Brazilian Legislation, %
01	3.61	0.30	13.90	3.7 mg·L^{-1}	03	33.33
02	3.20	0.10	17.00		10	34.48
03	1.16	0.10	5.50		02	11.76
04	3.28	0.00	8.70		07	43.75
05	0.67	0.30	0.90		00	00.00
06	2.85	0.30	14.30		02	16.66
07	1.24	0.10	4.20		01	10.00
08	7.87	1.40	30.60		09	50.00
09	10.70	0.10	36.23		01	14.28

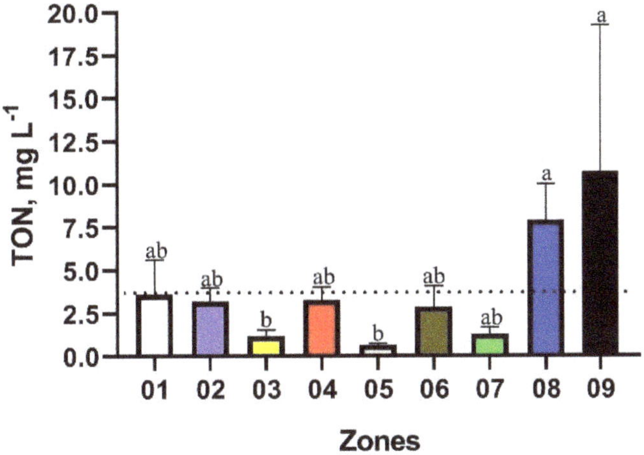

Figure 3. Total organic nitrogen (mg·L^{-1}) of all nine zones. The values are shown as mean ± SD. Different letters indicate significant statistical differences between the zones ($p < 0.05$). The horizontal dotted line shows the maximum limit per Brazilian legislation.

Phosphorus is a fundamental element for all living beings. However, excessively elevated phosphorus levels can result in eutrophication of surface water. Groundwater, on the other hand, can be an important diffuse source of phosphorus to surface aquatic environments. It is generally accepted that phosphate groundwater concentrations are negligible due to high rates of adsorption to the soil and sediment matrix. In the present study, recorded values of phosphate varied between 2.32 mg·L^{-1} and the detection limit (<0.01 mg·L^{-1}). The phosphorus concentrations indicated excess phosphorus, with at least 40% of the collection points of all zones being in disagreement with CONAMA 357 (Table 7). Approximately 69% of all zones had values above the CONAMA 357 maximum of 0.02 mg·L^{-1}. The ANOVA showed a statistical difference between zones 01 and 02 and zone 07 (Figure 4), with zone 07 presenting a mean phosphorus concentration that was significantly higher than those of zones 01 and 02 ($p < 0.05$).

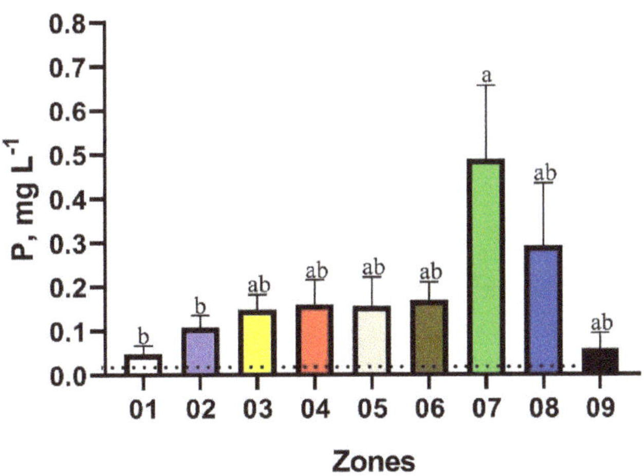

Figure 4. Phosphorus concentration (mg·L^{-1}) of all nine zones. The values are shown as mean ± SD. Different letters indicate significant statistical differences between the zones ($p < 0.05$). The horizontal dotted line shows the maximum limit per Brazilian legislation.

Table 7. Summary of the phosphorus concentrations of all areas sampled.

Zone	Mean	Minimum Value	Maximum Value	Brazilian Legislation Limits	N° of Points in Disagreement with Brazilian Legislation	Proportion of Points in Disagreement with Brazilian Legislation, %
01	0.05	0.01	0.19		04	44.44
02	0.11	0.01	0.61		16	55.17
03	0.15	0.01	0.44		13	76.47
04	0.16	0.00	0.89		14	87.5
05	0.16	0.02	0.46	$0.02\ mg{\cdot}L^{-1}$	05	83.33
06	0.17	0.01	0.49		11	91.66
07	0.35	0.01	1.23		07	70.00
08	0.26	0.01	2.32		12	66.66
09	0.06	0.01	0.26		03	42.85

Most pathogenic microorganisms found in water are of fecal origin and are spread via fecal–oral exposure. These pathogenic agents can produce relatively mild gastrointestinal disorders or serious disease such as infectious hepatitis, encephalitis, and myocarditis. Thus, the negative impacts resulting from the consumption of contaminated water on human health can range from low-level infections (mild diarrhea) lasting a few days to critical illnesses that require medical care or even hospitalization, potentially culminating in death [75]. Fecal coliforms constitute a group of relatively harmless microorganisms present in the intestinal tract of warm- and cold-blooded animals that are also involved in the digestive process [76]. Of these, *Esterichia coli* is the most common representative and indicates that the water is polluted with fecal material from humans or other animals [77].

On the other hand, the mere presence of these organisms, although harmless, may indicate that the water may have been contaminated by pathogenic organisms or disease-causing bacteria or viruses, which may also exist in fecal material. Runoff is the main carrier mechanism of pathogen transport to surface water bodies. During a rainfall event, the water distribution between surface runoff and soil percolation results from many factors: storm water flux and extension, soil physical features (e.g., porosity and permeability), land height gradients, and soil vegetal or urban cover [78]. If rainfall load exceeds the capacity of the soil to absorb water, overland flux occurs, and microorganisms can be transported in surface runoff [79,80].

The results found in the present study suggest significant important contamination of the groundwater with some source of industrial or, mainly, domestic sewage. Brazilian legislation allows the presence of up to 200 MPN (most probable number) of fecal coliforms in water. In the present study only zone 05 did not showed no collection points with values above the legal limit. However, even at concentrations within the permitted limit, all points in zone 05 showed the presence of fecal coliform cells, which suggests contamination with domestic sewage of the groundwater in that zone, even if small. At least 13% of the collected points of the other zones had coliform concentrations above 200 MPN; for example, 4 of the 29 monitoring stations in zone 02 had values above the Brazilian limit. Although in a relatively low proportion (compared to the other zones), the most worrying factor in zone 02 is the presence of highly contaminated samples, with concentrations up to 12 times higher (2400 MPN) than allowed by Brazilian legislation.

The other zones presented a critical scenario, with the proportion of contaminated points ranging from 35.29 to 83.33%. The minimum average value in these zones (01, 03, 04, 06, 07, 08, and 09) was 218 MPN, which is already above the maximum allowed by Brazilian legislation. The worrying maximum average value was 1455 MPN, found in zone 06. In addition to this worrying average, it is important to note that 83.33% of the sampled sites in zone 06 had contamination above the permitted level, further aggravating the magnitude of the problem in that area. Another important fact is that, of all 124 points analyzed, only 7 showed no presence of coliforms. With the exception of zone 05, all other zones had at least 1one point with at least four times more total coliforms than allowed by law. Table 8 presents the summary of the results found for total coliforms in all zones.

Table 8. Summary of the total coliform concentrations of all areas sampled.

Zone	Mean	Minimum Value	Maximum Value	Brazilian Legislation Limits	N° of Points in Disagreement with Brazilian Legislation	Proportion of Points in Disagreement with Brazilian Legislation, %
01	370	0	2400		04	44.44
02	218	2	2400		04	13.79
03	286	2	1600		06	35.29
04	433	0	1600		06	37.5
05	18	2	49	200 MPN	00	0.00
06	1455	2	1600		10	83.33
07	934	0	920		04	40.00
08	962	2	1600		09	50.00
09	291	0	1400		03	42.85

Pathogens are released from sewage in the soil; however, most of them remain associated with the fecal deposit. The amount depends on a number of factors, such as the source, age, type, and pathogen level in the fecal matter itself, as well as the survival characteristics of the pathogens. Pathogen survival in water depends on many factors, including water quality (e.g., turbidity, dissolved oxygen, pH, organic matter content) and environmental conditions (e.g., temperature, predation by zooplankton). Exposure to UV light is a key factor in bacterial, viral, and protozoan die-off in surface waters [78,81,82]. An aquifer environment also protects pathogens against UV exposure and facilitates their survival in groundwater. *E. coli* and fecal enterococci (FE) counts were suggested in some sampling sites. The results of the present study show a strong and positive correlation ($p < 0.05$) between total coliforms and the concentration of DOC, TON, and phosphorus in groundwater, corroborating possible contamination with domestic sewage. Figure 5 shows the three correlations.

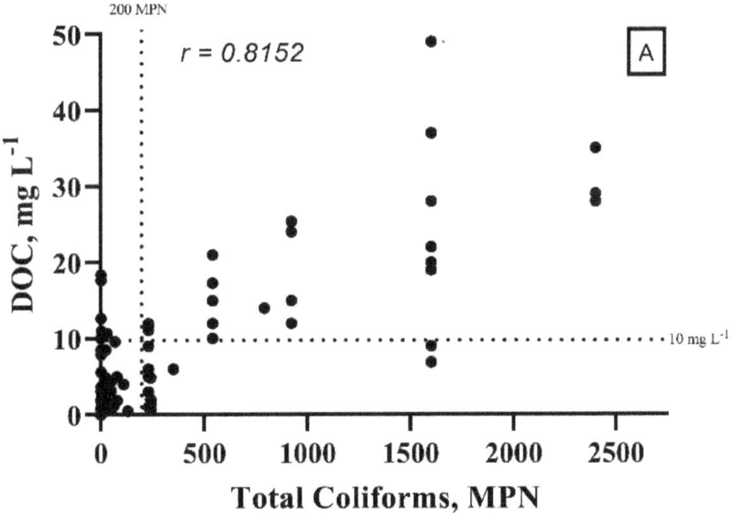

Figure 5. *Cont.*

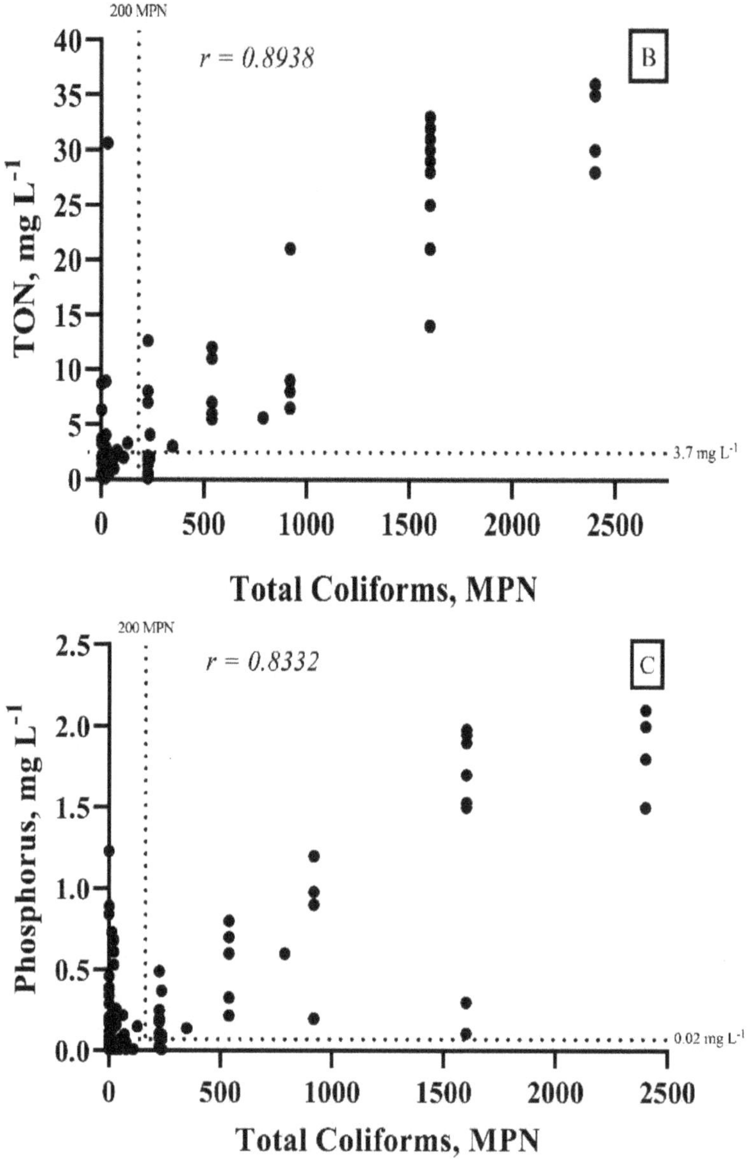

Figure 5. Pearson correlation coefficients between total coliforms and DOC (**A**), TON (**B**), and phosphorus (**C**) for all nine zones. The vertical dotted line shows the maximum limit per Brazilian legislation for total coliforms (200 MPN). The horizontal dotted line shows the maximum limit per Brazilian legislation for DOC, TON, and P.

5. Conclusions

Availability of groundwater of good quality is essential for the maintenance of human and animal health, especially in areas with scarce freshwater reservoir availability. This necessary quality can be ensured by regular monitoring and protection of water sources against potential contamination. In the present study, quality evaluation based on the chemical composition of groundwater revealed that the required level of some parameters for human health required by the legislation used around the world is not being met. None of the monitored points had the minimum quality required by Brazilian legislation; all points had at least one parameter in excess. The results of organic matter and fecal coliform evaluations suggest that one of the main sources of pollutants is domestic sewage resulting from a lack of basic sanitation. Additionally, with the exponential population growth observed in the municipality of Maricá and its plans for agro-industrial development, it is expected that the already established conditions will deteriorate significantly, with pollutants of a more recalcitrant and, therefore, more toxic nature being disseminated in the environment. Urgent measures of basic sanitation must be implemented, even before the implementation of an industrial park, as planned by the municipal government, to prevent even more severe impacts from being detected in the near future.

Although only one collection campaign was carried out, it was performed in the winter, when an improvement in groundwater quality is expected. The quality of the local water would be even worse in the summer, when the most intense rains promote greater transport of pollutants to the subterranean compartment.

Groundwater pollution represents a real danger to human communities. There is a great opportunity for the global scientific community to improve existing aquifer management, highlighting its importance to decision-makers and allowing them to understand how we can defend or rescue both the quality and the quantity of these essential resources.

Author Contributions: M.C.M.P.: experiment execution, analysis, and literature review. J.F.D.: experiment execution, analysis, literature review, and writing. B.S.P.: analysis, literature review, and writing. L.d.S.L.: analysis, literature review, and writing. C.C.G.: literature review, English review, and writing. J.A.B.N.: literature review and writing. C.V.N.: analysis and literature review. E.M.F.: supervisor researcher, literature review, and writing. All authors have read and agreed to the published version of the manuscript.

Funding: This research was funded by the Conselho Nacional de Desenvolvimento Científico e Tecnológico (CNPq) and Companhia de Desenvolvimento de Maricá (CODEMAR).

Institutional Review Board Statement: The Institutional Review Board Statement is not applicable to this study, as it does not involve human or animal testing.

Informed Consent Statement: The Informed Consent Statement is not applicable to this study, as it does not involve human or animal testing.

Data Availability Statement: The datasets generated during and/or analyzed during the current study are available from the corresponding author on reasonable request.

Conflicts of Interest: The authors declare that the research was conducted in the absence of any commercial or financial relationships that could be construed as potential conflicts of interest.

References

1. Ullah, Z.; Rashid, A.; Ghani, J.; Nawab, J.; Zeng, X.-C.; Shah, M.; Alrefaei, A.F.; Kamel, M.; Aleya, L.; Abdel-Daim, M.M.; et al. Groundwater Contamination through Potentially Harmful Metals and Its Implications in Groundwater Management. *Front. Environ. Sci.* **2022**, *10*, 1021596. [CrossRef]
2. International Association of Hydrogeologists. *Groundwater—More about the Hidden Resource*; International Association of Hydrogeologists: London, UK, 2020.
3. Li, P.; Tian, R.; Xue, C.; Wu, J. Progress, Opportunities, and Key Fields for Groundwater Quality Research under the Impacts of Human Activities in China with a Special Focus on Western China. *Environ. Sci. Pollut. Res.* **2017**, *24*, 13224–13234. [CrossRef]
4. Velis, M.; Conti, K.I.; Biermann, F. Groundwater and Human Development: Synergies and Trade-Offs within the Context of the Sustainable Development Goals. *Sustain. Sci.* **2017**, *12*, 1007–1017. [CrossRef]
5. Mazhar, S.; Pellegrini, E.; Contin, M.; Bravo, C.; De Nobili, M. Impacts of Salinization Caused by Sea Level Rise on the Biological Processes of Coastal Soils—A Review. *Front. Environ. Sci.* **2022**, *10*, 909415. [CrossRef]
6. Saalidong, B.M.; Aram, S.A.; Otu, S.; Lartey, P.O. Examining the Dynamics of the Relationship between Water PH and Other Water Quality Parameters in Ground and Surface Water Systems. *PLoS ONE* **2022**, *17*, e0262117. [CrossRef]
7. Li, P. To Make the Water Safer. *Expo. Health* **2020**, *12*, 337–342. [CrossRef] [PubMed]
8. Li, P.; Wu, J. Sustainable Living with Risks: Meeting the Challenges. *Human. Ecol. Risk Assess. Int. J.* **2019**, *25*, 1–10. [CrossRef]

9. Ugochukwu, U.C.; Ochonogor, A. Groundwater Contamination by Polycyclic Aromatic Hydrocarbon Due to Diesel Spill from a Telecom Base Station in a Nigerian City: Assessment of Human Health Risk Exposure. *Environ. Monit. Assess.* **2018**, *190*, 249. [CrossRef] [PubMed]
10. Marić, N.; Štrbački, J.; Mrazovac Kurilić, S.; Beškoski, V.P.; Nikić, Z.; Ignjatović, S.; Malbašić, J. Hydrochemistry of Groundwater Contaminated by Petroleum Hydrocarbons: The Impact of Biodegradation (Vitanovac, Serbia). *Environ. Geochem. Health* **2020**, *42*, 1921–1935. [CrossRef]
11. Nivetha, C.; Deepika, T.; Arjunan, A.; Sivalingam, P.; Revathi, N.; Muthuselvam, M. Antimicrobial and Antioxidant Activities of Streptomyces Sps Isolated from Muthupettai Mangrove Soil. *J. Pharm. Res. Int.* **2021**, *33*, 210–234. [CrossRef]
12. Li, P.; Karunanidhi, D.; Subramani, T.; Srinivasamoorthy, K. Sources and Consequences of Groundwater Contamination. *Arch. Environ. Contam. Toxicol.* **2021**, *80*, 1–10. [CrossRef] [PubMed]
13. Monteiro da Fonseca, E.; Machado Publio, M.C.; de Freitas Delgado, J. Segurança Hídrica. *Sist. Gestão* **2023**, *18*, 2. [CrossRef]
14. Basu, A.; Saha, D.; Saha, R.; Ghosh, T.; Saha, B. A Review on Sources, Toxicity and Remediation Technologies for Removing Arsenic from Drinking Water. *Res. Chem. Intermed.* **2014**, *40*, 447–485. [CrossRef]
15. Pandey, H.K.; Duggal, S.K.; Jamatia, A. Fluoride Contamination of Groundwater and It's Hydrogeological Evolution in District Sonbhadra (U.P.) India. *Proc. Natl. Acad. Sci. India Sect. A Phys. Sci.* **2016**, *86*, 81–93. [CrossRef]
16. Subba Rao, N.; Ravindra, B.; Wu, J. Geochemical and Health Risk Evaluation of Fluoride Rich Groundwater in Sattenapalle Region, Guntur District, Andhra Pradesh, India. *Human. Ecol. Risk Assess. Int. J.* **2020**, *26*, 2316–2348. [CrossRef]
17. Clement, M.; Meunie, A. Is Inequality Harmful for the Environment? An Empirical Analysis Applied to Developing and Transition Countries. *Rev. Soc. Econ.* **2010**, *68*, 413–445. [CrossRef]
18. Hayashi, A.; Akimoto, K.; Tomoda, T.; Kii, M. Global Evaluation of the Effects of Agriculture and Water Management Adaptations on the Water-Stressed Population. *Mitig. Adapt. Strateg. Glob. Change* **2013**, *18*, 591–618. [CrossRef]
19. Lam, S.; Nguyen-Viet, H.; Tuyet-Hanh, T.T.; Nguyen-Mai, H.; Harper, S. Evidence for Public Health Risks of Wastewater and Excreta Management Practices in Southeast Asia: A Scoping Review. *Int. J. Environ. Res. Public Health* **2015**, *12*, 12863–12885. [CrossRef]
20. Bertrand, G.; Hirata, R.; Pauwels, H.; Cary, L.; Petelet-Giraud, E.; Chatton, E.; Aquilina, L.; Labasque, T.; Martins, V.; Montenegro, S.; et al. Groundwater Contamination in Coastal Urban Areas: Anthropogenic Pressure and Natural Attenuation Processes. Example of Recife (PE State, NE Brazil). *J. Contam. Hydrol.* **2016**, *192*, 165–180. [CrossRef]
21. Gomes, O.V.O.; Marques, E.D.; Kütter, V.T.; Aires, J.R.; Travi, Y.; Silva-Filho, E.V. Origin of Salinity and Hydrogeochemical Features of Porous Aquifers from Northeastern Guanabara Bay, Rio de Janeiro, SE, Brazil. *J. Hydrol. Reg. Stud.* **2019**, *22*, 100601. [CrossRef]
22. Nicolodi, J.L.; Pettermann, R.M. Vulnerability of the Brazilian Coastal Zone in Its Environmental, Social, and Technological Aspects. *J. Coast. Res.* **2011**, *64*, 1372–1379.
23. Filgueira, J.M.; Pereira Júnior, A.O.; de Araújo, R.S.B.; da Silva, N.F. Economic and Social Impacts of the Oil Industry on the Brazilian Onshore. *Energies* **2020**, *13*, 1922. [CrossRef]
24. Da Silva, A.L.C.; da Silva, M.A.M.; Gambôa, L.A.P.; Rodrigues, A.R. Sedimentary Architecture and Depositional Evolution of the Quaternary Coastal Plain of Maricá, Rio de Janeiro, Brazil. *Braz. J. Geol.* **2014**, *44*, 191–206. [CrossRef]
25. Cruz, A.; da Silva, G.C., Jr.; de Almeida, G.M. Modelagem Hidrogeoquímica Do Aquífero Freático Da Restinga de Piratininga, Niterói-RJ. In *XIV Congresso Brasileiro de Águas Subterrâneas*; ABAS: Curitiba, Brazil, 2006; pp. 1–19.
26. *GB/T 14848-2017*; Standard for Groundwater Quality. Standardization Administration of the PRC: Beijing, China, 2017.
27. Van Hall, C.E.; Safranko, J.; Stenger, V.A. Rapid Combustion Method for the Determination of Organic Substances in Aqueous Solutions. *Anal. Chem.* **1963**, *35*, 315–319. [CrossRef]
28. Sharp, J.H.; Peltzer, E.T.; Alperin, M.J.; Cauwet, G.; Farrington, J.W.; Fry, B.; Karl, D.M.; Martin, J.H.; Spitzy, A.; Tugrul, S.; et al. Procedures Subgroup Report. *Mar. Chem.* **1993**, *41*, 37–49. [CrossRef]
29. EPA. *Method 1603: Escherichia Coli (E. Coli) in Water by Membrane Filtration Using Modified Membrane-Thermotolerant Escherichia Coli Agar (Modified MTEC)*; EPA: Washington, DC, USA, 2014.
30. Srivastav, A.L.; Ranjan, M. Inorganic Water Pollutants. In *Inorganic Pollutants in Water*; Elsevier: Amsterdam, The Netherlands, 2020; pp. 1–15. [CrossRef]
31. Block, S.S. *Desinfection, Sterilization, and Preservation*, 5th ed.; Lippincot Willians & Wilkins: Philadelphia, PA, USA, 2001.
32. Puri, A.; Kumar, M. A Review of Permissible Limits of Drinking Water. *Indian. J. Occup. Environ. Med.* **2012**, *16*, 40–44. [CrossRef]
33. Banks, D. *An Introduction to Thermogeology: Ground Source Heating and Cooling*, 2nd ed.; John Wiley & Sons, Ltd: Oxford, UK, 2008.
34. Bonte, M.; Stuyfzand, P.J.; Hulsmann, A.; Van Beelen, P. Underground Thermal Energy Storage: Environmental Risks and Policy Developments in the Netherlands and European Union. *Ecol. Soc.* **2011**, *16*, 1–22. [CrossRef]
35. Griffioen, J.; Appelo, A.J. Nature and Extent of Carbonate Precipitation during Aquifer Thermal Energy Storage. *Appl. Geochem.* **1993**, *8*, 161–176. [CrossRef]
36. Bonte, M.; Van Breukelen, B.M.; Stuyfzand, P.J. Environmental Impacts of Aquifer Thermal Energy Storage Investigated by Field and Laboratory Experiments. *J. Water Clim. Change* **2013**, *4*, 77–89. [CrossRef]
37. Bonte, M.; van Breukelen, B.M.; Stuyfzand, P.J. Temperature-Induced Impacts on Groundwater Quality and Arsenic Mobility in Anoxic Aquifer Sediments Used for Both Drinking Water and Shallow Geothermal Energy Production. *Water Res.* **2013**, *47*, 5088–5100. [CrossRef]

38. Bonte, M.; Röling, W.F.M.; Zaura, E.; van der Wielen, P.W.J.J.; Stuyfzand, P.J.; van Breukelen, B.M. Impacts of Shallow Geothermal Energy Production on Redox Processes and Microbial Communities. *Environ. Sci. Technol.* **2013**, *47*, 14476–14484. [CrossRef] [PubMed]
39. Jesußek, A.; Köber, R.; Grandel, S.; Dahmke, A. Aquifer Heat Storage: Sulphate Reduction with Acetate at Increased Temperatures. *Environ. Earth Sci.* **2013**, *69*, 1763–1771. [CrossRef]
40. Fraser, D.W.; Tsai, T.R.; Orenstein, W.; Parkin, W.E.; Beecham, H.J.; Sharrar, R.G.; Harris, J.; Mallison, G.F.; Martin, S.M.; McDade, J.E.; et al. Legionnaires' Disease. *N. Engl. J. Med.* **1977**, *297*, 1189–1197. [CrossRef] [PubMed]
41. Shands, K.N. Potable Water as a Source of Legionnaires' Disease. *JAMA J. Am. Med. Assoc.* **1985**, *253*, 1412. [CrossRef]
42. WHO. *WHO Housing and Health Guidelines*; WHO: Geneva, Switzerland, 2018.
43. Weiner, E.R. *Applications of Environmental Aquatic Chemistry*, 3rd ed.; CRC Press: Boca Raton, FL, USA, 2013.
44. Hayton, J. *Industrial Water Treatment Process Technology*; Butterworth-Heinemann: Oxford, UK, 2017.
45. Rose, S.; Long, A. Monitoring Dissolved Oxygen in Ground Water: Some Basic Considerations. *Groundw. Monit. Remediat.* **1988**, *8*, 93–97. [CrossRef]
46. Zhang, Z.; Xiao, C.; Adeyeye, O.; Yang, W.; Liang, X. Source and Mobilization Mechanism of Iron, Manganese and Arsenic in Groundwater of Shuangliao City, Northeast China. *Water* **2020**, *12*, 534. [CrossRef]
47. Jakobsen, R.; Postma, D. In Situ Rates of Sulfate Reduction in an Aquifer (Romo Denmark) and Implications for the Reactivity of Organic Matter. *Geology* **1994**, *22*, 1103–1106. [CrossRef]
48. Harvey, R.W.; Barber, L.B. Associations of Free-Living Bacteria and Dissolved Organic Compounds in a Plume of Contaminated Groundwater. *J. Contam. Hydrol.* **1992**, *9*, 91–103. [CrossRef]
49. Judd, K.E.; Crump, B.C.; Kling, G.W. Variation in Dissolved Organic Matter Controls Bacterial Production and Community Composition. *Ecology* **2006**, *87*, 2068–2079. [CrossRef]
50. Romera-Castillo, C.; Pinto, M.; Langer, T.M.; Álvarez-Salgado, X.A.; Herndl, G.J. Dissolved Organic Carbon Leaching from Plastics Stimulates Microbial Activity in the Ocean. *Nat. Commun.* **2018**, *9*, 1430. [CrossRef]
51. Richards, L.A.; Lapworth, D.J.; Magnone, D.; Gooddy, D.C.; Chambers, L.; Williams, P.J.; van Dongen, B.E.; Polya, D.A. Dissolved Organic Matter Tracers Reveal Contrasting Characteristics across High Arsenic Aquifers in Cambodia: A Fluorescence Spectroscopy Study. *Geosci. Front.* **2019**, *10*, 1653–1667. [CrossRef]
52. Baker, M.A.; Valett, H.M.; Dahm, C.N. Organic Carbon Supply and Metabolism in a Shallow Groundwater Ecosystem. *Ecology* **2000**, *81*, 3133–3148. [CrossRef]
53. Shen, Y.; Chapelle, F.H.; Strom, E.W.; Benner, R. Origins and Bioavailability of Dissolved Organic Matter in Groundwater. *Biogeochemistry* **2015**, *122*, 61–78. [CrossRef]
54. Longnecker, K.; Kujawinski, E.B. Composition of Dissolved Organic Matter in Groundwater. *Geochim. Cosmochim. Acta* **2011**, *75*, 2752–2761. [CrossRef]
55. Batiot-Guilhe, C.; Emblanch, C.; Blavoux, B. Total Organic Carbon (TOC) and Magnésium: Two Complementary Tracers of Residence Time in Karstic Systems. *C. R. Geosci.* **2003**, *335*, 205–214.
56. Barcelona, M.J. TOC Determinations in Ground Water. *Ground Water* **1984**, *22*, 18–24. [CrossRef]
57. Pernthaler, J. Predation on Prokaryotes in the Water Column and Its Ecological Implications. *Nat. Rev. Microbiol.* **2005**, *3*, 537–546. [CrossRef]
58. Goldscheider, N.; Hunkeler, D.; Rossi, P. Review: Microbial Biocenoses in Pristine Aquifers and an Assessment of Investigative Methods. *Hydrogeol. J.* **2006**, 926–941. [CrossRef]
59. Gopal, K.; Tripathy, S.S.; Bersillon, J.L.; Dubey, S.P. Chlorination Byproducts, Their Toxicodynamics and Removal from Drinking Water. *J. Hazard. Mater.* **2007**, *140*, 1–6. [CrossRef] [PubMed]
60. Chomycia, J.C.; Hernes, P.J.; Harter, T.; Bergamaschi, B.A. Land Management Impacts on Dairy-Derived Dissolved Organic Carbon in Ground Water. *J. Environ. Qual.* **2008**, *37*, 333–343. [CrossRef]
61. Rajendran, A.; Shimizu, G.K.H.; Woo, T.K. The Challenge of Water Competition in Physical Adsorption of CO_2 by Porous Solids for Carbon Capture Applications—A Short Perspective. *Adv. Mater.* **2023**. [CrossRef] [PubMed]
62. Gruau, G.; Dia, A.; Olivié-Lauquet, G.; Davranche, M.; Pinay, G. Controls on the Distribution of Rare Earth Elements in Shallow Groundwaters. *Water Res.* **2004**, *38*, 3576–3586. [CrossRef] [PubMed]
63. Makehelwala, M.; Wei, Y.; Weragoda, S.K.; Weerasooriya, R.; Zheng, L. Characterization of Dissolved Organic Carbon in Shallow Groundwater of Chronic Kidney Disease Affected Regions in Sri Lanka. *Sci. Total Environ.* **2019**, *660*, 865–875. [CrossRef]
64. Ahmed, A.U.; Hoddinott, J.F.; Md Shaiful Islam, K.; Mahbubur Rahman Khan, A.; Abedin, N.; Hossain, N.Z.; Ghostlaw, J.; Parvin, A.; Quabili, W.; Tahsin Rahaman, S.; et al. *Impacts of Bt Brinjal (Eggplant) Impacts In Bangladesh*; International Food Policy Research Institute: Washington, DC, USA, 2019.
65. McDonough, L.K.; Rutlidge, H.; O'Carroll, D.M.; Andersen, M.S.; Meredith, K.; Behnke, M.I.; Spencer, R.G.M.; McKenna, A.M.; Marjo, C.E.; Oudone, P.; et al. Characterisation of Shallow Groundwater Dissolved Organic Matter in Aeolian, Alluvial and Fractured Rock Aquifers. *Geochim. Cosmochim. Acta* **2020**, *273*, 163–176. [CrossRef]
66. Anawar, H.M.; Akai, J.; Komaki, K.; Terao, H.; Yoshioka, T.; Ishizuka, T.; Safiullah, S.; Kato, K. Geochemical Occurrence of Arsenic in Groundwater of Bangladesh: Sources and Mobilization Processes. *J. Geochem. Explor.* **2003**, *77*, 109–131. [CrossRef]

67. McDonough, L.K.; Andersen, M.S.; Behnke, M.I.; Rutlidge, H.; Oudone, P.; Meredith, K.; O'Carroll, D.M.; Santos, I.R.; Marjo, C.E.; Spencer, R.G.M.; et al. A New Conceptual Framework for the Transformation of Groundwater Dissolved Organic Matter. *Nat. Commun.* **2022**, *13*, 1–11. [CrossRef]
68. MacDonald, A.M.; Dochartaigh, B.É.Ó.; Kinniburgh, D.G.; Darling, W.G. *Baseline Scotland: Groundwater Chemistry of Southern Scotland*; British Geological Survey: Nottingham, UK, 2008.
69. Edwards, A.M.C.; Thornes, J.B. Annual Cycle in River Water Quality: A Time Series Approach. *Water Resour. Res.* **1973**, *9*, 1286–1295. [CrossRef]
70. Cun, C.; Vilagines, R. Time Series Analysis on Chlorides, Nitrates, Ammonium and Dissolved Oxygen Concentrations in the Seine River near Paris. *Sci. Total Environ.* **1997**, *208*, 59–69. [CrossRef]
71. Hunt, M.; Herron, E.; Green, L. *Chlorides in Fresh Water*; University of Rhode Island: Kingston, RI, USA, 2012.
72. Hong, Y.; Zhu, Z.; Liao, W.; Yan, Z.; Feng, C.; Xu, D. Freshwater Water-Quality Criteria for Chloride and Guidance for the Revision of the Water-Quality Standard in China. *Int. J. Environ. Res. Public. Health* **2023**, *20*, 2875. [CrossRef]
73. Chen, J.; Wu, H.; Qian, H. Groundwater Nitrate Contamination and Associated Health Risk for the Rural Communities in an Agricultural Area of Ningxia, Northwest China. *Expo. Health* **2016**, *8*, 349–359. [CrossRef]
74. Majumdar, D. The Blue Baby Syndrome. *Resonance* **2003**, *8*, 20–30. [CrossRef]
75. Schijven, J.F.; Imůnek, J.S. Kinetic Modeling of Virus Transport at the Field Scale. *J. Contam. Hydrol.* **2002**, *55*, 113–135. [CrossRef] [PubMed]
76. Mahmoud, M.E.; Shoaib, S.M.A.; Salam, M.A.; Elsayed, S.M. Efficient and Fast Removal of Total and Fecal Coliform, BOD, COD and Ammonia from Raw Water by Microwave Heating Technique. *Groundw. Sustain. Dev.* **2022**, *19*, 100847. [CrossRef]
77. Dayanti, M.P.; Fachrul, M.F.; Wijayanti, A. Escherichia Coli as Bioindicator of the Groundwater Quality in Palmerah District, West Jakarta, Indonesia. *IOP Conf. Ser. Earth Environ. Sci.* **2018**, *106*, 012081. [CrossRef]
78. Rosen, B.H. *Waterborne Pathogens in Agricultural Watersheds*; NRAES: New York, NY, USA, 2000.
79. Tyrrel, S.F.; Quinton, J.N. Overland Flow Transport of Pathogens from Agricultural Land Receiving Faecal Wastes. *J. Appl. Microbiol.* **2003**, *94*, 87–93. [CrossRef]
80. Unc, A.; Goss, M.J. Movement Of Faecal Bacteria Through The Vadose Zone. *Water Air Soil. Pollut.* **2003**, *149*, 327–337. [CrossRef]
81. Cotruvo, J.A.; Dufour, A.; Rees, G.; Bartram, J.; Carr, R.; Cliver, D.O.; Craun, G.F.; Fayer, R.; Gannon, V.P.J.; Dufour, A.; et al. *Waterborne Zoonoses*; World Health Organization: London, UK, 2004.
82. Fong, T.-T.; Lipp, E.K. Enteric Viruses of Humans and Animals in Aquatic Environments: Health Risks, Detection, and Potential Water Quality Assessment Tools. *Microbiol. Mol. Biol. Rev.* **2005**, *69*, 357–371. [CrossRef]

havior [1,2,24–35], in an attempt to obtain a method for determining the elastic critical moment, or sought to analyze the LDB strength of SCCB [1,4,8–11,26,27,29–31,36–38]. The authors showed that there are considerable divergences between the existing methodologies for determining the elastic critical moment. In addition, Rossi et al. [3] presented disagreements between experimental results and post-buckling numerical analysis when compared with standard procedures. Finally, the researchers present the need for further investigations on SCCB under the action of the hogging moment. Previous investigations by Rossi et al. [1,32,33] showed that LDB in steel-concrete composite beams is not fully resolved and there is a need for further investigations on the subject. One of the parameters to be investigated is the influence of the steel mechanical characteristics of the I-beams on the LDB behavior. The need to analyze the influence of the yield strength of different types of steel on the LDB phenomenon can be seen in Zhao, Li and Sun [34], and Sun et al. [35]. In these works, the authors verified that important standard procedures underestimate the resistance to global instability of singly symmetric I-sections with high-strength steels (HSS).

In this way, although there is a considerable number of investigations about the LDB phenomenon in SCCB, no study has investigated the possible influence of different types of steel and their mechanical properties (f_y and f_u) on the LDB strength of SCCB. Even though standard procedures consider indirectly, through the slenderness ratio ($\lambda = \sqrt{\frac{M_{cr}}{M_{pl}}}$), the yield strength of the steel sections, no comparative study was carried out between the LDB strength obtained for different types of steel in numerical analyses with the SSRC and ECCS design curves used to verify the LDB strength of SCCB. In addition, no study on the LDB strength of SCCB has thoroughly investigated the influence of the longitudinal reinforcement area variation in the concrete slab and the behavior of standard procedures in relation to this parameter. Therefore, this article aims to assess the LDB strength of SCCB through the development of post-buckling numerical analysis with the Dassault Systèmes ABAQUS 6.14 [38] software. In the post-buckling numerical analysis, the initial geometric imperfection, residual stress, real shear connector, geometric nonlinearity, and material nonlinearity were considered through the numerical model developed. The SCCB analyzed are simply supported, with restrictions on lateral slab displacement, and are subjected to a uniform hogging moment distribution. In the parametric study, four types of steel with different mechanical properties were analyzed. In addition, the I-section, the unrestrained length, and the reinforcement rate in the concrete slab were varied. The analyses shown in this work can provide a reference for future research and specification reviews.

2. LDB Strength in SCCB

Vlasov's assumption [39] that the cross-section remains undistorted is not applicable to the mode of distortional buckling, which is characterized by lateral and torsional buckling in the compression flange accompanied by out-of-plane distortion in the web [19].

Eurocode 4 (EN 1994-1-1) [13] deals with the lateral buckling of SCCB by reducing the section moment resistance at the internal support (plastic moment of the composite beam), $M_{pl\text{-}CB}$, to a lower value, $M_{u,dist}$, referred to the beam buckling strength (Equations (1)–(4)). Because the composite beam is one of several parallel members attached to the same concrete slab, the design is based on the inverted U-frame model to determine of the elastic critical moment.

$$M_{u,dist} = \chi_{LT} M_{pl-C.\,beam} \quad (1)$$

$$\chi_{LT} = \left[\phi_{LT} + \sqrt{\phi_{LT}^2 - \overline{\lambda}_{LT}^2}\right]^{-1} \leq 1 \quad (2)$$

$$\phi_{LT} = 0.5\left[1 + \alpha_{LT}\left(\overline{\lambda}_{LT} - 0.2\right) + \overline{\lambda}^2{}_{LT}\right] \quad (3)$$

$$\overline{\lambda}_{LT} = \sqrt{\frac{M_{pl-C.\,beam}}{M_{cr}}} \quad (4)$$

In the current version of EC4 [13], no expression is presented to determine the LDB elastic critical moment. However, in the previous version of EC4 [13] (EN 1994-1-1:1992), the methodology proposed by Roik et al. [14] to determine the elastic distortional critical moment was adopted (Equations (5)–(8)).

$$M_{cr} = \frac{\alpha_g C_{dist}}{L}\sqrt{\left(GJ + \frac{k_s L^2}{\pi^2}\right) EI_{af,y}} \quad (5)$$

$$k_s = \frac{k_1 k_2}{k_1 + k_2} \quad (6)$$

$$k_1 = \frac{\alpha E_c I_{c2}}{a} \quad (7)$$

$$k_2 = \frac{E_a t_w^3}{4(1 - \nu_a^2) h_s} \quad (8)$$

The Brazilian standard code, ABNT NBR 8800:2008 [15], is also based on the inverted U-frame model to determine of the elastic critical moment. However, differently from EC4 [13], the current version of the Brazilian code recommends the use of the procedure by Roik et al. [14] for determining the LDB elastic critical moment. To obtain the LDB strength the Brazilian standard uses the 2P design curve provided by the SSRC (Structural Stability Research Council).

Zhou and Yan [4], observing divergences between numerical results and the EC4 [13] procedure, proposed practical formulas (Equations (9)–(12)) to estimate the LDB strength of steel-concrete composite beams under the action of uniform hogging moment. The authors [4] numerically investigated the LDB behavior of SCCCs. Zhou and Yan [4] analyzed the I-section influence, the unrestricted length, and the shear interaction degree between the concrete slab and the steel I-section.

$$M_{u,dist} = \phi_{PR} M_{pl-C.\,beam} \quad (9)$$

$$\phi_{PR} = 1.56 - 0.58\lambda_{PR} \leq 1.0 \quad (10)$$

$$\lambda_{PR} = 0.09\left(\frac{L}{b_f}\right)^{0.2}\left(\frac{t_f}{b_f}\right)^{0.4}\left(\frac{h_s}{t_w}\right)^{0.6}(1.5+\rho)^{0.5} \quad (11)$$

$$\rho = \frac{A_r f_r}{\left(A_w f_w + 2A_f f_f\right)} \quad (12)$$

Bradford [29] also stated that the U-frame model, adopted by standard procedures like EC4 [13], has been conservative in most cases. Then, the author [29] investigated the LDB through special-purpose inelastic finite element method of analysis to study the buckling of beams with continuous and complete tension flange restraint. Bradford [29] suggests a new design proposal based on AS4100 (Equations (13) and (14)).

$$M_{u,dist} = 0.6\left\{\sqrt{\left(\frac{M_{pl-I-B}}{M_o}\right)^2 + 3} - \left(\frac{M_{pl-I-B}}{M_o}\right)\right\} M_{pl-I-B} \leq M_{pl-I-B} \quad (13)$$

$$\lambda_B = \sqrt{\frac{M_{pl-I-B}}{M_o}} = 0.02\left(\frac{L}{r_{fc,y}}\right)^{0.5}\left(\frac{h_w}{t_w}\right)^{1/3}\alpha_m^{-0.5} \quad (14)$$

3. Numerical Model

The numerical model showed in this paper is the same presented in the previous paper, Rossi et al. [1], where the LDB in SCCB was investigated by the authors. To carry out the numerical analyses the ABAQUS software [38] was used. With this software it is possible to develop elastic buckling analyzes and post-buckling analyzes (nonlinear inelastic analysis). For the elastic buckling analysis, the buckle linear perturbation method was used to estimate the critical elastic stability load by obtaining eigenvalues and their eigenvectors. In this method, the critical elastic stability load is obtained by the product of the first positive eigenvalue (lower energy) by the external load applied to the structure in the initial state. Importantly, this type of analysis does not consider any imperfections in the structure. After this, the post-buckling analysis is performed, considering the initial geometric imperfections. The structure shape in the buckling analysis, normalized to the initial imperfection value, was adopted as the shape at the beginning of the post-buckling analysis. Thus, the implementation of geometric imperfection was performed by the "IMPERFECTION" command. Residual stresses were included within the model as initial conditions. These initial conditions are included in a set of finite elements (shells, in this case) to which a given value of initial stress is provided. The set of elements can be defined by means of standard pre-processing tools. The procedure consists of defining several longitudinal partitions of the geometry. Each partition corresponds to a given set of elements that present the same value of initial stress. The residual stress patterns are assumed constant along the longitudinal direction regardless of the presence of transverse stiffeners (which might affect locally such pattern). The Static Riks method was used to solve the geometric nonlinearity problem. This method, also known as modified Riks algorithm, can obtain equilibrium solutions for instability problems.

3.1. Geometry, Element Type and Mesh

The numerical model developed faithfully represents the geometry of the problem. The composite behavior between the I-section and the concrete slab was guaranteed by means headed studs shear connectors. The finite elements used and other properties of the developed numerical model can be seen in (Figure 3).

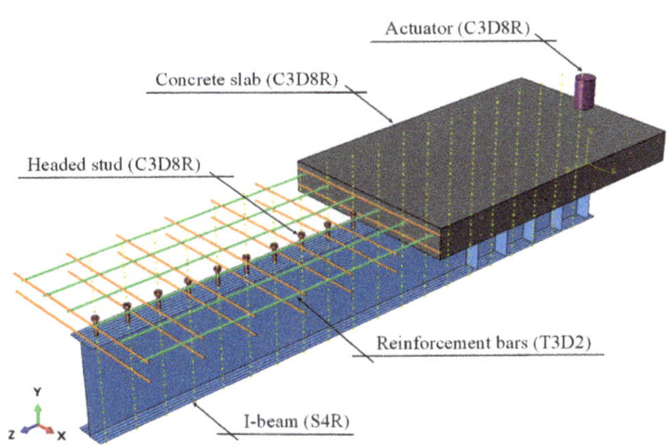

Figure 3. Numerical model.

The concrete slabs, the headed shear stud connectors and the actuator were modelled using the eight-node linear hexahedral solid elements with reduced integration and hourglass control (C3D8R). Elements with reduced-integration were adopted as they could reduce computer run time [40]. The mesh dimensions were verified by means of sensitivity analyzes, Rossi et al. [33]. For I-beams the quadrilateral element S4R (shell element) with four nodes and reduced integration was used. The S4R element has six degrees of freedom per node-three rotations and three translations. Rossi et al. [41,42] and Ferreira, Rossi, and

Martins [43] performed a sensitivity analysis for I beams discretized with S4R elements showing that the size of the 10 mm element provides good results with relatively low processing time. The reinforcing bars were performed with two-node linear three-dimensional truss elements (T3D2). The dimensions used in the discretization of each element are shown in Table 1 and can be seen in Figure 4.

Table 1. Element type and size.

Section Part	Element Type	Element Size
Reinforcement bars	T3D2	10 mm
Headed shear stud	C3D8R	2 mm
I-beam	S4R	10 mm
Concrete slab	C3D8R	20 mm
Actuator	C3D8R	20 mm

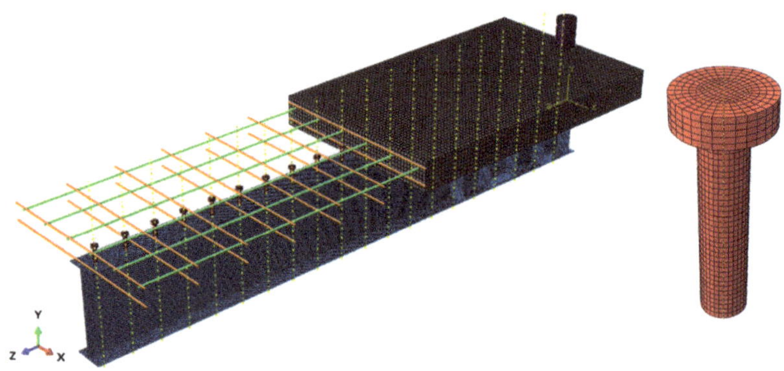

Figure 4. Discretized model.

3.2. Contact Properties

The interaction between the constituent parts of the numerical model was performed using interaction and constraint options available in ABAQUS, in Figure 5 the details of the model can be observed.

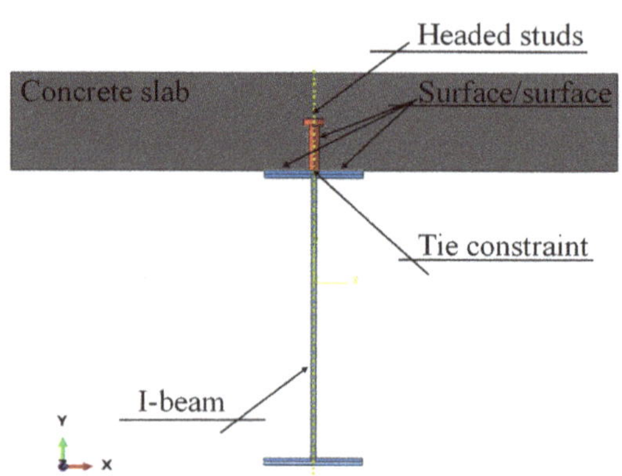

Figure 5. Interaction between contact surfaces.

For the interaction between the headed stud and the top flange, the option "Tie constrain" was used, which simulates the welded connection between these two elements, in which the relative displacement between both is totally prevented. The interactions between the following surfaces—headed stud/concrete slab and concrete slab/steel beam—was performed using the surface-to-surface interaction option (contact). The normal and tangential behavior between these contact surfaces was defined using the "Hard" and "Penalty" options, respectively. The value used for the friction coefficient was 0.4 [40,44,45].

The interaction between the reinforcement bars and concrete slab was performed using the embedded element technique. In this technique perfect bond between embedded elements and host elements was chosen. This bonding will constrain the translational degree of freedom of the embedded nodes and will also avoid slip between the reinforcing bars and concrete [40]. Finally, the interaction between the actuator and the concrete slab was also carried out through the "Tie constrain" option.

3.3. Boundary Conditions and Initial Imperfections

The SCCB analyzed are simply supported with lateral bracing (Figure 6). The lateral bracing simulates the condition of a composite floor, which has infinite stiffness in the slab plane (transversal stiffness). This situation prevents any lateral displacement of the composite beam, restrictions similar to those developed by Tong et al. [28] and Zhou and Yan [4]. The boundary conditions used try to represent the restrictions performed in the tests by Tong et al. [28], investigated in the present paper. The forces can be applied at one or both ends (cantilever) of the beam according to the desired negative moment distribution configuration. The possible modes of instability, LDB or LB (local buckling) that govern the strength of these elements are analyzed in the internal span (L), between supports. To prevent any form of instability from occurring in the cantilevered regions, avoiding interferences in the developed analysis, web stiffeners were placed at both ends, as shown in Figure 6. Details of the boundary conditions developed in the ABAQUS software can be seen in Figure 7.

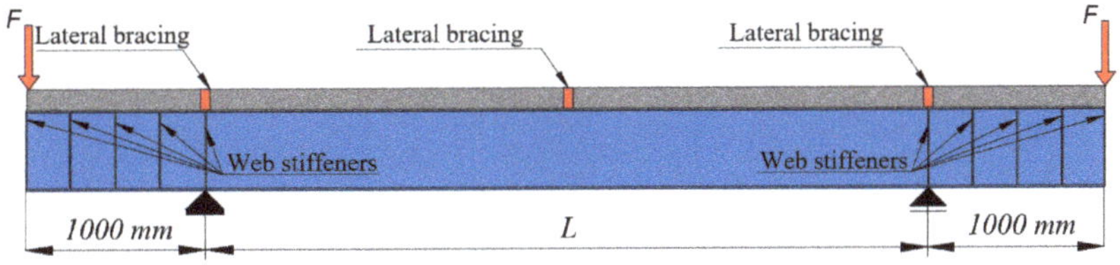

Figure 6. Boundary condition of the experimental tests by Tong et al. [35].

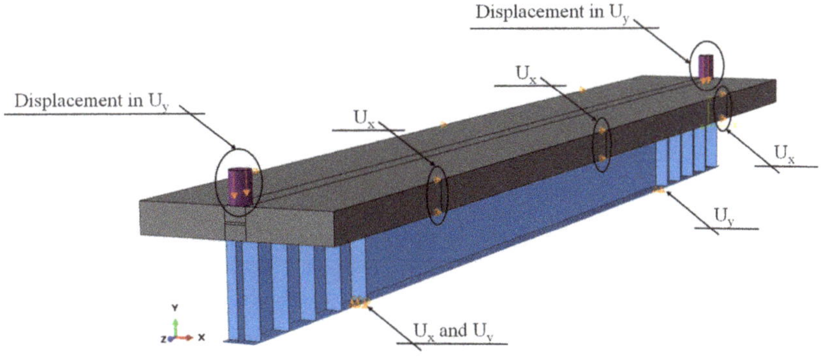

Figure 7. Details of the boundary conditions in ABAQUS software.

Regarding the geometric and structural imperfections, the initial lateral bending and the residual stresses were considered. In the former imperfection, the steel member is not ideally straight, but has some initial geometric deformation such as the initial bending and initial distortion. The latter imperfection is caused by the residual stress due to welding and hot rolling [46]. A nonlinear post-buckling analysis should consider both the initial geometric imperfection and residual stress. For the geometric imperfection an initial bending with magnitude of *L/1000* was adopted. To consider residual stresses, the Galambos and Ketter model [47] (Equations (15) and (16)) was adopted, as shown in Figure 8. The geometric and structural imperfections were determined through a sensitivity study published in a previous article, Rossi et al. [33]. In this study [33] the authors found that the magnitude of *L/1000* for geometric imperfections and the residual stress model by Galambos and Ketter [47] provided results closer to the experimental test.

$$\sigma_c = 0.3 f_y \tag{15}$$

$$\sigma_t = \left[\frac{b_f t_f}{b_f t_f + t_w \left(d - 2t_f \right)} \right] 0.3 f_y \tag{16}$$

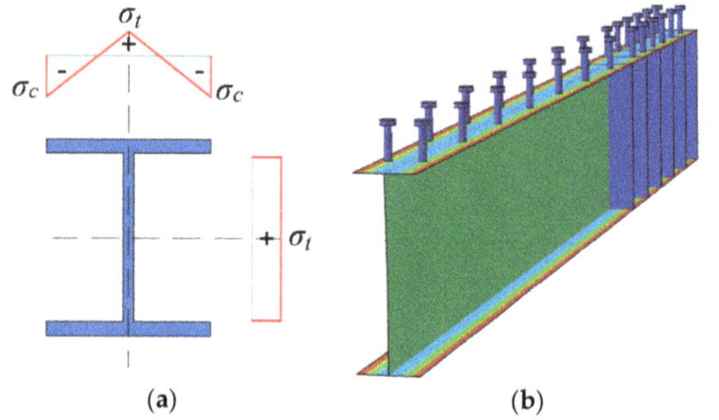

Figure 8. Residual stresses. (**a**) Galambos and Ketter model [47]; (**b**) residual stresses in the numerical model.

The residual stresses were included within the model as initial conditions. These initial conditions are included in a set of finite elements (shells, in this case) to which a given value of initial stress is provided. The set of elements can be defined by means of standard pre-processing tools. The procedure consists of defining several longitudinal partitions of the geometry. Each partition corresponds to a given set of elements that present the same value of initial stress. The residual stress patterns are assumed constant along the longitudinal direction regardless of the presence of transverse stiffeners (which might affect locally such patterns).

3.4. Material Properties

3.4.1. Steel

For the steel sections, the adopted constitutive relationship was proposed by Earls [48,49]. The stress–strain relationships presented by Earls is an elastoplastic model where the consideration of the steel hardening effect is performed through a tri-linear region (Figure 9a). For longitudinal and transverse reinforcement bars, an elastoplastic model with positive linear hardening was adopted, as shown in Figure 9b. Finally, the model assigned to the headed stud shear connectors was the perfect elastoplastic, as shown in Figure 9c.

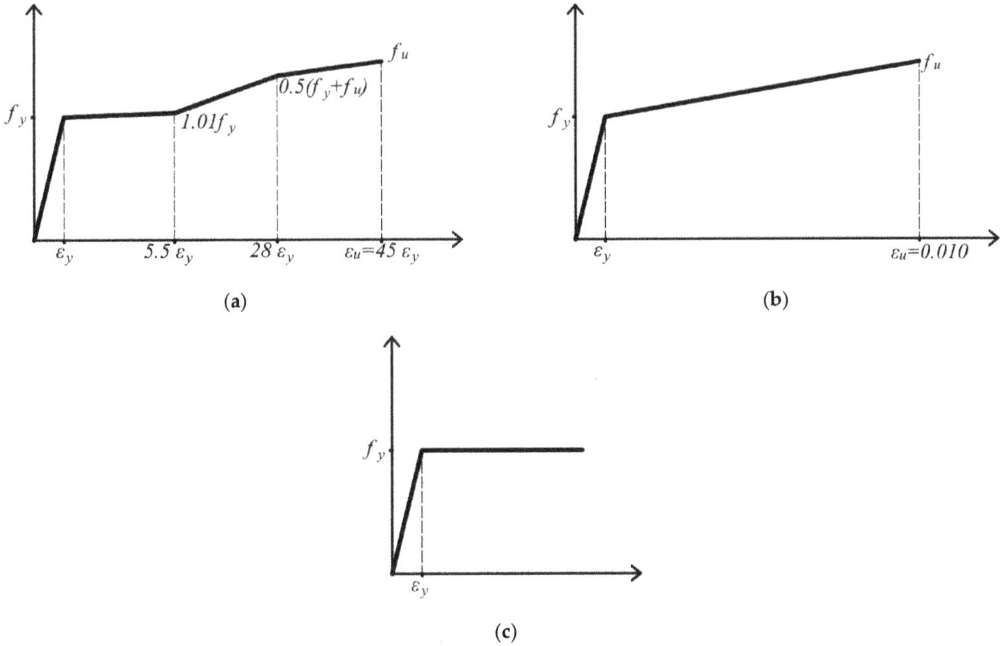

Figure 9. Stress–strain relationships of steel. (**a**) tri-linear hardening, (**b**) positive linear hardening, (**c**) perfect elastoplastic.

3.4.2. Concrete

The concrete damage plasticity (CDP) model was used to represent the concrete material. The CDP model is a criterion based on the models proposed by some authors [50–52], and has been widely used in modeling concrete and other quasi-brittle materials. The model considers three hypotheses based on the plasticity theory: The initial yield surface determines when plastic deformation begins; the flow rule determines the direction of plastic deformation; and the softening/hardening rule defines how the surface flow evolves with plastic deformation [53].

The stress–strain relationships used to represent the behavior of concrete in compression was proposed by Carreira and Chu [54] (Equations (17) and (18)).

$$\frac{\sigma}{f_{cm}} = \frac{\beta_c(\varepsilon/\varepsilon_c)}{\beta_c - 1 + (\varepsilon/\varepsilon_c)^{\beta_c}} \tag{17}$$

$$\beta_c = \left(\frac{f_{cm}}{32.4}\right)^3 + 1.55 \quad (MPa) \tag{18}$$

For the representation of the concrete behavior in tensile, the Carreira and Chu [55] proposal (Equation (19)) was also used.

$$\frac{\sigma}{f_{tm}} = \frac{\beta_c(\varepsilon/\varepsilon_t)}{\beta_c - 1 + (\varepsilon/\varepsilon_t)^{\beta_c}} \tag{19}$$

4. Numerical Model Verification

The numerical model developed to analyze the LDB behavior of SCCB under hogging moment was validated in the previous paper, Rossi et al. [1]. The numerical validation was developed considering the experimental models analyzed by Tong et al. [28]. The comparison between the experimental results of Tong et al. [28] and the results of the post-buckling numerical analysis can be seen in Table 2 and Figure 10.

Table 2. Calibration results.

Test Number	Test		FE		M_{test}/M_{FE}
	M_{test} (kNm)	Failure	M_{FE} (kNm)	Failure	
B3.0-350-1	205.1	LDB	207.8	LDB	0.987
B4.2-350-1	206.9	LDB	206.9	LDB	1.000
B4.2-350-1-C	203.7	LDB	203.8	LDB	0.999
B4.2-400-1	210.5	LDB	215.1	LDB	0.979

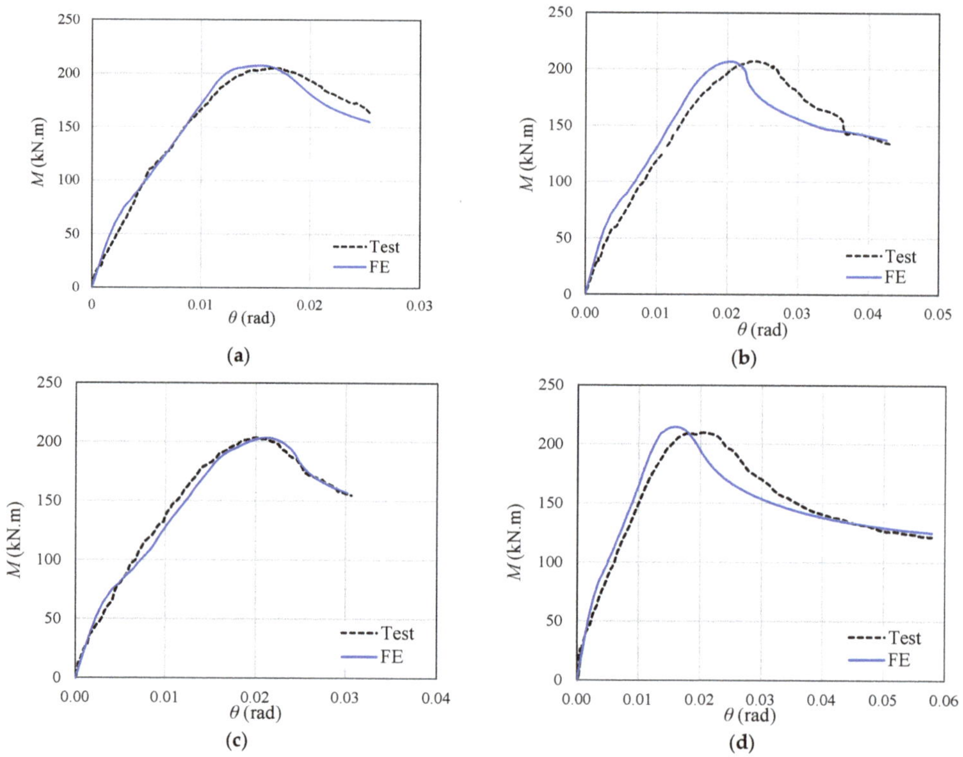

Figure 10. Test vs. FE results. (**a**) B3.0-350-1; (**b**) B4.2-350-1; (**c**) B4.2-350-1-C; (**d**) B4.2-400-1.

The comparison between the results of the numerical model and the experimental tests showed proximity. Therefore, the numerical model is validated given the conformity of the results and will be used for the development of the parametric study.

5. Parametric Study

The purpose of this article is to investigate the LDB behavior of SCCB. For this, 180 post-buckling numerical analyzes were developed with the ABAQUS [38] software. In the parametric study, the influence on the LDB behavior of the steel type and its mechanical properties was investigated. In addition, it was also varied the I-sections dimensions: the unrestricted span length and the reinforcement rate in the concrete slab. The analyses were developed considering SCCB with three different I-sections, whose geometric properties can be found in Table 3.

Table 3. I-Sections geometric properties.

I-Section	d (mm)	b_f (mm)	t_f (mm)	t_w (mm)	L (mm)	S_{stud} (mm)
CB350	349	127	8.5	5.8	4000; 5000; 6000; 7000; 8000	200
CB450	450	152	10.8	7.6	4000; 5000; 6000; 7000; 8000	200
CB600	599	178	12.8	10	4000; 5000; 6000; 7000; 8000	200

Four different steel types were considered for the I-sections, which have a yield strength (f_y) of 250 MPa, 290 Mpa, 350 Mpa, and 450 Mpa. Regarding the unrestricted span length (L), according to Figure 6, five situations were analyzed: 4000 mm; 5000 mm; 6000 mm; 7000 mm; and 8000 mm. Finally, it was considered three different reinforcement rates distributed in the slab width, as shown in Figure 11, by means of bars with diameters of 8 mm, 16 mm, and 25 mm. This reinforcement distribution was chosen with the objective of allowing the neutral line to rise towards the concrete slab; in this way, almost the entire I-section is under compression, thus allowing for the LDB occurrence. Furthermore, according to Zhou and Yan [4] the ratio between the total area of the longitudinal reinforcement bars and the I-sections cross-sectional area must be less than 0.5, which represents usual values, which were respected. The cross-section characteristics of the SCCB analyzed can be seen in Figure 11. The composite beams were subjected a uniform distribution of the hogging moment (Figure 12).

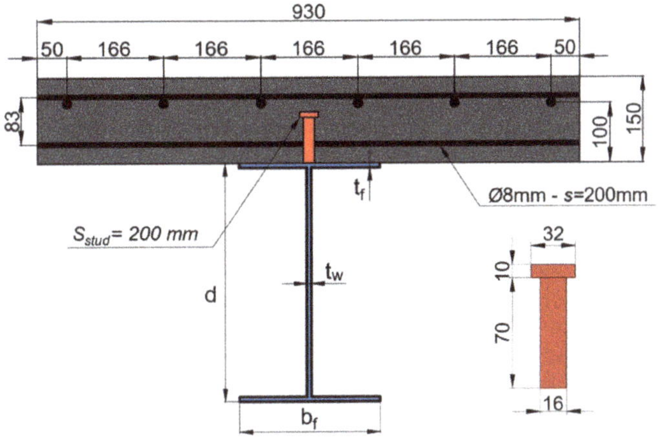

Figure 11. Cross-section details of the SCCB (dimensions in mm).

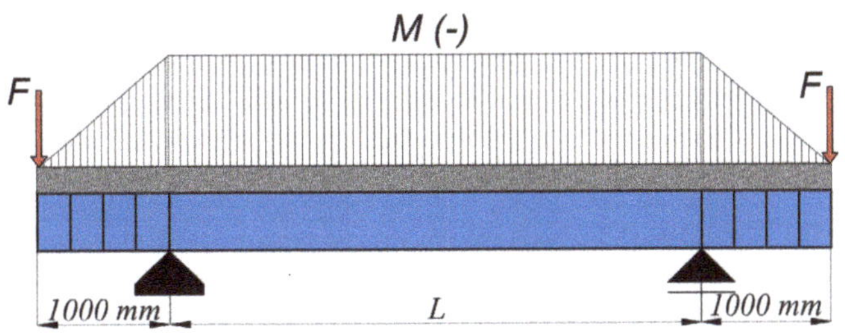

Figure 12. Hogging moment distribution in the parametric study.

The representation of the 180 numerical analyses developed can be seen in Figure 13.

compared with the reference values. The average value of the ratio ($\chi/\chi_{ref.}$) for A572-GR42 steel is 0.98, that is, only 2% lower than the values of A36 steel. This proximity occurs due to the yield strength (f_y) of the two types of steels are close: 250 MPa for A36 steel and 290 MPa for A572-GR42 steel.

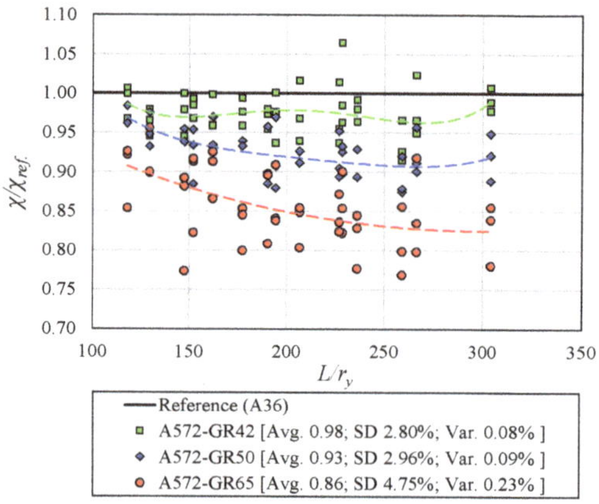

Figure 15. LDB reduction factor for different steel type.

In order to compare the values of the reduction factor (χ), obtained through the FE analyses, with the values obtained through the standard procedures (ABNT NBR 8800:2008 [15] and EC4 [13]), Tables 5–7 are shown. In standard procedures, the reduction factor (χ) is obtained as a function of the relative slenderness factor ($\lambda = \sqrt{\frac{M_{cr}}{M_{pl-CB}}}$) calculated as a function of the LDB elastic critical moment and the plastic moment of the composite section, which is dependent on the steel yield strength (f_y).

Table 5 presents the results for the models with 8 mm reinforcement bars, comparisons are made between the values of the reduction factor (χ) obtained by the standard procedures and through the FE analyses for each type of steel used in the I-sections. In Tables 6 and 7 the same comparisons are made, however, for models with 16 mm and 25 mm reinforcement bars, respectively. More information about the results obtained for each beam can be found on Appendix A.

It is observed in Table 5 that the values of the reduction factor (χ) obtained by the standard procedures (NBR 8800:2008 [15] and EC4 [13]) are lower than the values obtained with the FE analyses. This situation shows a very conservative behavior of the Brazilian [15], for models with 8 mm reinforcement bars, a situation also observed by Rossi et al. [1], Zhou and Yan [4], and Bradford [29], in relation to the European procedure [13]. Table 5 also presents the percentage error values of the standard procedures calculated in relation to the results obtained with the FE analyses. Percentage error values were calculated for models with different steel types. For the Brazilian standard procedure [15], the average values of the percentage error for steels with a yield strength (f_y) of 250 MPa, 290 Mpa, 350 Mpa, and 450 MPa were −31.4%, −32.1%, −39.8%, and −35.1%, respectively. These results show that the LDB ultimate moment values obtained by the Brazilian standard procedure [15] are inferior to the values obtained by the FE analyses, verifying a considerably conservative situation. When the results are compared with the values obtained by the European standard procedure [13], there is an intensification of this conservative situation. The percentage error values for the European standard [13] are −39.7%, −42.2%, −51.7%, and −51.5% for steels with a yield strength of 250 MPa, 290 MPa, 350 MPa, and 450 MPa, respectively. Therefore, the ultimate moment values obtained by the European procedure [13] are up to 50% lower than the values obtained with the FE analyses for the models with reinforcement bars of 8mm.

Table 5. Comparison between the reduction factor values (χ) calculated by standard procedures and obtained by FE analyses for models with 8 mm bars.

			f_y = 250 MPa						f_y = 290 MPa						f_y = 350 MPa						f_y = 450 MPa					
			λ	χ			Percent Error (%)		λ	χ			Percent Error (%)		λ	χ			Percent Error (%)		λ	χ			Percent Error (%)	
I-Section	ø (mm)	L (m)		FE	NBR	EC3	$(\chi_{NBR} - \chi_{FE})/\chi_{NBR}$	$(\chi_{EC3} - \chi_{FE})/\chi_{EC3}$		FE	NBR	EC3	$(\chi_{NBR} - \chi_{FE})/\chi_{NBR}$	$(\chi_{EC3} - \chi_{FE})/\chi_{EC3}$		FE	NBR	EC3	$(\chi_{NBR} - \chi_{FE})/\chi_{NBR}$	$(\chi_{EC3} - \chi_{FE})/\chi_{EC3}$		FE	NBR	EC3	$(\chi_{NBR} - \chi_{FE})/\chi_{NBR}$	$(\chi_{EC3} - \chi_{FE})/\chi_{EC3}$
CB350	8 mm	4	0.76	1.05	0.79	0.75	−33.8	−40.3	0.81	1.02	0.76	0.72	−33.7	−41.4	0.87	1.00	0.73	0.68	−38.1	−48.0	0.98	0.96	0.67	0.61	−43.2	−57.0
		5	0.76	1.01	0.78	0.75	−28.6	−34.9	0.81	0.98	0.76	0.72	−29.3	−36.8	0.88	0.96	0.72	0.67	−33.5	−43.2	0.98	0.90	0.67	0.61	−35.6	−49.0
		6	0.77	1.04	0.78	0.74	−33.1	−39.7	0.82	1.00	0.76	0.71	−32.4	−40.2	0.89	0.93	0.72	0.67	−29.6	−39.3	0.99	0.89	0.66	0.60	−33.8	−47.1
		7	0.77	1.04	0.78	0.74	−33.2	−39.9	0.82	1.00	0.75	0.71	−33.0	−40.9	0.89	0.95	0.72	0.67	−31.8	−41.6	0.99	0.87	0.66	0.60	−31.1	−44.3
		8	0.77	1.01	0.78	0.74	−30.0	−36.6	0.82	1.00	0.75	0.71	−32.9	−40.9	0.89	0.96	0.72	0.67	−34.0	−44.1	1.00	0.85	0.66	0.60	−28.7	−41.7
CB450	8 mm	4	0.81	1.00	0.76	0.72	−31.7	−39.3	0.86	0.98	0.73	0.68	−34.1	−43.3	0.94	0.95	0.69	0.64	−37.5	−49.5	1.06	0.90	0.63	0.56	−43.7	−60.4
		5	0.82	1.00	0.76	0.71	−32.8	−40.7	0.88	0.96	0.73	0.68	−32.5	−42.1	0.95	0.93	0.68	0.63	−36.6	−49.0	1.07	0.87	0.62	0.55	−40.3	−57.1
		6	0.83	1.05	0.75	0.71	−40.1	−48.7	0.88	1.03	0.72	0.67	−42.4	−52.9	0.96	0.96	0.68	0.62	−40.7	−53.7	1.08	0.89	0.61	0.55	−44.3	−61.8
		7	0.83	0.95	0.75	0.71	−27.2	−35.1	0.89	0.97	0.72	0.67	−34.4	−44.4	0.97	0.85	0.68	0.62	−25.9	−37.7	1.09	0.79	0.61	0.54	−28.6	−44.4
		8	0.83	0.98	0.75	0.70	−31.4	−39.6	0.89	0.90	0.72	0.67	−25.3	−34.7	0.97	0.86	0.67	0.62	−27.4	−39.4	1.09	0.76	0.61	0.54	−24.1	−39.4
CB600	8 mm	4	0.89	0.93	0.72	0.67	−28.9	−38.5	0.95	0.90	0.69	0.63	−30.9	−42.6	0.92	1.12	0.70	0.65	−59.2	−72.3	1.17	0.79	0.56	0.49	−40.5	−60.1
		5	0.90	0.93	0.71	0.66	−31.4	−41.7	0.97	0.88	0.68	0.62	−30.7	−43.0	0.94	1.06	0.69	0.64	−52.4	−65.6	1.19	0.72	0.55	0.48	−31.0	−49.8
		6	0.91	0.94	0.70	0.65	−33.3	−44.1	0.98	0.90	0.67	0.61	−34.6	−47.6	0.95	1.08	0.69	0.63	−56.6	−70.6	1.21	0.75	0.54	0.47	−38.2	−58.3
		7	0.92	0.94	0.70	0.65	−34.5	−45.6	0.99	0.89	0.67	0.61	−33.1	−46.3	0.95	1.07	0.68	0.63	−57.1	−71.4	1.21	0.76	0.54	0.47	−40.6	−61.2
		8	0.92	0.85	0.70	0.65	−21.1	−31.2	0.99	0.82	0.66	0.60	−23.1	−35.4	0.96	0.93	0.68	0.62	−36.7	−49.3	1.22	0.66	0.54	0.47	−22.7	−40.8
						Avg.	−31.4	−39.7				Avg.	−32.1	−42.2				Avg.	−39.8	−51.7				Avg.	−35.1	−51.5
						SD (%)	4.2	4.4				SD (%)	4.4	4.7				SD (%)	11.1	12.3				SD (%)	7.1	8.2
						Var. (%)	17.4	19.1				Var. (%)	19.0	21.8				Var. (%)	124.1	152.2				Var. (%)	51.1	67.9

Table 6. Comparison between the reduction factor values (χ) calculated by standard procedures and obtained by FE analyses for models with 16 mm bars.

			f_y = 250 MPa						f_y = 290 MPa						f_y = 350 MPa						f_y = 450 MPa					
			λ	χ			Percent Error (%)		λ	χ			Percent Error (%)		λ	χ			Percent Error (%)		λ	χ			Percent Error (%)	
I-Section	ø (mm)	L (m)		FE	NBR	EC3	$(\chi_{NBR} - \chi_{FE})/\chi_{NBR}$	$(\chi_{EC3} - \chi_{FE})/\chi_{EC3}$		FE	NBR	EC3	$(\chi_{NBR} - \chi_{FE})/\chi_{NBR}$	$(\chi_{EC3} - \chi_{FE})/\chi_{EC3}$		FE	NBR	EC3	$(\chi_{NBR} - \chi_{FE})/\chi_{NBR}$	$(\chi_{EC3} - \chi_{FE})/\chi_{EC3}$		FE	NBR	EC3	$(\chi_{NBR} - \chi_{FE})/\chi_{NBR}$	$(\chi_{EC3} - \chi_{FE})/\chi_{EC3}$
CB350	16 mm	4	0.62	0.82	0.85	0.83	3.3	0.4	0.66	0.81	0.83	0.81	2.8	−0.5	0.71	0.77	0.81	0.78	5.3	1.5	0.78	0.75	0.78	0.74	2.8	−2.2
		5	0.63	0.81	0.85	0.82	4.7	1.8	0.66	0.77	0.83	0.80	7.2	4.0	0.71	0.72	0.81	0.78	10.4	6.7	0.78	0.72	0.77	0.73	6.2	1.2
		6	0.63	0.80	0.85	0.82	6.1	3.2	0.67	0.78	0.83	0.80	5.6	2.3	0.72	0.74	0.81	0.77	8.7	4.9	0.79	0.72	0.77	0.73	7.1	2.1
		7	0.63	0.79	0.85	0.82	7.0	4.2	0.67	0.75	0.83	0.80	9.9	6.7	0.72	0.75	0.81	0.77	6.5	2.6	0.79	0.72	0.77	0.73	6.2	1.1
		8	0.63	0.79	0.85	0.82	6.6	3.7	0.67	0.77	0.83	0.80	6.8	3.5	0.72	0.73	0.80	0.77	9.6	5.8	0.79	0.68	0.77	0.73	12.2	7.4
CB450	16 mm	4	0.69	0.83	0.82	0.79	−0.7	−4.5	0.73	0.80	0.80	0.77	0.4	−3.9	0.78	0.81	0.77	0.74	−4.2	−9.7	0.87	0.79	0.73	0.68	−8.2	−15.7
		5	0.69	0.80	0.82	0.79	1.6	−2.2	0.74	0.80	0.80	0.76	−0.6	−5.1	0.79	0.78	0.77	0.73	−1.4	−6.9	0.88	0.74	0.72	0.68	−2.6	−10.1
		6	0.70	0.83	0.81	0.78	−2.0	−6.0	0.74	0.83	0.79	0.76	−4.6	−9.4	0.80	0.80	0.77	0.73	−5.1	−11.0	0.88	0.75	0.72	0.67	−4.7	−12.4
		7	0.70	0.80	0.81	0.78	2.2	−1.7	0.75	0.76	0.79	0.76	4.3	−0.2	0.80	0.76	0.76	0.72	0.9	−4.6	0.89	0.69	0.72	0.67	3.6	−3.6
		8	0.71	0.79	0.81	0.78	3.3	−0.6	0.75	0.76	0.79	0.76	4.2	−0.3	0.80	0.72	0.76	0.72	5.3	−0.1	0.89	0.67	0.72	0.67	6.3	−0.8
CB600	16 mm	4	0.78	0.79	0.78	0.74	−1.6	−6.9	0.82	0.79	0.76	0.71	−5.0	−11.2	0.89	0.78	0.72	0.67	−7.6	−15.6	0.99	0.73	0.67	0.61	−9.1	−19.8
		5	0.79	0.79	0.77	0.73	−3.0	−8.7	0.83	0.79	0.75	0.70	−5.8	−12.4	0.90	0.76	0.71	0.66	−6.1	−14.3	1.00	0.70	0.66	0.60	−6.4	−17.3
		6	0.80	0.81	0.76	0.72	−5.9	−11.9	0.84	0.80	0.74	0.70	−8.2	−15.2	0.91	0.76	0.71	0.65	−7.5	−16.1	1.01	0.69	0.65	0.59	−6.2	−17.4
		7	0.81	0.80	0.76	0.72	−5.4	−11.4	0.85	0.82	0.74	0.70	−10.0	−17.3	0.92	0.73	0.70	0.65	−3.9	−12.3	1.02	0.68	0.65	0.58	−5.3	−16.5
		8	0.81	0.73	0.76	0.72	3.6	−2.0	0.85	0.73	0.74	0.69	1.7	−4.8	0.92	0.68	0.70	0.65	3.1	−4.9	1.02	0.61	0.64	0.58	5.8	−4.3
						Avg.	1.3	−2.8				Avg.	0.6	−4.2				Avg.	0.9	−4.9				Avg.	0.5	−7.2
						SD (%)	4.2	5.3				SD (%)	6.1	7.4				SD (%)	6.5	8.1				SD (%)	6.8	8.8
						Var. (%)	17.7	27.6				Var. (%)	37.0	55.0				Var. (%)	41.8	65.2				Var. (%)	46.4	77.5

Table 7. Comparison between the reduction factor values (χ) calculated by standard procedures and obtained by FE analyses for models with 25 mm bars.

			f_y = 250 MPa						f_y = 290 MPa						f_y = 350 MPa						f_y = 450 MPa					
				χ			Percent Error (%)			χ			Percent Error (%)			χ			Percent Error (%)			χ			Percent Error (%)	
I-Section	ø (mm)	L (m)	λ	FE	NBR	EC3	$(\chi_{NBR} - \chi_{FE})/\chi_{NBR}$	$(\chi_{EC3} - \chi_{FE})/\chi_{EC3}$	λ	FE	NBR	EC3	$(\chi_{NBR} - \chi_{FE})/\chi_{NBR}$	$(\chi_{EC3} - \chi_{FE})/\chi_{EC3}$	λ	FE	NBR	EC3	$(\chi_{NBR} - \chi_{FE})/\chi_{NBR}$	$(\chi_{EC3} - \chi_{FE})/\chi_{EC3}$	λ	FE	NBR	EC3	$(\chi_{NBR} - \chi_{FE})/\chi_{NBR}$	$(\chi_{EC3} - \chi_{FE})/\chi_{EC3}$
CB350	25 mm	4	0.48	0.81	0.91	0.89	10.5	9.0	0.50	0.81	0.90	0.88	10.5	8.8	0.56	0.72	0.88	0.86	18.0	16.0	0.64	0.67	0.84	0.82	20.6	18.1
		5	0.49	0.79	0.91	0.89	13.2	11.7	0.50	0.77	0.90	0.88	14.3	12.7	0.57	0.70	0.87	0.85	20.3	18.3	0.65	0.64	0.84	0.81	24.2	21.7
		6	0.49	0.76	0.91	0.89	16.2	14.7	0.51	0.81	0.90	0.88	10.0	8.3	0.57	0.71	0.87	0.85	18.9	16.9	0.65	0.62	0.84	0.81	25.5	23.0
		7	0.49	0.78	0.90	0.89	14.3	12.8	0.51	0.79	0.90	0.88	11.6	9.9	0.57	0.70	0.87	0.85	19.9	17.9	0.65	0.62	0.84	0.81	26.1	23.6
		8	0.49	0.78	0.90	0.89	13.6	12.1	0.51	0.79	0.90	0.88	12.4	10.7	0.57	0.69	0.87	0.85	20.4	18.4	0.65	0.61	0.84	0.81	27.2	24.7
CB450	25 mm	4	0.57	0.74	0.87	0.85	15.0	12.9	0.61	0.70	0.86	0.83	17.9	15.6	0.66	0.69	0.83	0.81	16.8	13.9	0.73	0.67	0.80	0.76	16.1	12.4
		5	0.57	0.72	0.87	0.85	17.2	15.1	0.62	0.69	0.85	0.83	18.9	16.5	0.67	0.67	0.83	0.80	18.6	15.8	0.74	0.66	0.79	0.76	16.9	13.1
		6	0.58	0.79	0.87	0.85	9.6	7.3	0.62	0.74	0.85	0.83	13.4	10.9	0.68	0.69	0.83	0.80	16.3	13.3	0.75	0.66	0.79	0.76	16.7	12.8
		7	0.58	0.72	0.87	0.85	17.4	15.3	0.62	0.67	0.85	0.83	20.9	18.5	0.68	0.65	0.82	0.80	21.3	18.4	0.75	0.60	0.79	0.75	23.9	20.3
		8	0.58	0.71	0.87	0.85	17.8	15.7	0.63	0.66	0.85	0.82	22.2	19.9	0.68	0.63	0.82	0.79	23.9	21.2	0.75	0.57	0.79	0.75	27.7	24.3
CB600	25 mm	4	0.67	0.69	0.83	0.80	16.9	14.0	0.71	0.69	0.81	0.78	15.0	11.5	0.77	0.66	0.78	0.75	15.4	11.1	0.85	0.64	0.74	0.70	13.9	8.3
		5	0.68	0.70	0.82	0.79	14.9	11.8	0.72	0.69	0.80	0.77	14.6	11.0	0.78	0.66	0.78	0.74	15.3	10.8	0.86	0.63	0.73	0.69	14.8	8.9
		6	0.69	0.73	0.82	0.79	11.6	8.3	0.73	0.71	0.80	0.77	11.5	7.6	0.79	0.68	0.77	0.73	12.4	7.7	0.87	0.61	0.73	0.68	15.9	9.9
		7	0.69	0.73	0.82	0.79	11.3	7.9	0.73	0.70	0.80	0.76	12.0	8.0	0.79	0.67	0.77	0.73	12.6	7.8	0.87	0.62	0.73	0.68	14.6	8.5
		8	0.69	0.66	0.82	0.79	19.4	16.3	0.74	0.64	0.80	0.76	19.0	15.4	0.80	0.59	0.77	0.73	23.4	19.2	0.88	0.56	0.72	0.68	23.2	17.6
						Avg.	14.6	12.3				Avg.	14.9	12.3				Avg.	18.2	15.1				Avg.	20.5	16.5
						SD (%)	2.9	3.0				SD (%)	3.9	3.9				SD (%)	3.5	4.2				SD (%)	5.1	6.2
						Var. (%)	8.6	9.0				Var. (%)	15.3	15.3				Var. (%)	12.1	17.4				Var. (%)	26.0	38.7

Still in relation to Table 5, observing the percentage error evolution for the different steel types, a growth tendency of the errors is verified with the yield strength (f_y) increase, except for the steel with a yield strength of 450 MPa. This situation shows that the use of steels with different mechanical characteristics has an impact on the LDB strength that is not captured by the design curves of SSRC (used by NBR 8800:2008 [15]) and ECCS (used by EC3), which use the relative slenderness factor ($\lambda = \sqrt{\frac{M_{cr}}{M_{pl-CB}}}$).

Table 6 presents the results for models with 16 mm reinforcement bars. The results of Table 6 show a greater proximity between the LDB ultimate moment values obtained by the standard procedures and by the FE analyses. This situation can be explained by the plastic moment calculation of the composite section, which is dependent on the longitudinal reinforcement area present in the effective width of the concrete slab. Therefore, with the increase in the steel area, there is an increase in the plastic moment, and, therefore, a reduction in the relative slenderness factor ($\lambda = \sqrt{\frac{M_{cr}}{M_{pl-CB}}}$), and, consequently, an increase in the reduction factor (χ). Thus, with the increase in the plastic moment and the reduction factor, there is an increase in the value of the LDB ultimate moment, which leads to a closer approximation with the standard values. It is observed in Table 6 that the average values of the percentage error for the Brazilian standard [15] are 1.3%, 0.6%, 0.9%, and 0.5%, and for the European standard [13] −2.8%, −4.2%, −4.9%, and −7.2%, for steels with a yield strength of 250 MPa, 290 MPa, 350 MPa, and 450 MPa, respectively. Regarding the steel type, it is verified, for the Brazilian procedure [15] (curve 2P of the SSRC), that there is no great variation in the average values of the percentage errors for the different steel types. However, for the European standard [13], there is an increase in divergences with the increase in the I-section steel yield strength.

Finally, Table 7 shows the results for models with reinforced bars with 25 mm diameter. Contrary to the results observed in Table 5, Table 7 shows a non-conservative situation of the standard procedures; that is, the results of the FE analyses were lower than the ultimate moment values obtained by the standard procedures. The explanation for this situation is the same as that presented in the previous paragraph; that is, with the increase of the steel area in the effective width of the concrete slab, there is an increase in the plastic moment, and, therefore, a reduction in the relative slenderness factor ($\lambda = \sqrt{\frac{M_{cr}}{M_{pl-CB}}}$), and, consequently, an increase in the reduction factor (χ). Thus, with the increase in the plastic moment and the reduction factor, there is an increase in the LDB ultimate moment value, higher than those obtained in the FE analyses. In Table 7, it is observed that the average values of the percentage error in relation to the Brazilian standard [15] are 14.6%, 14.9%, 18.2%, and 20.5%, for the European standard [13] these values are 12.3%, 12.3%, 15.1%, and 16.5%, for steels with a yield strength of 250 MPa, 290 MPa, 350 MPa, and 450 MPa, respectively. Analyzing the percentage error evolution, it is also observed a small influence of the yield strength variation on the LDB ultimate moment values that are not captured by the classic SSRC and ECCS curves.

Figure 16 shows the deformed shape and von Mises stresses, for the CB450 model with a length of 6.0 m and reinforcement bars with 8 mm diameter for the four steel types analyzed. For the model with steel yield strength of 250 MPa (Figure 16a), the maximum stress in the I-section web (middle of the span) varies from 230.30 to 258.98 MPa. For the model with 290 MPa yield strength (Figure 16b), the maximum stress in the I-section web varies from 287.60 to 316.31 MPa. For the model in Figure 16c, with a yield strength of 350 MPa, the maximum web stress ranges from 332.94 to 363.16 MPa, and finally, for the model with a yield strength of 450 MPa (Figure 16d) the maximum web stress varies from 422.35 to 460.67 MPa.

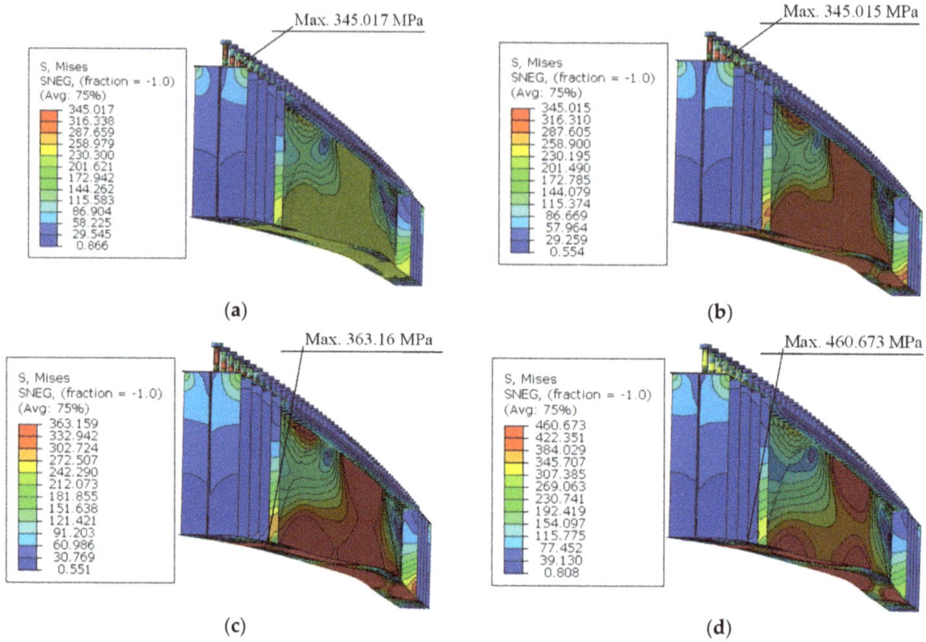

Figure 16. von Mises stress distribution (Unit = MPa). (**a**) f_y = 250 MPa; (**b**) f_y = 290 MPa; (**c**) f_y = 350 MPa; (**d**) f_y = 450 MPa.

6.2. Analytical Procedures and Other Parameters

The increase in the area of reinforcement bars present in the effective width of the concrete slab causes the increase in the plastic moment of the composite section calculated by plastic theory, which consequently leads to an increase in the LDB ultimate moment calculated by standard procedures such as EC4 [13] and NBR 8800:2008 [15]. However, the numerical analyses developed in this article show a small variation in LDB strength due to the increase in the area of reinforcement bars. Figure 17 shows the LDB ultimate moment results for models with reinforcement bars with a diameter of 8 mm, 16 mm, and 25 mm. In Figure 17 the reference values are those obtained by the models with 8 mm bars.

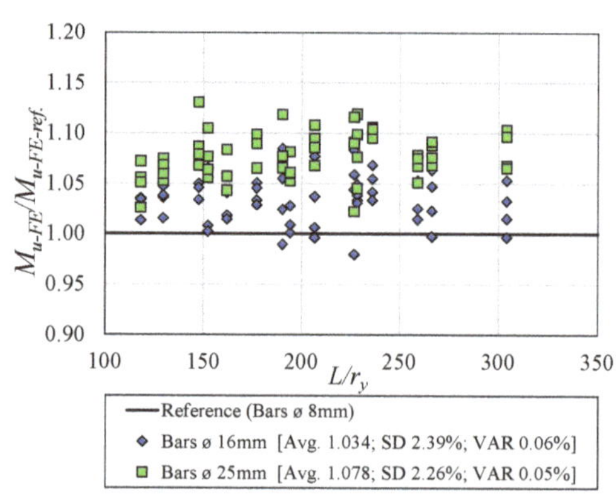

Figure 17. Influence of longitudinal reinforcement.

It is observed in Figure 17 that the average value of the ratio ($M_{u\text{-}FE}/M_{u\text{-}FE\text{-}ref.}$) for models with reinforcement bars with 16 mm diameter is 1.034; that is, there is an average gain in the ultimate moment value of about 3.4%. Regarding the models with 25 mm bars, the average value of the ratio ($M_{u\text{-}FE}/M_{u\text{-}FE\text{-}ref.}$) is 1.078. This shows that the ultimate moment has an average gain of 7.8% compared with the models with 8 mm bars. This situation shows that, despite a considerable increase in the longitudinal reinforcement area, there is an insignificant increase in the value of the LDB ultimate moment.

Figure 18 shows the deformed shape and the distribution of the absolute maximum stresses in the principal plane for models CB350 with a length of 6.0 m, A572-GR50 steel and with longitudinal reinforcements of 8 mm (Figure 18a), 16 mm (Figure 18b), and 25 mm (Figure 18c). It is observed in Figure 18 that the increase in the longitudinal reinforcement area in the effective width of the concrete slab causes an increase in the maximum tension and compression stresses, reflecting the increase in the LDB ultimate moment, a situation verified in the numerical models analyzed.

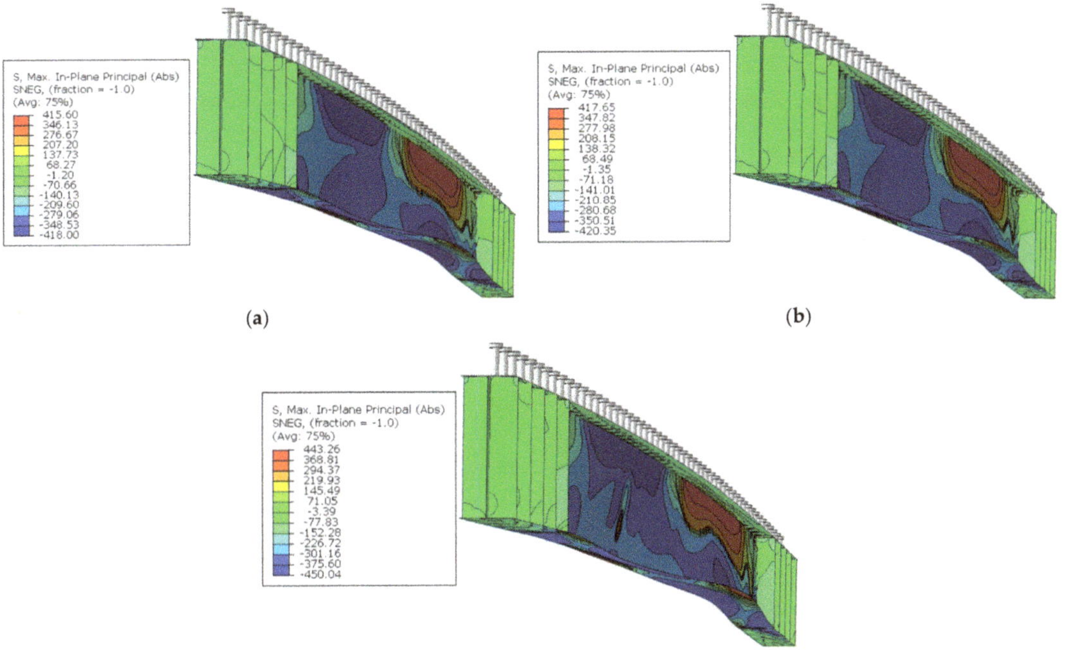

(c)

Figure 18. Stress distribution for the CB350 model with 6 m span (Unit = MPa). (**a**) Model with 8 mm longitudinal reinforcement bars; (**b**) model with 16 mm longitudinal reinforcement bars; (**c**) model with 25 mm longitudinal reinforcement bars.

As seen in Figure 17, the increase in the longitudinal reinforcement area in the effective width of the concrete slab leads to an insignificant increase in the LDB ultimate moment. However, this increase in the longitudinal reinforcement area causes a considerable increase in the plastic moment value calculated according to EC4 [13] by plastic theory. This situation causes an increase in the LDB ultimate moment value, calculated according to the procedures of EC4 [13] and NBR 8800:2008 [15], as the longitudinal reinforcement area increases. Figure 19 shows the comparison between the European [13] and Brazilian [15] standard design curves with the results of the FE analyses for models with 8 mm (Figure 19a), 16 mm (Figure 19b), and 25 mm (Figure 19c). In addition, Figure 19 also shows the values of the experimental analysis of four beams tested by Tong et al. [28], which served to validate the numerical model of this article.

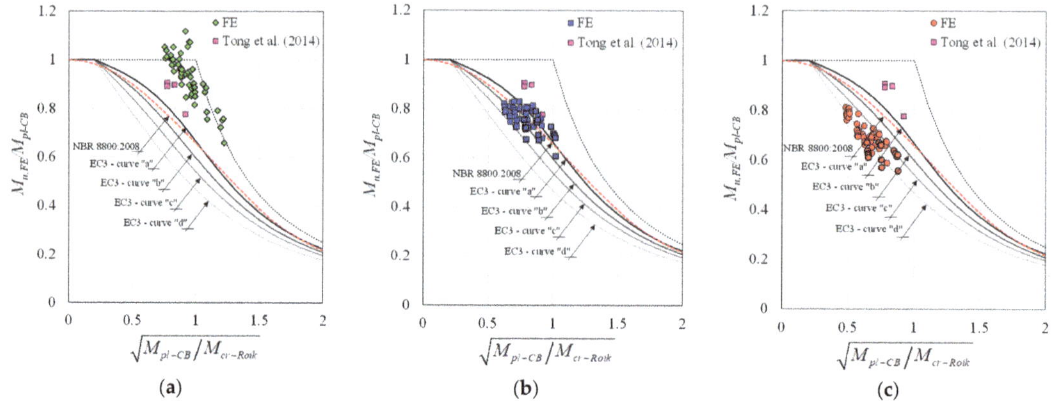

Figure 19. LDB strength for different steel type and comparison with [28]. (**a**) Models with 8 mm bars; (**b**) models with 16 mm bars; (**c**) models with 25 mm bars.

It is observed in Figure 19 three different situations. In Figure 19a, the results of the FE analyses provided LDB ultimate moment values higher than those obtained by the Brazilian [15] and European [13] standard procedures. The average percentage error of the FE analyses results when compared to the European procedure [13] was −46.28%, which shows considerable conservatism in the EC4 procedure [13]. Regarding the Brazilian standard [15], which is based on the SSRC 2P curve, the average percentage error was −34.60%, also showing conservatism. Figure 19a also presents the results of four beams tested experimentally by Tong et al. [28]. The results of these four beams are close to the values of the FE analyses developed in this article, since the longitudinal reinforcement area of the beams by Tong et al. [28] is similar to the models with 8 mm bars analyzed in the present work.

In relation to Figure 19b, which presents the results for the models with 16 mm diameter longitudinal reinforcement, a close proximity is observed between the results of the FE analyses and the standard procedures of EC4 [13] and NBR 8800:2008 [15]. The increase in the plastic moment of the composite section is the reason that leads to the greater proximity between the results of the standard procedures and those obtained through the FE analyses. For the results presented in Figure 19b, the average percentage error of the numerical results compared with the standard procedures is 0.83% for the Brazilian standard [15] and −4.78% for the European standard [13]. Finally, in Figure 19c, the results of the FE analyses are compared with the standard procedures for the models with 25 mm diameter longitudinal reinforcement. It is observed in Figure 19c that the LDB ultimate moment values obtained by means of the FE analyses are inferior to the standard results, which leads to an unsafe situation of the European [13] and Brazilian [15] standards. As verified, the increase in the longitudinal reinforcement area does not cause a considerable gain in the LDB strength. However, for the EC4 [13] and NBR 8800:2008 [15] procedures, the increase in the longitudinal reinforcement area leads to a considerable gain from the LDB strength of SCCB, which leads to this unsafe situation for the European [13] and Brazilian [15] procedures. In Figure 19c the average percentage error of the numerical results compared with the standard procedures is 17.05% for the Brazilian standard [15] and 14.05% for the European standard [13].

The results of the FE analyses were also compared with the analytical procedures presented by Zhou and Yan [4] and Bradford [29]. In Figure 20, the results of the FE analyses are compared with the Zhou and Yan procedure [4].

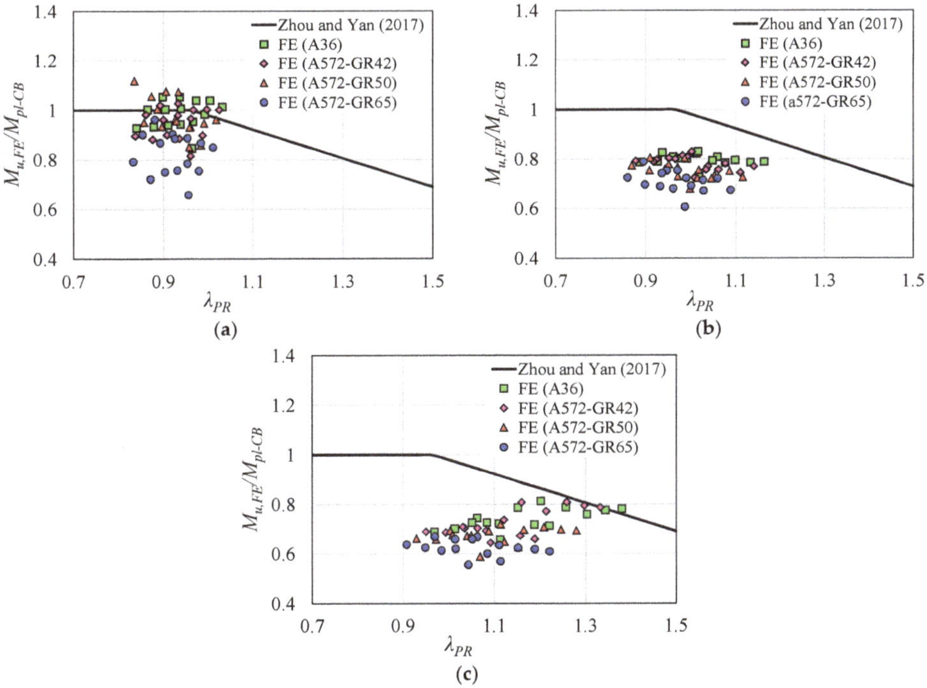

Figure 20. Comparison of FE results with the Zhou and Yan procedure [4]. (**a**) Models with 8 mm bars; (**b**) models with 16 mm bars; (**c**) models with 25 mm bars.

It is observed in Figure 20 that the longitudinal reinforcement area variation in the concrete slab is also responsible for three different situations when the results of the FE analyses are compared with the Zhou and Yan procedure [4]. For the models with 8 mm diameter longitudinal reinforcement (Figure 20a) the results of the FE analyses show greater agreement with the Zhou and Yan [4] procedure. The results analysis of the LDB ultimate moment shows that the average value of the ratio ($M_{Zhou\ and\ Yan}/M_{u,FE}$) is 1.08, that is, the values obtained through the analytical procedure of Zhou and Yan [4] are on average 8.0% higher than those obtained with the FE analyses. Figure 20b presents the results obtained with the FE analyses for the models with 16 mm bars. It is verified that with the increase in the longitudinal reinforcement area there is also an increase in the divergence between the results of the FE analyses and the results of the Zhou and Yan procedure [4]. The average value of the ratio ($M_{Zhou\ and\ Yan}/M_{u,FE}$) is 1.28, that is, the results of the procedure developed by the authors [4] are on average 28% higher than the results obtained in the numerical analysis, an unsafe situation. Finally, Figure 20c shows the results for the models with 25 mm diameter longitudinal reinforcement bars. There is also a tendency to increase the divergence between the numerical results and the Zhou and Yan [4] procedure as the longitudinal reinforcement area is increased. For the results in Figure 20c, the average value of the ratio ($M_{Zhou\ and\ Yan}/M_{u,FE}$) is 1.34, which shows an unsafe situation in the Zhou and Yan procedure [4]. Figure 20 also presents the results for the models with the different steel types analyzed, it is verified that, as for the standard procedures, in the Zhou and Yan procedure [4] the influence of the steel type is not captured properly.

Finally, Figure 21 presents the comparison between the results obtained with the FE analyses and the Bradford [29] procedure.

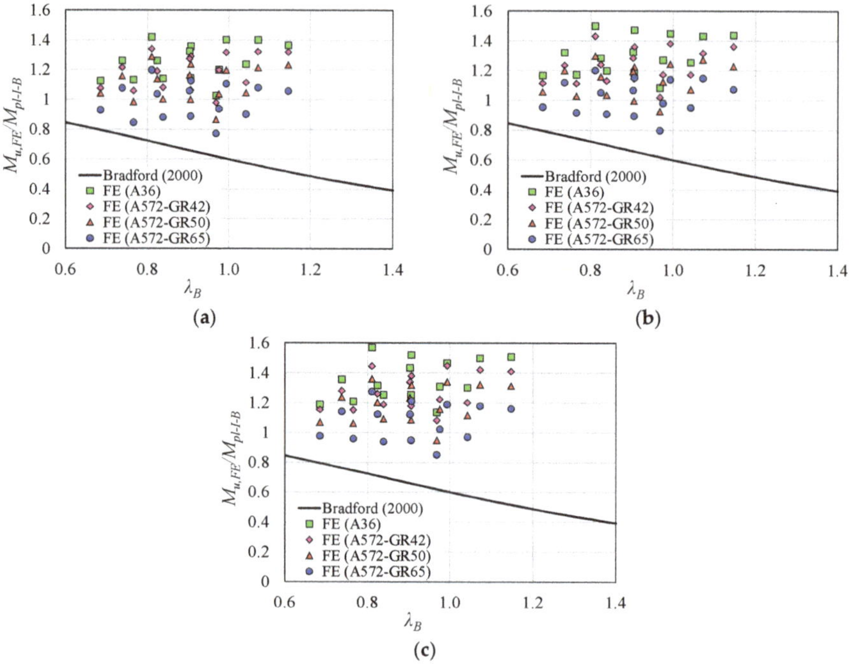

Figure 21. Comparison of FE results with the Bradford [29] procedure. (**a**) Models with 8 mm bars; (**b**) models with 16 mm bars; (**c**) models with 25 mm bars.

The results in Figure 21 show a similar behavior of the Bradford procedure [29] in comparison with the results of the FE analysis between the models with longitudinal reinforcement of 8 mm (Figure 21a), 16 mm (Figure 21b), and 25 mm (Figure 21c). This situation is due to the fact that the Bradford procedure [29] does not consider the plastic moment of the composite section, but only the I-section plastic moment to determine the LDB ultimate moment. This situation can be confirmed with the analysis of the average value of the ratio ($M_{Bradford}/M_{u,FE}$) which provided values of 0.59, 0.57, and 0.55 for the models with reinforcement bars with a diameter of 8 mm, 16 mm, and 25 mm, respectively. However, despite providing safe results, the Bradford procedure [29] is considerably conservative. Regarding the results for the different steel types analyzed, it is verified that for the models with greater steel yield strength, there is a greater proximity of the FE results with the author's procedure [29]. On the other hand, with the reduction in the steel yield strength, there is a tendency to increase the divergences between the FE results and the Bradford procedure [29].

Finally, Figure 22 shows the LDB ultimate moment values for models with A572-GR50 steel and 8 mm reinforcement bars. The results show that the unrestricted length variation does not significantly influence the LDB ultimate moment value, a situation also observed by Rossi et al. [1]. It is verified in Figure 22 that the preponderant factor in the LDB ultimate moment values is the I-section geometric properties.

The results presented show that the standard procedures are still flawed in determining the LDB strength of SCCB under the action of the hogging moment. The influence of parameters not previously investigated was presented, such as the I-section steel type and a considerable variation in the longitudinal reinforcement area in the effective width of the concrete slab. The investigation of these factors showed flaws until then not observed, in the standard procedures such as EC4 [13] and NBR8800:2008 [15]. The comparison between the FE results and the analytical procedures, such as Zhou and Yan [4] and Bradford [29], shows that further investigations are still needed to fully understand the LDB phenomenon in SCCB. Thus, the results presented in this article can provide a reference for future research and specification reviews.

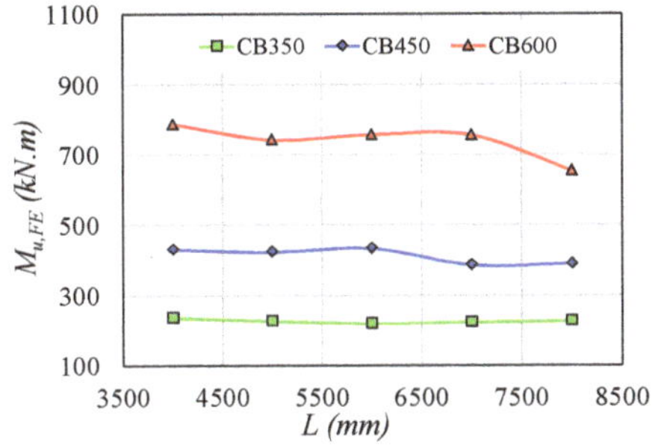

Figure 22. Influence of unrestricted length.

7. Conclusions

In this article, the LDB strength of SCCB was investigated through the development of post-buckling numerical analyses with the ABAQUS software. The SCCB analyzed are simply supported with lateral bracing. In the parametric study, 180 analyses were developed, varying the following factors: the mechanical properties of the steel I-section, the dimensions of the I-sections, the unrestricted length, and the longitudinal reinforcement area present in the effective width of the concrete slab. Finally, the results were compared with current standard procedures and analytical proposals. In this way, it was concluded:

- The increase in the I-section steel yield strength causes an increase in the LDB ultimate moment values. Taking as reference the steel with a yield strength of 250 MPa, it is verified that for an increase of 16%, 40%, and 80% in the yield strength (f_y = 290; f_y = 350; f_y = 450), the variation in the LDB ultimate moment was of 10.12%, 23.60%, and 42.43%, respectively.
- It was verified that for the models with a higher steel yield strength (f_y), the LDB occurs for ultimate moment values lower than the plastic moment of the composite section, that is, the greater the steel yield strength, the lower the chances of the composite section reaching the plastic moment, with the LDB being the failure mode responsible for defining the ultimate moment value;
- There was a tendency to increase divergences between the numerical results and the curves of ECCS (EC4) and SSRC (NBR 8800:2008), with the increase in the steel yield strength (f_y), that is, for steels with greater yield strength, the greater the divergences between the FE results and standard procedures.
- The increase in the longitudinal reinforcement area present in the concrete slab causes a small increase in the LDB strength. For an increase of 244% in the longitudinal reinforcement area, there was only an increase of about 7.8% in the value of the LDB ultimate moment;
- Different behavior of the standard procedures was verified, due to the variation of the longitudinal reinforcement area, when compared to the ultimate moment obtained by the FE analyses. For a longitudinal reinforcement rate that approximates practical situations (8 mm bars), the standard procedures proved to be considerably conservative. As the longitudinal reinforcement area in the effective width of the concrete slab is increased, the behavior of the standard procedures is changed. For the longitudinal reinforcement with 16 mm bars, the standard results are close to the numerical values, showing a good behavior of the standard procedures. However, for models with 25 mm bars, the European [13] and Brazilian [15] standard procedures provided ultimate moment results superior to those obtained by FE analyses;

- The Zhou and Yan [4] procedure showed a behavior similar to that observed in the standard procedures, that is, with the increase in the longitudinal reinforcement rate, there is a variation of the divergences with the results of the FE analyses. This situation is caused by the fact that the procedures of Zhou and Yan [4], the European standard [13] and the Brazilian standard [15], define the LDB ultimate moment value due to the reduction of the plastic moment of the composite section, and, any change in this value was shown to be sensitive to the behavior of the mentioned procedures;
- It was found that the Bradford procedure [29] is not sensitive to the variation in the longitudinal reinforcement rate, since the definition of the LDB ultimate moment is a function of the I-section plastic moment, and not of the composite section plastic moment. The comparison between the results of the FE analyses and Bradford's procedure [29] showed that this provides safe results, however, considerably conservative;
- It was verified that the LDB ultimate moment is not very sensitive to the unrestricted length variation of the SCCB. The predominant factor is the dimensions of the I-section.

In general, it was verified that the standard procedures are still flawed in determining the LDB strength of SCCB. It was found that the European [13] and the Brazilian [15] standard procedures show different behaviors due to the longitudinal reinforcement area variation. In addition, they are sensitive to the variation of the steel mechanical properties, tending to an increase or reduction of the divergences depending on the yield strength variation (f_y). The comparison between the FE results and the analytical procedures, such as Zhou and Yan [4] and Bradford [29], shows that further investigations are still needed to fully understand the LDB phenomenon in SCCB. Thus, the results presented in this article can provide a reference for future research and specification reviews.

Author Contributions: Conceptualization, A.R., C.H.M. and A.S.C.d.S.; methodology, A.R., A.S.d.C. and V.M.d.O.; software, A.R., A.S.d.C. and V.M.d.O.; validation, A.R., A.S.d.C. and V.M.d.O.; formal analysis, A.R.; investigation, A.R., A.S.d.C. and V.M.d.O.; resources, A.R.; data curation, A.R., C.H.M. and A.S.C.d.S.; writing—original draft preparation, A.R,.; writing—review and editing, A.R., A.S.d.C. and V.M.d.O.; visualization, A.R.; supervision, C.H.M. and A.S.C.d.S.; project administration, C.H.M. and A.S.C.d.S.; funding acquisition, C.H.M. and A.S.C.d.S. All authors have read and agreed to the published version of the manuscript.

Funding: This study was financed by the Coordenação de Aperfeiçoamento de Pessoal de Nível Superior—Brasil (CAPES)—Finance Code 001 and by the Conselho Nacional de Desenvolvimento Científico e Tecnológico (CNPq).

Data Availability Statement: The data is available upon request.

Acknowledgments: The authors acknowledge CAPES and CNPq for the financial support.

Conflicts of Interest: The authors declare no conflict of interest.

Notation

The following symbols are used in this paper:

A	sectional area of the steel I-beam
A_f	sectional area of the steel beam flange
A_r	sectional area of the longitudinal reinforcement in concrete slab
A_w	sectional area of the steel beam web
a	spacing between parallel beams
b_f	flange width
C_{dist}	property associated with the distribution of bending moments
d	section depth
E	modulus of elasticity
E_a	elasticity modulus of structural steel
E_cI_{c2}	flexural stiffness per unit width of the slab
$f_{ck\text{-}cubic}$	characteristic compressive cubic strength of concrete
$f_{ck\text{-}cylindrical}$	characteristic compressive cylinder strength of concrete

f_{cm}	compressive strength of concrete
f_f	yield strength of the flange I-beams
f_r	yield strength of the reinforcement bars
f_{tm}	concrete tension strength
f_u	ultimate strength
f_y	yield strength
G	shear modulus
h_s	distance between the centroids of the flanges
h_w	web depth
$I_{af,y}$	moments of inertia of area of the structural steel section about the weak axis
J	torsional constant
k_s	rotational stiffness
k_1	bending stiffness of the concrete slab
k_2	bending stiffness of the web of I-beam
L	unrestrained length of composite beam
M_{cr}	elastic distortional critical moment
M_{FE}	ultimate moment obtained from the finite element analysis
$M_{pl\text{-}CB}$	plastic moment obtained by I-beam more reinforcement bars
$M_{pl\text{-}I\text{-}B}$	plastic moment of I-beam
M_{test}	experimental test ultimate moment
$M_{u,dist}$	resistant moment obtained by analytical procedures
$r_{fc,y}$	radius of gyration of the compression flange
S_{stud}	spacing between headed stud connectors
t_f	flange thickness
t_w	web thickness
U	displacement
UR	rotational displacement
α	coefficient equal to 2 for an edge beam, with or without a cantilever, or 3 for an inner beam
α_g	factor related to the cross-section geometry of the composite beam by Roik et al.
α_{LT}	imperfection factor according EC3
α_m	modification factor of the moment distribution according AS4100
β_c	stress–strain relationship form factor of concrete in compression
δ	lateral displacement
ε	strain
ε_c	compressive strain
ε_t	tensile strain
ε_u	ultimate strain
ε_y	yield strain
θ	rotation
$\overline{\lambda}$	slenderness ratio
$\overline{\lambda}_{LT}$	slenderness ratio according EC3
λ_{PR}	slenderness ratio according Zhou and Yan
ν_a	Poisson's ratio of structural steel
ρ	force ratio according Zhou and Yan
σ	stress
σ_c	residual compression stress
σ_t	residual tensile stress
ϕ_{PR}	stability coefficient according Zhou and Yan
ϕ_{LT}	stability coefficient according EC3
χ_{LT}	reduction factor according EC3

Appendix A

Geometric Parameters			f_y = 250 MPa			f_y = 290 MPa			f_y = 350 MPa			f_y = 450 MPa		
I-Section	*ø* (mm)	*L* (m)	M_{FE} (kNm)	[a] $M_{pl\text{-}I}$(kNm)	[b] $M_{pl\text{-}C.beam}$(kNm)	M_{FE} (kNm)	[a] $M_{pl\text{-}I}$(kNm)	[b] $M_{pl\text{-}C.beam}$(kNm)	M_{FE} (kNm)	[a] $M_{pl\text{-}I}$(kNm)	[b] $M_{pl\text{-}C.beam}$(kNm)	M_{FE} (kNm)	[a] $M_{pl\text{-}I}$(kNm)	[b] $M_{pl\text{-}C.beam}$(kNm)
CB350	8 mm	4	187.32	131.85	177.87	205.39	152.95	201.53	237.75	184.59	236.86	284.13	237.33	295.46
		5	179.13	131.85	177.87	197.51	152.95	201.53	228.29	184.59	236.86	267.04	237.33	295.46
		6	184.87	131.85	177.87	201.63	152.95	201.53	221.00	184.59	236.86	262.23	237.33	295.46
		7	184.77	131.85	177.87	202.16	152.95	201.53	224.12	184.59	236.86	256.30	237.33	295.46
		8	180.19	131.85	177.87	201.82	152.95	201.53	227.54	184.59	236.86	251.10	237.33	295.46
CB450	8 mm	4	336.68	266.93	335.81	376.44	309.63	383.59	432.31	373.70	455.13	517.47	480.47	574.18
		5	336.71	266.93	335.81	368.87	309.63	383.59	425.02	373.70	455.13	498.42	480.47	574.18
		6	353.75	266.93	335.81	394.32	309.63	383.59	434.81	373.70	455.13	508.65	480.47	574.18
		7	320.11	266.93	335.81	370.79	309.63	383.59	387.73	373.70	455.13	451.14	480.47	574.18
		8	330.19	266.93	335.81	345.09	309.63	383.59	391.20	373.70	455.13	433.87	480.47	574.18
CB600	8 mm	4	607.05	539.40	654.24	674.08	625.70	751.44	787.23	755.16	704.23	903.03	970.92	1139.82
		5	610.78	539.40	654.24	663.06	625.70	751.44	743.05	755.16	704.23	822.99	970.92	1139.82
		6	614.81	539.40	654.24	676.83	625.70	751.44	757.06	755.16	704.23	856.43	970.92	1139.82
		7	617.29	539.40	654.24	665.84	625.70	751.44	755.31	755.16	704.23	863.88	970.92	1139.82
		8	554.04	539.40	654.24	613.21	625.70	751.44	655.03	755.16	704.23	749.43	970.92	1139.82
CB350	16 mm	4	197.70	131.85	240.13	218.83	152.95	269.88	239.58	184.59	311.71	284.72	237.33	377.33
		5	194.28	131.85	240.13	208.33	152.95	269.88	225.82	184.59	311.71	273.47	237.33	377.33
		6	191.05	131.85	240.13	211.47	152.95	269.88	229.44	184.59	311.71	270.26	237.33	377.33
		7	188.96	131.85	240.13	201.59	152.95	269.88	234.63	184.59	311.71	272.49	237.33	377.33
		8	189.71	131.85	240.13	208.38	152.95	269.88	226.70	184.59	311.71	254.79	237.33	377.33
CB450	16 mm	4	352.54	266.93	426.17	382.13	309.63	478.59	447.85	373.70	555.12	537.13	480.47	679.50
		5	342.56	266.93	426.17	383.88	309.63	478.59	432.69	373.70	555.12	505.44	480.47	679.50
		6	353.97	266.93	426.17	397.65	309.63	478.59	446.80	373.70	555.12	513.02	480.47	679.50
		7	338.85	266.93	426.17	363.14	309.63	478.59	419.95	373.70	555.12	470.91	480.47	679.50
		8	334.80	266.93	426.17	363.04	309.63	478.59	400.93	373.70	555.12	456.89	480.47	679.50
CB600	16 mm	4	628.58	539.40	797.48	697.16	625.70	879.14	797.42	755.16	1028.62	926.56	970.92	1275.36
		5	631.46	539.40	797.48	695.93	625.70	879.14	777.01	755.16	1028.62	890.53	970.92	1275.36
		6	645.84	539.40	797.48	707.56	625.70	879.14	782.25	755.16	1028.62	881.05	970.92	1275.36
		7	640.07	539.40	797.48	716.77	625.70	879.14	752.17	755.16	1028.62	868.96	970.92	1275.36
		8	584.30	539.40	797.48	638.74	625.70	879.14	699.76	755.16	1028.62	774.36	970.92	1275.36
CB350	25 mm	4	206.91	131.85	254.63	221.16	152.95	274.12	251.04	184.59	349.45	302.23	237.33	452.10
		5	200.34	131.85	254.63	211.23	152.95	274.12	243.32	184.59	349.45	287.65	237.33	452.10
		6	193.26	131.85	254.63	221.52	152.95	274.12	247.26	184.59	349.45	282.26	237.33	452.10
		7	197.41	131.85	254.63	217.50	152.95	274.12	243.84	184.59	349.45	279.76	237.33	452.10
		8	198.84	131.85	254.63	215.55	152.95	274.12	242.41	184.59	349.45	275.42	237.33	452.10
CB450	25 mm	4	361.81	266.93	486.42	396.20	309.63	563.41	461.84	373.70	666.25	548.33	480.47	818.91
		5	351.22	266.93	486.42	389.96	309.63	563.41	449.23	373.70	666.25	539.93	480.47	818.91
		6	382.57	266.93	486.42	415.02	309.63	563.41	460.75	373.70	666.25	539.66	480.47	818.91
		7	349.24	266.93	486.42	378.96	309.63	563.41	432.70	373.70	666.25	491.72	480.47	818.91
		8	347.02	266.93	486.42	372.13	309.63	563.41	417.53	373.70	666.25	466.47	480.47	818.91
CB600	25 mm	4	641.11	539.40	930.85	722.59	625.70	1050.25	807.54	755.16	1219.74	948.98	970.92	1487.96
		5	652.32	539.40	930.85	720.41	625.70	1050.25	801.63	755.16	1219.74	930.19	970.92	1487.96
		6	675.12	539.40	930.85	743.66	625.70	1050.25	824.83	755.16	1219.74	912.33	970.92	1487.96
		7	675.68	539.40	930.85	737.68	625.70	1050.25	819.96	755.16	1219.74	922.46	970.92	1487.96
		8	612.75	539.40	930.85	677.31	625.70	1050.25	717.36	755.16	1219.74	827.37	970.92	1487.96

[a] I-section plastification moment; [b] Composite section plastification moment.

References

1. Rossi, A.; Nicoletti, R.S.; de Souza, A.S.C.; Martins, C.H. Numerical Assessment of Lateral Distortional Buckling in Steel-Concrete Composite Beams. *J. Constr. Steel Res.* **2020**, *172*, 106192. [CrossRef]
2. Vrcelj, Z.; Bradford, M.A. Inelastic Restrained Distortional Buckling of Continuous Composite T-Beams. *J. Constr. Steel Res.* **2009**, *65*, 850–859. [CrossRef]
3. Rossi, A.; Nicoletti, R.S.; de Souza, A.S.C.; Martins, C.H. Lateral Distortional Buckling in Steel-Concrete Composite Beams: A Review. *Structures* **2020**, *27*, 1299–1312. [CrossRef]
4. Zhou, W.-B.; Yan, W.-J. Refined Nonlinear Finite Element Modelling towards Ultimate Bending Moment Calculation for Concrete Composite Beams under Negative Moment. *Thin-Walled Struct.* **2017**, *116*, 201–211. [CrossRef]
5. Moghbeli, A.; Sharifi, Y. New Predictive Equations for Lateral-Distortional Buckling Capacity Assessment of Cellular Steel Beams. *Structures* **2021**, *29*, 911–923. [CrossRef]
6. Hosseinpour, M.; Sharifi, Y.; Sharifi, H. Neural Network Application for Distortional Buckling Capacity Assessment of Castellated Steel Beams. *Structures* **2020**, *27*, 1174–1183. [CrossRef]
7. Dietrich, M.Z.; Calenzani, A.F.G.; Fakury, R.H. Analysis of Rotational Stiffness of Steel-Concrete Composite Beams for Lateral-Torsional Buckling. *Eng. Struct.* **2019**, *198*, 109554. [CrossRef]
8. Chen, S.; Wang, X. Finite Element Analysis of Distortional Lateral Buckling of Continuous Composite Beams with Transverse Web Stiffeners. *Adv. Struct. Eng.* **2012**, *15*, 1607–1616. [CrossRef]
9. Yang, L.; Lewei, T.; Bo, S.; Yiyi, C.; Feng, Z.; Tian Hal, S.X. FEA and Bending Capacity Calculation for Mechanical Behavior of Steel-Concrete Composite Beams under Negative Bending. *J. Build. Struct.* **2014**, *35*, 10–21.
10. Vasdravellis, G.; Uy, B.; Tan, E.L.; Kirkland, B. The Effects of Axial Tension on the Hogging-Moment Regions of Composite Beams. *J. Constr. Steel Res.* **2012**, *68*, 20–33. [CrossRef]
11. Chen, S.; Jia, Y. Numerical Investigation of Inelastic Buckling of Steel–Concrete Composite Beams Prestressed with External Tendons. *Thin-Walled Struct.* **2010**, *48*, 233–242. [CrossRef]
12. Bradford, M.; Johnson, R. Inelastic Buckling of Composite Bridge Girders Near Internal Supports. *Proc. Inst. Civ. Eng.* **1987**, *83*, 143–159. [CrossRef]
13. *EN 1994-1-1*; Eurocode 4: Design of Composite Steel and Concrete Structures—Part 1-1: General Rules and Rules for Buildings. Standard. Comite Europeen de Normalisation: Brussels, Belgium, 2004.
14. Roik, K.; Hanswille, G.; Kina, J. Solution for the Lateral Torsional Buckling Prob Lem of Composite Beams. *Stahlbau* **1990**, *59*, 327–332. (In German)
15. *ABNT NBR 8800*; Projeto de Estruturas de Aço e de Estruturas Mistas de Aço e Concreto de Edifícios, Standard. Associação Brasileira de Normas Técnicas: Rio de Janeiro, Brazil, 2008.
16. *ANSI/AISC 360-16*; ANST/AISC AISC 360-16: Specification for Structural Steel Buildings. ANSI: Washington, DC, USA, 2016; Volume 676.
17. *AS 4100*; Steel Structures. Standards Association of Australia: Sydney, Australia.
18. Dias, J.V.F.; Oliveira, J.P.S.; Calenzani, A.F.G.; Fakury, R.H. Elastic Critical Moment of Lateral-Distortional Buckling of Steel-Concrete Composite Beams under Uniform Hogging Moment. *Int. J. Struct. Stab. Dyn.* **2019**, *19*, 1950079. [CrossRef]
19. Zhou, W.B.; Jiang, L.Z.; Li, S.J.; Kong, F. Elastic Distortional Buckling Analysis of I-Steel Concrete Composite Beam Considering Shear Deformation. *Int. J. Struct. Stab. Dyn.* **2016**, *16*, 1–22. [CrossRef]
20. Ye, J.H.; Chen, W. Elastic Restrained Distortional Buckling of Steel-Concrete Composite Beams Based on Elastically Supported Column Method. *Int. J. Struct. Stab. Dyn.* **2013**, *13*, 1–29. [CrossRef]
21. Svensson, S.E. Lateral Buckling of Beams Analysed as Elastically Supported Columns Subject to a Varying Axial Force. *J. Constr. Steel Res.* **1985**, *5*, 179–193. [CrossRef]
22. Williams, F.W.; Jemah, A.K. Buckling Curves for Elastically Supported Columns with Varying Axial Force, to Predict Lateral Buckling of Beams. *J. Constr. Steel Res.* **1987**, *7*, 133–147. [CrossRef]
23. Hanswille, G.; Lindner, J.; Munich, D. Lateral Torsional Buckling of Composite Beams. *Stahlbau* **1998**, *67*, 525–535. (In German) [CrossRef]
24. Chen, W.; Ye, J. Elastic Lateral and Restrained Distortional Buckling of Doubly Symmetric I-Beams. *Int. J. Struct. Stab. Dyn.* **2010**, *10*, 983–1016. [CrossRef]
25. Dekker, N.W.; Kemp, A.R.; Trinchero, P. Factors Influencing the Strength of Continuous Composite Beams in Negative Bending. *J. Constr. Steel Res.* **1995**, *34*, 161–185. [CrossRef]
26. Johnson, R.P.; Chen, S. Stability of Continuous Composite Plate Girders with U-Frame Action. *Proc. Inst. Civ. Eng. Struct. Build.* **1993**, *99*, 187–197. [CrossRef]
27. Gizejowski, M.A.; Salah, W.A. Stability and Ductility of Castellated Composite Beams Subjected to Hogging Bending. In Proceedings of the SDSS' Rio 2010: International Colloquium Stability and Ductility of Steel Structures, Rio de Janeiro, Brazil, 8–10 September 2010; Volume 2, pp. 839–846.
28. Tong, L.; Liu, Y.; Sun, B.; Chen, Y.; Zhou, F.; Tian, H.; Sun, X. Experimental Investigation on Mechanical Behavior of Steel-Concrete Composite Beams under Negative Bending. *J. Build. Struct.* **2014**, *35*, 1–10. [CrossRef]
29. Bradford, M.A. Strength of Compact Steel Beams with Partial Restraint. *J. Constr. Steel Res.* **2000**, *53*, 183–200. [CrossRef]

30. Bradford, M.A.; Ronagh, H.R. Generalized Elastic Buckling of Restrained I-Beams by FEM. *J. Struct. Eng.* **1997**, *123*, 1631–1637. [CrossRef]
31. Johnson, R.P.; Fan, C.K.R. Distortional Lateral Buckling of Continuous Composite Beams. *Proc.—Inst. Civ. Eng. Part 2. Res. theory* **1991**, *91*, 131–161. [CrossRef]
32. Rossi, A.; Souza, A.S.C.d.; Nicoletti, R.S.; Martins, C.H. Stability Behavior of Steel–Concrete Composite Beams Subjected to Hogging Moment. *Thin-Walled Struct.* **2021**, *167*, 108193. [CrossRef]
33. Rossi, A.; de Souza, A.S.C.; Nicoletti, R.S.; Martins, C.H. The Influence of Structural and Geometric Imperfections on the LDB Strength of Steel–Concrete Composite Beams. *Thin-Walled Struct.* **2021**, *162*, 107542. [CrossRef]
34. Zhao, J.; Li, J.; Sun, Y. Experimental and Numerical Study on Overall Buckling Behavior of Q460 High-Strength Steel Continuous Beams with Welded Singly Symmetric I-Section. *Eng. Struct.* **2023**, *280*, 115678. [CrossRef]
35. Sun, Y.; He, A.; Liang, Y.; Zhao, O. In-Plane Bending Behaviour and Capacities of S690 High Strength Steel Welded I-Section Beams. *J. Constr. Steel Res.* **2019**, *162*, 105741. [CrossRef]
36. Bradford, M.A.; Kemp, A.R. Buckling in Continuous Composite Bems. *Prog. Struct. Eng. Mater.* **2000**, *2*, 169–178. [CrossRef]
37. Weston, G.; Nethercot, D.A.; Crisfield, M.A. Lateral Buckling in Continuous Composite Bridge Girders. *Struct. Eng. London* **1991**, *69*, 79–87.
38. Dassault Systemes Simulia, Inc. Abaqus Cae, User's Guide, 2014. Available online: http://130.149.89.49:2080/v6.14/pdf_books/CAE.pdf (accessed on 15 July 2023).
39. Vlasov, V.Z. *Thin-Walled Elastic Beams*, 2nd ed.; Israel Program for Scientific Translation: Jerusalem, Israel, 1961.
40. Wijesiri Pathirana, S.; Uy, B.; Mirza, O.; Zhu, X. Flexural Behaviour of Composite Steel–Concrete Beams Utilising Blind Bolt Shear Connectors. *Eng. Struct.* **2016**, *114*, 181–194. [CrossRef]
41. Rossi, A.; Martins, C.H.; Nicoletti, R.S.; Souza, A.S.C. Reassesment of Lateral Torsional Buckling in Hot-Holled I-Beams. *Structures* **2020**, ISTRUC801. [CrossRef]
42. Rossi, A.; Ferreira, F.P.V.; Martins, C.H.; Mesacasa Júnior, E.C. Assessment of Lateral Distortional Buckling Resistance in Welded I-Beams. *J. Constr. Steel Res.* **2020**, *166*. [CrossRef]
43. Ferreira, F.P.V.; Rossi, A.; Martins, C.H. Lateral-Torsional Buckling of Cellular Beams According to the Possible Updating of EC3. *J. Constr. Steel Res.* **2019**, *153*, 222–242. [CrossRef]
44. Nicoletti, R.S.; Rossi, A.; Souza, A.S.C.d.; Martins, C.H. Numerical Assessment of Effective Width in Steel-Concrete Composite Box Girder Bridges. *Adv. Struct. Eng.* **2020**, 136943322097174. [CrossRef]
45. Nicoletti, R.S.; Rossi, A.; de Souza, A.S.C.; Martins, C.H. Numerical Assessment of Effective Width in Steel-Concrete Composite Box Girder Bridges with Partial Interaction. *Eng. Struct.* **2021**, *239*, 112333. [CrossRef]
46. Rossi, A.; Hideyuki Saito, D.; Humberto Martins, C.; Sander Clemente de Souza, A. The Influence of Structural Imperfections on the LTB Strength of I-Beams. *Structures* **2021**, *29*, 1173–1186. [CrossRef]
47. Galambos, T.V.; Ketter, R.L. Columns under Combined Bending and Thrust. *J. Eng. Mech. Div.* **1959**, *85*, 1–30. [CrossRef]
48. Earls, C.J. On the Inelastic Failure of High Strength Steel I-Shaped Beams. *J. Constr. Steel Res.* **1999**, *49*, 1–24. [CrossRef]
49. Earls, C.J. Effects of Material Property Stratification and Residual Stresses on Single Angle Flexural Ductility. *J. Constr. Steel Res.* **1999**, *51*, 147–175. [CrossRef]
50. Hillerborg, A.; Modéer, M.; Petersson, P.E. Analysis of Crack Formation and Crack Growth in Concrete by Means of Fracture Mechanics and Finite Elements. *Cem. Concr. Res.* **1976**, *6*, 773–781. [CrossRef]
51. Lubliner, J.; Oliver, J.; Oller, S.; Oñate, E. A Plastic-Damage Model for Concrete. *Int. J. Solids Struct.* **1989**, *25*, 299–326. [CrossRef]
52. Lee, J.; Fenves, G.L. Plastic-Damage Model for Cyclic Loading of Concrete Structures. *J. Eng. Mech.* **1998**, *124*, 892–900. [CrossRef]
53. Yu, T.; Teng, J.G.; Wong, Y.L.; Dong, S.L. Finite Element Modeling of Confined Concrete-I: Drucker–Prager Type Plasticity Model. *Eng. Struct.* **2010**, *32*, 665–679. [CrossRef]
54. Carreira, D.; Chu, K. Stress-Strain Relationship for Plain Concrete in Compression. *ACI J. Proc.* **1985**, *82*. [CrossRef]
55. Carreira, D.J.; Chu, K.H. Stress-Strain Relatonship for Reinforced Concrete in Tension. *J. Am. Concr. Inst.* **1986**, *83*, 21–28. [CrossRef]

eng

MDPI

Article

Insight into the Optimization of Implementation Time in Cob Construction: Field Test and Compressive Strength Versus Drying Kinetics

Karim Touati [1,2,*], Baraa Al Sahmarany [2], Malo Le Guern [2], Yassine El Mendili [3], François Streiff [4] and Steve Goodhew [5]

1 EPF Ecole d'Ingénieurs, 21 Boulevard Berthelot, 34000 Montpellier, France
2 ComUE Normandie Université, Builders Ecole d'Ingénieurs, 1 Rue Pierre et Marie Curie, 14610 Epron, France; baraa.alsahmarany96@gmail.com (B.A.S.); malo.leguern@builders-ingenieurs.fr (M.L.G.)
3 Institut de Recherche en Constructibilité IRC, Ecole Spéciale des Travaux Publics, 28 Avenue du Président Wilson, 94234 Cachan, France; yelmendili@estp-paris.eu
4 Parc Naturel Régional des Marais du Cotentin et du Bessin, 50500 Carentan les Marais, France; fstreiff@parc-cotentin-bessin.fr
5 School of Art, Design and Architecture, University of Plymouth, Plymouth PL4 8AA, UK; s.goodhew@plymouth.ac.uk
* Correspondence: karim.touati@epf.fr; Tel.: +33-4-99-65-99-55

Abstract: Mastering construction times is of paramount importance in making vernacular earth construction techniques attractive to modern clients. The work presented here is a contribution towards the optimization of the construction time of cob buildings. Therefore, this paper follows the evolution of a cob's mechanical properties during its drying process in the case of a double-walling CobBauge system. Laboratory tests and in situ measurements were performed, and further results were described. Volumetric water content sensors were immersed in the walls of a CobBauge prototype building during its construction. The evolution of the cob layer's compressive strength and Clegg Impact Value (CIV) as a function of its water content has been experimentally studied and discussed. These studies showed that compressive strength and CIV are correlated with water content, and both properties decrease exponentially with time. In this study, a new tool to evaluate cob's mechanical performances in situ has been proposed, Clegg Impact Soil Tester. This was linked to compressive strength, and a linear relationship between these two properties was found. Finally, appropriate values of compressive strength and CIV to satisfy before formwork stripping and re-lifting were proposed. For this study's conditions, these values are reached after approximately 27 days.

Keywords: implementation time; cob; water content; compressive strength; Clegg Impact Value

Citation: Touati, K.; Al Sahmarany, B.; Le Guern, M.; El Mendili, Y.; Streiff, F.; Goodhew, S. Insight into the Optimization of Implementation Time in Cob Construction: Field Test and Compressive Strength Versus Drying Kinetics. *Eng* **2023**, *4*, 2075–2089. https://doi.org/10.3390/eng4030117

Academic Editor: Alessio Cascardi

Received: 24 June 2023
Revised: 19 July 2023
Accepted: 19 July 2023
Published: 25 July 2023

1. Introduction

Climate change has been observed worldwide over recent decades. This is in major part due to greenhouse gas emissions (GHG). In France, the sector of the manufacturing and construction industry, in particular, related to the use and construction of residential/tertiary buildings contributed to 152.7 $MtCO_2eq$ in 2021, representing 36.5% of total GHG emissions [1]. Thus, construction is one of the major target sectors that should be focused on to reduce carbon footprints. This can be achieved by specifying less processed materials, using locally sourced natural materials, and low environmental impact processes. In the context of global climate change, the development of earthen construction is a real alternative to reduce the CO_2 emissions from the construction sector. However, this age-old material must be able to demonstrate good mechanical resistance and adequate water resistance whilst complying with conditions imposed by the current building regulations.

Earthen construction is gaining popularity as a potential means of establishing local value chains with minimal environmental impact. However, the growth of this historically existing building technology, concentrated in a number of locations of developed countries is still limited due to the high cost, labour intensity, and construction periods because of the material drying time. One of the most popular earthen construction techniques in the Northwest of Europe is cob. Natural fibers, water, and silty-clayey soils are typically used to make cob. Water is added to the mixture to cause it to transform into a plastic state, which enables the efficient production of reasonably thick load-bearing building walls. Cob has gained less interest since the beginning of the 20th century in favor of industrial materials thought to be more effective and contemporary with a high degree of standardization. When compared to modern construction methods, cob has actually numerous perceived drawbacks, including low insulating capabilities, long construction durations, high labor demands, etc. [2]. Actually, buildings using cob (in its original conception) do not adhere to global thermal construction rules. To overcome this issue and allow building designers to use cob as a walling material, some aspects of the properties of the finished walling need to be improved in order to enhance cob housings' thermal performances. Alongside this need, the implementation of new methods is being examined to reduce construction times and associated costs.

Cob- and earthen-based materials, in general, have seen a rise in attention since the 1980s due to climatic concerns, especially in France and the United Kingdom [3–11]. This is partly because of its many environmental benefits, namely its durability, minimal environmental effect, and occupational thermal comfort [12,13]. Understanding and improving cob's geotechnical, thermal, and mechanical properties has been the subject of several studies [7,12,14,15].

Accordingly, the CobBauge EU Interreg project is concentrated on creating, implementing, and testing a novel low-carbon technology employing regional soils and plant fibers. This earth fiber-based technique attempts to create a hybrid walling system by combining a load-bearing (cob) and an insulating (light earth) layer, as shown in Figure 1. In order to create 100% earth-based walling that ensures structural resistance and thermal insulation, cob and light earth are naturally blended. In the first phase of the project, many formulations have been examined and evaluated to identify the best earth/fiber mixtures that would enable the construction of the building with a double walling system that complies with French and United Kingdom standards. In this first phase, studies focused on the geotechnical properties of soils, followed by the mechanical and thermal properties of soil–fiber composites. Two prototype buildings have been built in France and the United Kingdom as part of the project's second phase in order to evaluate how the CobBauge hybrid walling affects the behavior of those buildings on the site.

When compared to other modern construction techniques, a building with a cob represents an issue possibly limiting its large-scale rollout. This concerns construction times. With cob, construction times are usually long and sometimes weather dependent. These long construction times are mainly due to the leakage of knowledge regarding the drying kinetics of the mixtures. Thus, the first aim of this present work is to understand the cob's drying process in situ and propose a methodology to optimize the walling's construction time. This objective can be achieved by studying and understanding the impact of water content on the mechanical performances of cob. The second aim of this work is to propose a field test allowing the knowledge of in situ performances (drying kinetics and compressive strength). The results will make it possible to evaluate cob drying and compressive strength progress in situ. Since a double walling system is implemented within formwork, this contribution can be useful in the optimization of form-stripping times, particularly, influencing cob construction times, in general.

Figure 1. In situ implementation of a CobBauge double walling system.

2. Materials and Methods

2.1. Soils Particle Size and Geotechnical Characterization

Three soils were used in the construction of a prototype building in France, designed to allow the research team to assess the construction process performances and energy efficiency of the CobBauge technique. The cob layer (subject of this present paper) is constituted of two different soils. The soils used in this study were collected in Lieusaint quarry (Société des Sablières du Cotentin, SABCO Normandy). These soils are locally sourced from a geographical area that is associated with existing cob buildings. Their geotechnical characterizations were performed. The clay activity was evaluated using the methylene blue value test according to standard NF P94-068 [16] and Atterberg's limit according to standard NF EN ISO 17892-12 [17]. Retrieved soil properties constituting cob mixture and classification can be found in Table 1.

Table 1. Atterberg limits, methylene blue values, and soil classification.

Soil	Liquid Limit [%]	Plasticity Index [%]	Methylene Blue Value [g/100 g]	USCS Classification
Soil 1	22.8	2.3	1.35	Low plasticity silt (ML)
Soil 2	28.5	4.2	2.31	Silty sand with gravel (SM)

The mineralogical composition of soil 1 reveals the presence of the following major phases: quartz (54.8%), muscovite (26.2%), montmorillonite (6.9%), and albite (4.2%), with small occurrences of illite, kaolinite, goethite, rutile, and huntite (see Table 2) [18]. Subsequently, soil 1 is typical of silty soil. It is composed of quartz grains and silicates (feldspars, micas, serpentines, and smectites). Silt particles are intermediate sand and clay in size and have similar properties.

Table 2. Mineralogical composition of soil 1 with refined values of unit cell volume and average diameter. One standard deviation is indicated in parenthesis on the last digit.

Phases	V (%)	⟨D⟩ (nm)
Quartz SiO_2	54.8 (5)	492 (10)
Muscovite $KAl_2(AlSi_3O_{10})(F,OH)_2$	26.2 (5)	35 (5)
Montmorillonite $(Na,Ca)_{0\cdot3}$ $(Al,Mg)_2Si_4O_{10}(OH)_2$	6.9 (2)	111 (6)
Albite $NaAlSiO_3$	4.2 (2)	43 (5)
Kaolinite $Al_2Si_2O_5(OH)_4$	2.1 (3)	78 (5)
Goethite α-FeO(OH)	2.0 (3)	21 (1)
Rutile TiO_2	1.6 (3)	92 (5)
Illite $(K,H_3O)(Al,Mg,Fe)_2(Si,Al)_4O_{10}[(OH)_2,(H_2O)]$	1.1 (2)	100 (5)
Huntite $Mg_3Ca(CO_3)_4$	1.1 (2)	123 (5)

Soil 2 is made of natural quartz (99% silica) [18].

2.2. Flax Straw

The selection of flax straw was based on the results of an earlier study [14], which established that flax straw offered earth–fiber mixes optimum compressive strength when dried while retaining good workability during mixing. Flax straw incorporated into the soil represents theoretically a proportion of 2.5% of the mixed dry mass. In laboratory, the flax straw is cut to a length of 7 $\pm$ 1 cm. On site, flax straw was introduced into the mix in its raw state. This latter presents an absolute density equal to 1266 $\pm$ 55 $kg{\cdot}m^{-3}$ and an absorption coefficient at 24 h equal to 350 $\pm$ 11%.

2.3. Sample Preparation

To study cob's compressive strength as a function of water content, 24 specimens were prepared at the same time with same proportion of soil, straw, and water. To be representative of what can be encountered in situ after cob implementation, 8 different water contents are considered: 19%, 17%, 15%, 13%, 11%, 9%, 5%, and 0%. Three samples are made for each water content. Cylindrical molds with the following dimensions were used: Ø110 mm $\times$ H220 mm. Before starting the production of the specimens, the mold's inner face was oiled. Afterwards, the mixture was compacted in several layers (a new layer is added when the previous one is considered completely compact) with a wooden tamper with dimensions equal to 30 mm $\times$ 30 mm $\times$ 410 mm (see Figure 2). Then, 24 h after filling the molds, the samples were turned upside down in order to ensure a good distribution of the water in them. In addition, the molds are slightly opened to accelerate the drying of the mixtures. Then, these samples are demolded after two or three days and left to dry until reaching their required water content.

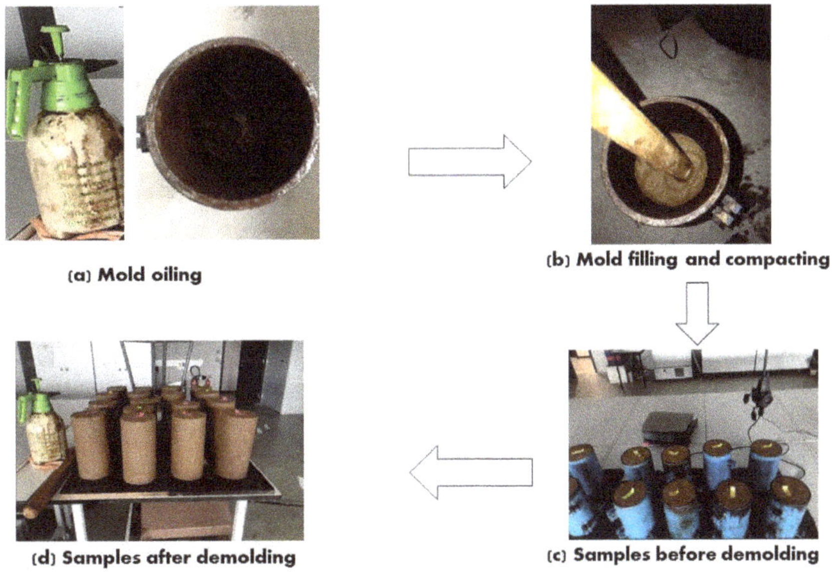

Figure 2. Preparation of cob samples at different water contents for the compressive strength study.

2.4. *Water Content Control and Bulk Density*

To better control the drying process, the samples were placed in the laboratory at a temperature of 21 ± 2 °C and a relative humidity of 50 ± 2%. When the required moisture content was reached, the samples were covered for 48 h to homogenize the water content within them, and their bulk density was measured before performing the compressive strength test (see Figure 3). The bulk density was determined by following the standard NF X31-501 [19]. Successive measurements of the sample's weight are made until reaching and stabilizing at the corresponding required water content. Samples with following water content have been prepared: 19, 17, 15, 13, 11, 9, 5, and 0%. These percentages have been chosen to approach values encountered on site during the cob's drying course.

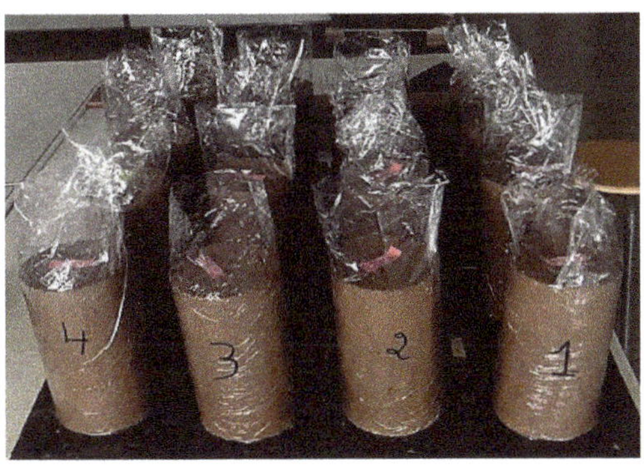

Figure 3. Samples coverage to homogenize water content (the samples are numbered from 1 to 24).

2.5. *Compressive Strength*

To maintain a measure of consistency between the samples, enabling the research team to relate the structural performance of the samples to the needs of the walling system, compressive strength measurements were undertaken on samples with different water contents. Compressive strength tests are performed on the prepared cylindrical samples

(∅110 mm × H220 mm) in accordance with the NF EN 13286-41 standard [20]. An IGM press with a load capacity of 250 kN was used. The tests are performed with an imposed loading rate of 0.40 kN/s. The deformation of the specimens is measured with a vertical displacement sensor in contact with the lower plate of the press (see Figure 4).

Figure 4. Compressive strength test on cob samples at different water content under an IGM press.

2.6. Clegg Impact Value

The Clegg Impact Soil Tester (CIST) was proposed by Dr. Bade Clegg in the 1970s as an alternative to the CBR test. The output of the CIST device is called the Clegg Impact Value (CIV). In this present study, this value will be linked to the compressive strength in order to propose a site control test. To study Clegg Impact Value, 10 specimens were prepared. To be representative of what can be encountered in situ, 5 different water contents have been considered: 19%, 15%, 11%, 5%, and 0%. For mold availability considerations, two samples are made for each water content. CBR molds with the following dimensions were used: Ø152 mm × H117 mm. Preparation method is quite similar to the one described in the precedent paragraph (see Figure 5).

Figure 5. Preparation of cob samples at different water contents for the Clegg Impact Value measurement.

The mobility of the CIST makes it possible to perform in situ tests relatively quickly. Initially, the CIST device was developed for non-cohesive backfill materials. Clegg impact soil tester is implemented in accordance with ASTM D5874-02. This method using a hammer of 4.5 kg is suitable for soils or aggregates with particle sizes less than 37.5 mm.

The soils used in this present study have particle sizes less than 20 mm (see Figure 6). The impact height is equal to 0.45 m. The hammer diameter is 0.05 m. The CIV value is obtained using the knowledge of gravitational acceleration ($g = 9.81\ m \cdot s^{-2}$) and the deceleration measured during the hammer drop (a in $m \cdot s^{-2}$) is expressed as the following equation [21]:

$$CIV = \frac{a}{10 \cdot g} \quad (1)$$

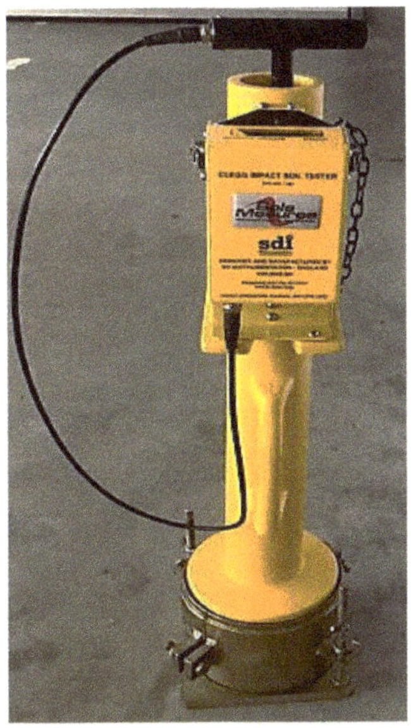

Figure 6. Clegg Impact Soil Tester on cob samples at different water contents.

The Clegg Impact Value can be used to calculate quantities such as the modulus of elasticity or the CBR (California Bearing Ratio) value using correlations. In this study, CIV is linked to cob's compressive strength. This can represent an easy way for craftsmen to control cob's resistance in situ. This can be interesting to know if a cob lift is sufficiently dry and resistant to receive a new lift on its top.

2.7. In Situ Implementation and Monitoring

The main focus of this research is the drying rates of CobBauge dual walls. It is, therefore, important to undertake moisture measurements of representative samples. In this present work, Campbell scientific CS655 sensors, based on reflectometry principle, have been used to locally measure Volumetric Water Content (VWC). This type of sensor (with an accuracy of ±3%) has shown its efficiency when measuring VWC in soil materials [22–24].

The theory behind the CS655 water content reflectometer is based on the speed at which an electromagnetic wave travels through the sensor's two rods. The latter depends on the material's dielectric permittivity surrounding the two rods. Dielectric permittivity is then converted to volumetric water content using the Topp equation [25]. For the sake of greater contact between mixes and rods, sensors were positioned horizontally in cob at the same heights and depths. Probes were positioned parallel to the west wall surfaces at two different heights, 25 and 50 cm, from the lift basis. Volumetric water content considered in this present work is an average of these two considered heights. The thickness of cob in this west wall is equal to 25 cm. Additionally, information gathered via CS 655 probes was logged in the CR1000X data logger every 15 min during the course of more than a year (from May 2020 to November 2021). The following weather conditions are presented in

Figure 7: temperature, relative humidity, wind speed and direction, radiation, and rainfall recovered every 15 min. Instrumentation implemented on site is shown in Figure 8.

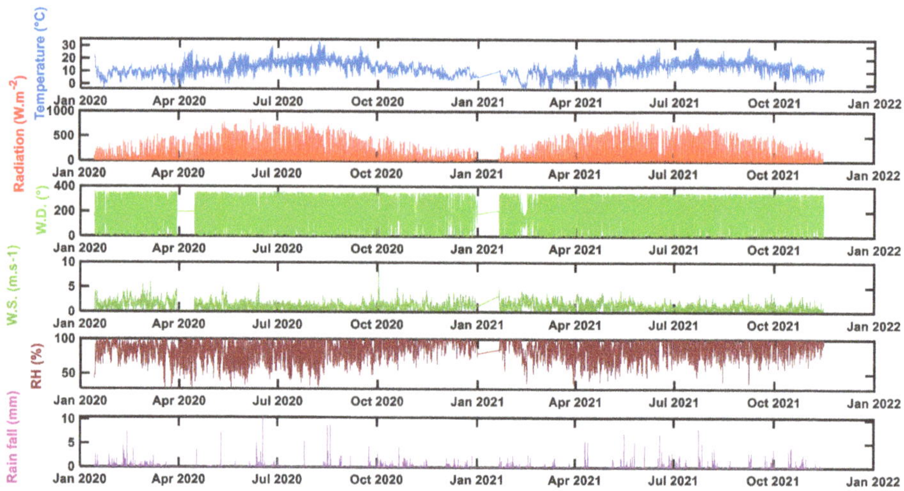

Figure 7. Weather conditions (temperature, RH, W.D., W.S., radiation, rainfall) recovered via a WS-GP1 weather station near the prototype building.

Figure 8. Instrumentation (weather station and volumetric water content sensors) implemented in situ.

3. Results and Discussion

3.1. Cob's Compressive Strength and Density as Function of Water Content

Figure 9 shows the different types of failure (deformation) that occurred in the samples after the compressive loading. When the water content is above 11%, there is crushing of the sample, and plastic behavior is predominant. However, when the water content is equal to or less than 11%, the sample breaks, and cracks are present throughout the sample.

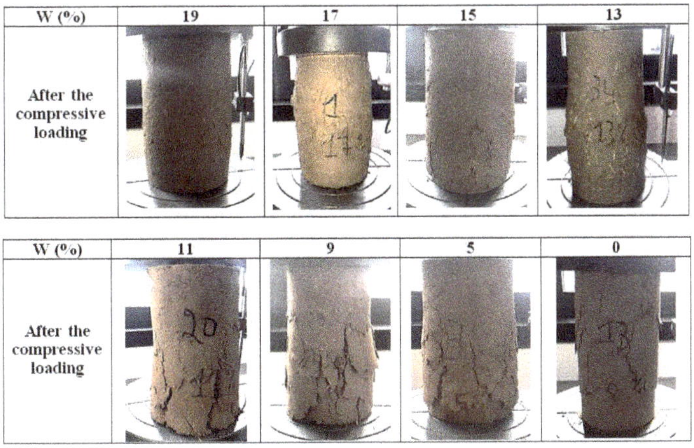

Figure 9. Cob's samples facies after compressive loading at different water content.

The evolution of the cob's compressive strength and density as a function of water content is reported in Table 3 and Figure 10. On this latter, maximum compressive strength and one at 1% deformation are presented. A deformation of 1% is considered in this present work for considerations regarding the ease of straightening wall surfaces before the application of plasters or renders.

Table 3. Evolution of compressive strength and Clegg Impact Value as function of water contents usually encountered in situ.

WC [wt.%]	VWC [$m^3 \cdot m^{-3}$]	Density [$kg \cdot m^{-3}$]	$\sigma_{\varepsilon = 1\%}$ [MPa]	σ_{max} [MPa]	CIV [-]
19	0.3672738	1933.02 ± 15.65	0.04 ± 0.03	0.08 ± 0.01	04.0 ± 0.0
17	0.3315272	1950.16 ± 09.67	0.06 ± 0.02	0.15 ± 0.01	-
15	0.2822145	1881.43 ± 35.80	0.12 ± 0.02	0.20 ± 0.00	08.0 ± 1.4
13	0.2425423	1865.71 ± 05.37	0.20 ± 0.02	0.29 ± 0.03	-
11	0.2002693	1820.63 ± 05.64	0.37 ± 0.03	0.42 ± 0.01	12.5 ± 0.7
9	0.1625859	1806.51 ± 02.25	0.58 ± 0.02	0.65 ± 0.02	-
5	0.0868015	1736.03 ± 10.24	1.19 ± 0.03	1.35 ± 0.03	20.0 ± 1.4
0	0	1656.98 ± 08.11	2.27 ± 0.30	2.60 ± 0.03	34.0 ± 1.4

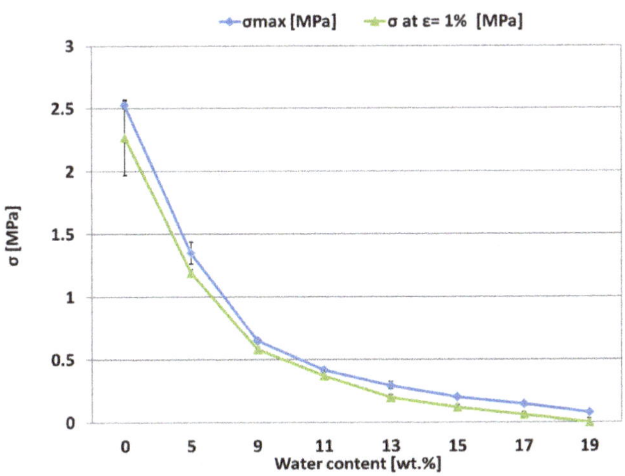

Figure 10. Evolution of cob's compressive strength as function of water content.

The clay content is critical in earth building because it maintains the larger particles connected. However, soils containing more than 30% clay have very high shrinkage/swelling ratios, which, along with their proclivity to absorb moisture, can result in large fissures in the finished cob product and thus impact its mechanical performance. The soil 1 is composed of quartz (54.8%), muscovite (26.2%), montmorillonite (6.9%), and albite (4.2%), with minor traces of kaolinite, goethite, rutile, illite, and huntite. Some clays have the ability to widen the interfoliar spaces between their leaves. The incorporation of hydrated cations (Na, Ca, etc.) provides it with this property, allowing it to compensate for chronic charge shortages (Andrade et al., 2011). The phenomenon disappears if the clay charge is too high (e.g., micas or muskovite in our sample: total clay charge of −1 entirely counterbalanced by the dehydrated cations (e.g., pyrophyllite, talc: total clay charge of 0, no interfoliar cation). With a charge ranging from 0.3 to 0.8, the smectites subclass is among the expandable species. The crystalline structure might expand due to the water injected via the hydrated cations [26]. The swelling increased as a result of the high humidity. Montmorillonite is the only expandable species identified in our soil 1, at a rate of 6.9%. The amount of muscovite, albite, kaolinite, and illite in our soil will impact its shrinkage characteristics. These crystals have few water molecules between their layers due to their tiny interfoliar space [26]. As a result, they have negligible intercrystalline swelling when immersed in water [27]. As a result, when dried, these four species shrink significantly less than smectite clays like montmorillonite [28]. Considering the number of smectites, the cob layer shrinkage properties are mainly affected by the quantity of muscovite, albite, kaolinite, and illite. These structures have few water molecules between their layers, and they exhibit negligible intercrystalline swelling and shrink substantially less than smectite clays like montmorillonite.

To explain the influence of the drying processes, the phase composition of soil 1 after drying was determined via XRD (Table 4). The corresponding XRD pattern is shown in Figure 11. The XRD analysis shows the disappearance of huntite combined with the decrease in the montmorillonite content in favor of the formation of carbonated calcium hemicarboaluminate (see Tables 2 and 3). Montmorillonite is the only expandable species found in the structural cob. The presence of kaolinite, muscovite, and illite, as well as the reduction in montmorillonite, will decrease the shrinking behavior. Indeed, these crystals have a weak intercrystalline swelling behavior and contain a minor amount of water [26]. The formation of carbonated calcium hemicarboaluminate leads to the enhancement of compressive strength (see Figure 10).

Table 4. Mineralogical composition of soil 1 after the drying process.

Phases	V (%)	⟨D⟩ (nm)
Quartz SiO_2	61.2 (5)	492 (10)
Muscovite $KAl_2(AlSi_3O_{10})(F,OH)_2$	26.0 (5)	34 (5)
Montmorillonite $(Na,Ca)_{0\cdot3}$ $(Al,Mg)_2Si_4O_{10}(OH)_2$	0.9 (2)	67 (5)
Albite $NaAlSiO_3$	1.7 (2)	46 (5)
Kaolinite $Al_2Si_2O_5(OH)_4$	1.5 (3)	78 (5)
Illite $(K,H_3O)(Al,Mg,Fe)_2(Si,Al)_4O_{10}[(OH)_2,(H_2O)]$	2.1 (2)	100 (5)
Rutile TiO_2	3.4 (3)	92 (5)
Carbonated calcium hemicarboaluminate Al $Ca_2C_{0\cdot4}O_{9\cdot2}$	3.2 (1)	61 (5)

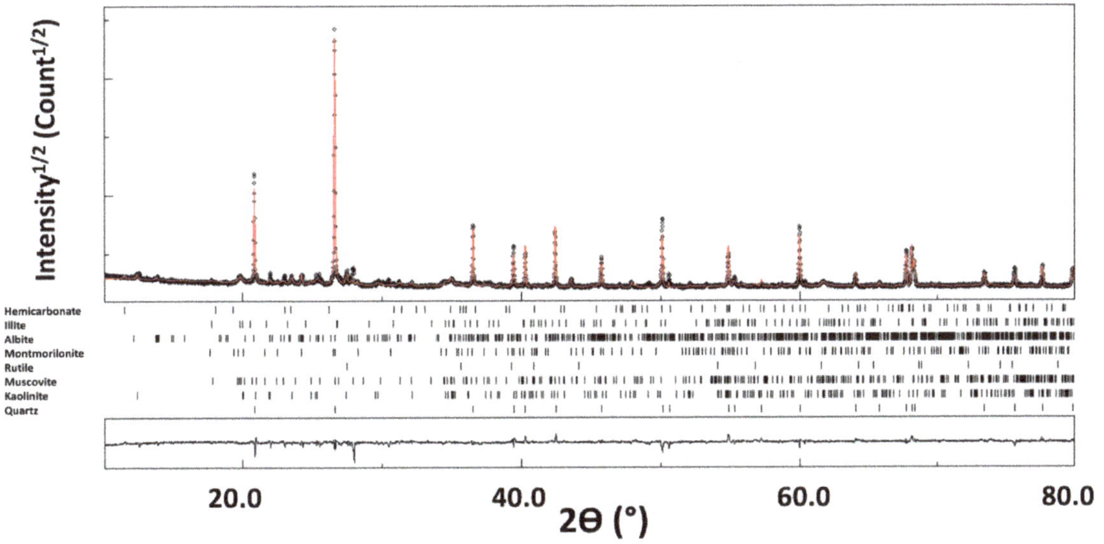

Figure 11. XRD pattern of soil 1. The computed pattern (red line) is superimposed on the experimental pattern (black dots). In the center, the fitted phases, and at the bottom, the difference curve (Iobs−Icalc).

3.2. Clegg Impact Value as Function of Water Content

Results from the Clegg Impact Soil Tester are reported in Table 3. Thus, the evolution of the Clegg Impact Value as a function of water content is represented in Figure 12. On this latter, it can be seen that CIV increases with the decrease in cob's water content. For water content ranging between 19 wt.% and 0 wt.%, Clegg Impact Value varies from 4 to 35 following a quasi-exponential law. As announced in the introduction, CIV measurements will allow us to propose a field test, allowing us to deduce the in situ cob's compressive strength and water content and in-fine to know if a lift is sufficiently dry and resistant to receive a new lift on its upper side.

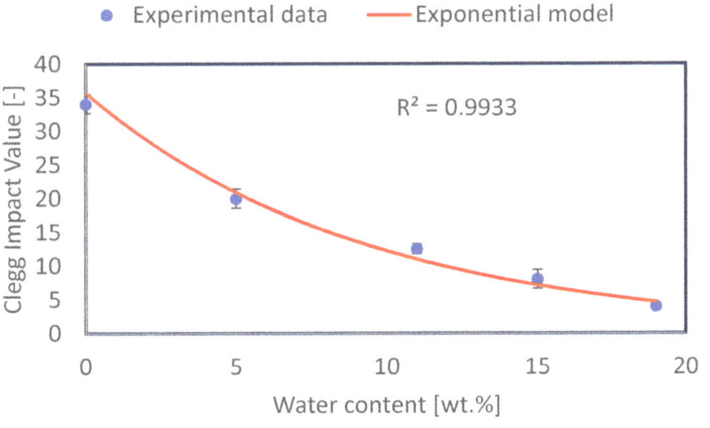

Figure 12. Evolution of Clegg Impact Value as function of cob's water content.

3.3. In Situ Water Content and Compressive Strength Evolution

Figure 13 shows the course of the volumetric water content and associated compressive strength (at 1% deformation). As observed in the figure, cob's volumetric water content has decreased gradually with time and tends to a practical value after several months. Moreover, it can be shown that the drying speed becomes slower with the decrease in the water content in the wall. The drying presents two phases: a first one that is faster and a second one slower. With water content getting lower, the drying process gets slower. This drying is consistent with the typical behavior of construction materials.

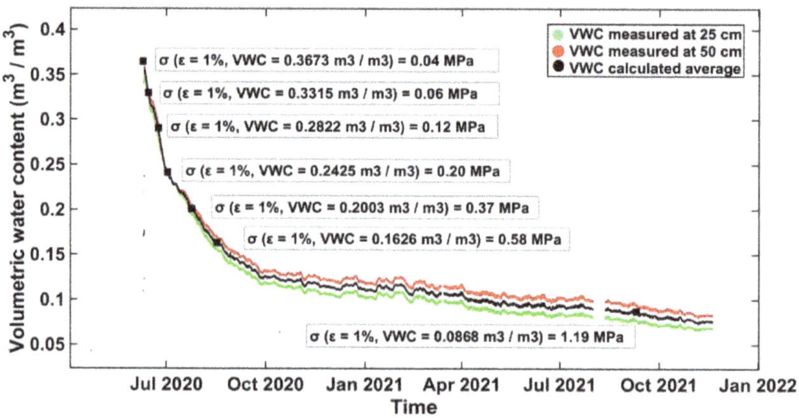

Figure 13. Evolution of cob's compressive strength with the water content course.

With the decrease in water content, the cob's compressive strength increases. In this figure, it can be observed that compressive strength passes from 0.04 to 0.06, 0.12, 0.20, 0.37, and 0.58 MPa after 04, 15, 22, 44, and 68 days, respectively. After approximately one year and two months, compressive strength is approaching 1.2 MPa.

In Figure 13, the reported water content is recovered in situ, but mechanical performances are obtained on samples produced in the laboratory and tested at determined water contents considering those encountered in situ.

In the studied prototype, formworks are released after 27 days, and a new lift was raised directly. The choice of this delay is based on Norman craftsmen practice. At that time, the VWC of the existing lift was approx. 0.2278 m^3/m^3 corresponding to a compressive strength of approx. 0.28 MPa. From this experience, it can be affirmed that 0.28 MPa (reached after approx. 27 days in this typical case and weather conditions) is sufficient to raise up a new lift. However, this value can be optimized by performing judicious calculations.

In this sense, calculations were undertaken by considering the stress imposed by the new wet lift (layer of wet cob) on the existing lift. These calculations are based on the knowledge of the wet cob density, its volume, gravitational acceleration, and existing lift horizontal surface (see Equation (2)).

$$\sigma = \frac{N}{S} = \frac{m \cdot g}{S} = \frac{\rho \cdot V \cdot g}{S} \tag{2}$$

where N is the normal stress, m is the wet cob mass, ρ is the wet cob density, V is the wet cob volume, g is the gravitational acceleration, and S is the surface of the existing lift.

Considering the measured density of the wet cob, it was found that a new wet cob lift (height = 70 cm) exerts a stress of 0.013 MPa directly after its implementation. This value seems to be too low compared to that of compressive strength measured at 0.3673 m^3/m^3 (water content of the newly implemented wet cob), which is equal to 0.04 MPa (for a deformation of 1%). This would imply that a cob lift could support a new wet lift from the first day after its implementation. But, this seems to be too optimistic. Consequently, when trying to calculate the minimum mechanical resistance that a cob lift should reach before raising up a new wet lift, other methods than the one presented here should be identified and considered.

3.4. Clegg Impact Value and Compressive Strength

In the previous subsections, it was seen that both cob's compressive strength and CIV increase exponentially with the decrease in water content. Thus, the evolution of Clegg Impact Value is plotted as a function of compressive strength. Figure 14 shows that the evolution of CIV as a function of σ is best fitted via a linear relationship. On this latter, it can be seen that the value of compressive strength obtained after 27 days of drying (approx. 0.28 MPa) correspond to a CIV value of approx. 9. As it was already reported that this compressive strength is sufficient to raise up a new lift, it can be stated here that in situ measurement of a CIV value equal to or larger than 9 is sufficient for raising up a new lift on the existing one. More on-site investigations could help to optimize this CIV value. However, when dealing with soils different from those reported in this present study, the absolute values of compressive strength and Clegg Impact Value should be considered carefully.

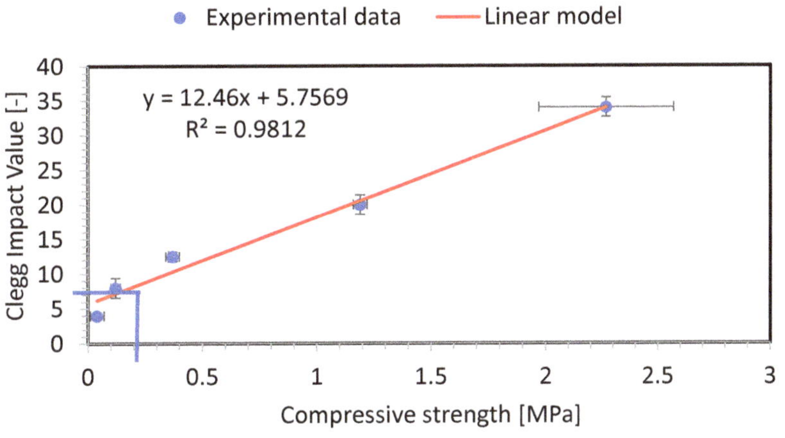

Figure 14. Relationship between cob's compressive strength (at 1% deformation) and Clegg Impact Value.

4. Conclusions

The first aim of this present work was to understand the link between cob's drying process and its mechanical performances in a CobBauge double-walling system. This is consistent with the objective of better controlling construction times. In this regard, laboratory and in situ studies were performed. First, the water content was recovered during the construction of a cob wall (under real conditions). These measurements showed that water content decreases exponentially with time. Then, samples were produced in the laboratory at the water contents encountered in situ. Afterward, these samples were subjected to compressive strength tests at different water contents. When considering the decrease in cob's water content, as shown in the in situ measurement, compressive strength was found to follow an exponential increase.

The second aim of this present work was to propose a simple tool allowing evaluation of cob's hydromechanical performances in situ. Thus, the Clegg Impact Soil Tester (CIST) was proposed. Indeed, this can represent an easy way for craftsmen to control in situ cob drying and readiness to receive a new lift. In this sense, samples were also subjected to the CIST in order to obtain the Clegg Impact Value (CIV) at different water contents. The obtained results showed that the evolution of CIV as a function of water content can be best fitted via an exponential model.

From there, the evolution of Clegg Impact Value is plotted as a function of compressive strength (σ). It was found that the evolution of CIV as a function of σ is best fitted via a linear relationship.

In the studied wall, a new lift was raised up on the monitored one after 27 days. In this present study, it was found that after this duration, volumetric water content (VWC) in the wall was approximately equal to 0.2278 m^3/m^3. At this VWC, the cob presents a compressive strength of approx. 0.28 MPa. When CIV was plotted as a function of compressive strength, it was found that 0.28 MPa corresponds to a CIV value of approximately 9. As the form striping and the raise up of a new lift at this compressive strength went well, it can be stated that measuring a CIV value of at least 9 can allow the rise of a new CobBauge lift safely.

Author Contributions: Conceptualization, K.T., M.L.G., F.S. and S.G.; methodology, K.T., B.A.S. and M.L.G.; investigation, data curation, and formal analysis, K.T., B.A.S., M.L.G. and Y.E.M.; writing—original draft preparation, K.T., B.A.S., M.L.G. and Y.E.M.; writing—review and editing, K.T., Y.E.M., F.S. and S.G.; funding acquisition, S.G. and F.S. All authors have read and agreed to the published version of the manuscript.

Funding: The results presented in this article were obtained in the framework of the collaborative project CobBauge, funded by the European cross-border cooperation program INTERREG V France (Manche/Channel) England.

Informed Consent Statement: Not applicable.

Data Availability Statement: The experimental and computational data presented in this present paper are available from the corresponding author upon request.

Conflicts of Interest: The authors declare that they have no known competing financial interest or personal relationship that could have appeared to influence the work reported in this paper.

References

1. INSEE, Émissions de Gaz à Effet de Serre par Activité. 2022. Available online: https://www.insee.fr/fr/statistiques/2015759#tableau-figure1 (accessed on 30 January 2023).
2. Watson, L.; McCabe, K. La técnica constructiva del cob. Pasado, presente y futuro. *Inf. Constr.* **2011**, *63*, 59–70. [CrossRef]
3. Laborel-Préneron, A.; Aubert, J.; Magniont, C.; Tribout, C.; Bertron, A. Plant aggregates and fibers in earth construction materials: A review. *Constr. Build. Mater.* **2016**, *111*, 719–734. [CrossRef]
4. Soudani, L.; Woloszyn, M.; Fabbri, A.; Morel, J.-C.; Grillet, A.-C. Energy evaluation of rammed earth walls using long term in-situ measurements. *Sol. Energy* **2017**, *141*, 70–80. [CrossRef]
5. Hamard, E.; Lemercier, B.; Cazacliu, B.; Razakamanantsoa, A.; Morel, J.-C. A new methodology to identify and quantify material resource at a large scale for earth construction—Application to cob in Brittany. *Constr. Build. Mater.* **2018**, *170*, 485–497. [CrossRef]
6. Colinart, T.; Vinceslas, T.; Lenormand, H.; De Menibus, A.H.; Hamard, E.; Lecompte, T. Hygrothermal properties of light-earth building materials. *J. Build. Eng.* **2020**, *29*, 101134. [CrossRef]
7. Goodhew, S.; Boutouil, M.; Streiff, F.; Le Guern, M.; Carfrae, J.; Fox, M. Improving the thermal performance of earthen walls to satisfy current building regulations. *Energy Build.* **2021**, *240*, 110873. [CrossRef]
8. Alassaad, F.; Touati, K.; Levacher, D.; Sebaibi, N. Impact of phase change materials on lightened earth hygroscopic, thermal and mechanical properties. *J. Build. Eng.* **2021**, *41*, 102417. [CrossRef]
9. Alassaad, F.; Touati, K.; Levacher, D.; Sebaibi, N. Effect of latent heat storage on thermal comfort and energy consumption in lightweight earth-based housings. *Build. Environ.* **2023**, *229*, 109915. [CrossRef]
10. Alassaad, F.; Touati, K.; Levacher, D.; El Mendili, Y.; Sebaibi, N. Improvement of cob thermal inertia by latent heat storage and its implication on energy consumption. *Constr. Build. Mater.* **2022**, *329*, 127163. [CrossRef]
11. Giuffrida, G.; Caponetto, R.; Nocera, F.; Cuomo, M. Prototyping of a Novel Rammed Earth Technology. *Sustainability* **2021**, *13*, 11948. [CrossRef]
12. Hamard, E.; Cazacliu, B.; Razakamanantsoa, A.; Morel, J.-C. Cob, a vernacular earth construction process in the context of modern sustainable building. *Build. Environ.* **2016**, *106*, 103–119. [CrossRef]
13. Hall, M.R.; Lindsay, R.; Krayenhoff, M. *Modern Earth Buildings: Materials, Engineering, Construction and Applications*; Woodhead Publishing: Sawston, UK, 2012.
14. CobBauge Project, Soil and Fibre Characterisation, Mixes Choice and Mixes Characteristics, n.d. Available online: http://www.cobbauge.eu/wp-content/uploads/2019/05/Technical-report_ESITC_feb2019.pdf (accessed on 30 January 2023).

15. Azil, A.; Le Guern, M.; Touati, K.; Sebaibi, N.; Boutouil, M.; Streiff, F.; Goodhew, S.; Gomina, M. Earth construction: Field variabilities and laboratory reproducibility. *Constr. Build. Mater.* **2022**, *314*, 125591. [CrossRef]
16. *NF P94-068*; Soils: Investigation and Testing. Measuring of the Methylene Blue Adsorption Capacity of à Rocky Soil. Deter-Mination of the Methylene Blue of à Soil by Means of the Stain Test. GlobalSpec: Albany, NY, USA, 1998. Available online: https://sagaweb.afnor.org/fr-FR/sw/Consultation/Notice/1261353?directFromSearch=true (accessed on 11 April 2022).
17. *NF EN ISO 17892-12*; Geotechnical Investigation and Testing—Laboratory Testing of Soil—Part 12: Determination of Liquid and Plastic Limits. ISO: Geneva, Switzerland, 2018. Available online: https://sagaweb.afnor.org/fr-FR/sw/Consultation/Notice/1446382?directFromSearch=true (accessed on 11 April 2022).
18. Touati, K.; Le Guern, M.; El Mendili, Y.; Azil, A.; Streiff, F.; Carfrae, J.; Fox, M.; Goodhew, S.; Boutouil, M. Earthen-based building: In-situ drying kinetics and shrinkage. *Constr. Build. Mater.* **2023**, *369*, 130544. [CrossRef]
19. *NF X31-501*; Qualité des Sols-Méthodes Physiques-Mesure de la Masse Volumique Apparente d'un échantillon de sol non Remanié-Méthode du Cylindre. Afnor EDITIONS: Saint-Denis, France, 1992. Available online: https://www.boutique.afnor.org/fr-fr/norme/nf-x31501/qualite-des-sols-methodes-physiques-mesure-de-la-masse-volumique-apparente-/fa024711/355 (accessed on 11 April 2022).
20. *NF EN 13286-41*; Mélanges Traités et Mélanges non Traités aux Liants Hydrauliques-Partie 41: Méthode d'essai pour la Détermination de la Résistance à la Compression des Mélanges Traités aux Liants Hydrauliques. Afnor EDITIONS: Saint-Denis, France, 2003. Available online: https://www.boutique.afnor.org/fr-fr/norme/nf-en-1328641/melanges-traites-et-melanges-non-traites-aux-liants-hydrauliques-partie-41-/fa101317/21720 (accessed on 11 April 2022).
21. Vlček, J.; Valašková, V. Analysis of Applicability of Clegg Impact Soil Tester for Clayey Soils. *MATEC Web Conf.* **2018**, *196*, 02031. [CrossRef]
22. Chabriac, P.-A.; Fabbri, A.; Morel, J.-C.; Laurent, J.-P.; Blanc-Gonnet, J. A Procedure to Measure the in-Situ Hygrothermal Behavior of Earth Walls. *Materials* **2014**, *7*, 3002–3020. [CrossRef] [PubMed]
23. Caldwell, T.G.; Bongiovanni, T.; Cosh, M.H.; Halley, C.; Young, M.H. Field and Laboratory Evaluation of the CS655 Soil Water Content Sensor. *Vadose Zone J.* **2018**, *17*, 1–16. [CrossRef]
24. INSTRUCTION MANUAL, CS650 and CS655 Water Content Reectometers, Revision: 07/2021, (n.d.). Available online: https://s.campbellsci.com/documents/es/manuals/cs650.pdf (accessed on 11 April 2022).
25. Topp, G.C.; Davis, J.L.; Annan, A.P. Electromagnetic determination of soil water content: Measurements in coaxial transmission lines. *Water Resour. Res.* **1980**, *16*, 574–582. [CrossRef]
26. Andrade, F.; Al-Qureshi, H.; Hotza, D. Measuring the plasticity of clays: A review. *Appl. Clay Sci.* **2011**, *51*, 1–7. [CrossRef]
27. Zhang, X.; Chen, B.; Ahmad, M.R. Characterization of a novel bio-insulation material for multilayer wall and research on hysteresis effect. *Constr. Build. Mater.* **2021**, *290*, 123162. [CrossRef]
28. Meinhold, G. Rutile and its applications in earth sciences. *Earth-Science Rev.* **2010**, *102*, 1–28. [CrossRef]

eng

MDPI

Article

Seismic Resilience and Design Factors of Inline Seismic Friction Dampers (ISFDs)

Ali Naghshineh [1,*], Ashutosh Bagchi [1] and Fariborz M. Tehrani [2]

[1] Department of Building, Civil and Environmental Engineering Concordia University, Montreal, QC H3G 1M8, Canada
[2] Department of Civil & Geomatics Engineering, California State University, Fresno, CA 93740-8030, USA
* Correspondence: ali_nag@encs.concordia.ca

Abstract: While damping devices can provide supplemental damping to mitigate building vibration due to wind or earthquake effects, integrating them into the design is more complex. For example, the Canadian code does not provide building designs with inline friction dampers. The objective of this present article was to study the overstrength, ductility, and response modification factors of concrete frame buildings with inline friction dampers in the Canadian context. For that purpose, a set of four-, eight-, and fourteen-story ductile concrete frames with inline seismic friction dampers, designed based on the 2015 National Building Code of Canada (NBCC), was considered. The analyses included pushover analysis in determining seismic characteristics and dynamic response history analysis using twenty-five ground motion records to assess the seismic performance of the buildings equipped with inline seismic friction dampers. The methodology considered diagonal braces, including different 6 m and 8 m span lengths. The discussion covers the prescribed design values for overstrength, ductility, and response modification factors, as well as the performance assessment of the buildings. The results revealed that increasing the height of the structure and reducing the span length increases the response modification factors.

Keywords: seismic; overstrength; ductility; resilience; inline seismic friction damper

Citation: Naghshineh, A.; Bagchi, A.; Tehrani, F.M. Seismic Resilience and Design Factors of Inline Seismic Friction Dampers (ISFDs). *Eng* **2023**, *4*, 2015–2033. https://doi.org/10.3390/eng4030114

Academic Editor: Alessio Cascardi

Received: 12 June 2023
Revised: 11 July 2023
Accepted: 14 July 2023
Published: 18 July 2023

1. Introduction

Utilizing the inherent ductility of a building to prevent catastrophic failure has led to the development of seismic design codes and inelastic design response spectra. The relationship between seismic load and damage is defined by equilibrium energy, in which the total seismic inertia equals the sum of the elastic vibration energy, the cumulative inelastic strain energy, and the energy absorbed by damping. Consequently, the energy induced by structural damage equals the total seismic inertia minus the energy absorbed by damping. Simply put, the damping of and damage to a system directly affects one another. In conventional structural systems, seismic energy is dissipated through lateral load-resisting systems. However, supplemental damping is employed in modern structural systems to mitigate and limit structural damage [1,2].

Consequently, modern structures consider the actual dynamic behavior of systems, which can be categorized into isolation systems and energy dissipation devices, including passive, semi-active, active, and hybrid systems [3,4]. Friction dampers (seismic brakes) are examples of passive energy dissipation devices that dissipate energy through the friction caused by sliding two solid bodies against one another. When a structure is subjected to a major earthquake, the friction dampers attached to a typical brace dissipate a significant proportion of the energy and slip instead of yielding before yielding the primary members of a structure. Therefore, the initial cost could be significantly reduced with a high capacity for energy dissipation. There are several approaches to using friction brace frames as an energy dissipation mechanism in steel frames. Friction dampers have been used in combination with a frame to capture the global structural response with the tri-linear lateral elastic-perfect-plastic shape. These methods are based on an equivalent one degree of freedom and a multi-degree of freedom to determine the appropriate design of structural components and slip loads to reduce structural damage and control drifts [5–8]. In another instance, in an experimental setting, a full-scale Moment-Resisting Frame (MRF) system with inline friction dampers exhibited enhanced performance and resilience via decreased acceleration and increased damping. The friction-damped frame reduced acceleration by an average of 25% across different ground and artificial motions. The friction-damped frame enhanced structural damping by 6.74%, preserving strength and stiffness. The calculated story drift demonstrated a resilient system with minimal deflection.

Additionally, the MRF effectively corrected residual elongation in the inline friction dampers [9]. Some methods have been developed to capture the plastic mechanism of friction dampers [10]. Alternatively, an energy-based method has been proposed to determine the slip resistance of friction dampers in reinforced concrete structures [11].

Structural analysis under an earthquake in the elastic region can create reverse strength in the structures that surpasses their structural response. Seismic codes take advantage of the inherent capabilities of structures, including overstrength and ductility, to dissipate significant amounts of earthquake energy. The ductility factor represents the ability of a structure to dissipate energy in the inelastic range. It can be determined by dividing the maximum displacement by the displacement at the yield point, which is influenced by factors such as soil type and the fundamental period of the structure [2,12,13]. The National Building Code of Canada [14] divides the minimum earthquake lateral force by the Seismic Force-Resisting System (SFRS) reduction factor. This factor, known as the response modification factor, can be calculated by multiplying the overstrength factor (R_0) by the ductility-related force modification factor (R_d). However, the 2015 National Building Code of Canada does not explicitly include the overstrength factor (R_0) or the ductility-related force modification factor (R_d). In the 2015 National Building Code of Canada [14], several types of Seismic Force-Resisting Systems (SFR_S) are considered for ductility, with a minimum value of 1.0 for brittle systems such as unreinforced masonry and a maximum value of 5.0 for ductile moment-resisting frames. The overstrength factor ranges from 1.0 to 1.7, signifying the level of overstrength in the SFRS [14]. For friction brace frames, limited to 10 stories in Canada, proposed values of 5.0 and 1.1 have been suggested for the ductility and overstrength factors, respectively [15]. The seismic performance of fourteen-story concrete moment-resisting frames (CMRFs) with and without friction dampers was studied. The models with integrated friction dampers showed improved performance compared to similar models without dampers. When friction dampers were integrated, the moment and shear were reduced by approximately 75%, 69%, and 56% for ductile, moderately ductile, and elastic CMRFs. This integration enhanced the building's performance and reduced potential damage to the primary frame members.

Moreover, it offset the cost of the damping system, resulting in an average cost saving of around 11.5%. The closest equivalent system in the 2015 National Building Code of Canada (NBCC) is for ductile buckling-restrained braced frames (R_d = 4, R_0 = 1.2). These factors are already conservative, primarily because the non-damage-based modification factor for Inline Seismic Friction Dampers (ISFDs) is substantially higher [16,17]. Furthermore, the system can be tested with Maximum Considered Earthquake (MCE) ground motion forces and at MCE displacement levels, contrasting with the equivalent systems that cannot avoid uncertainty in their actual behavior.

Unlike yielding steel elements, the activation load and stiffness of the friction damper are decoupled, meaning there is a wide range of energy dissipation capacities/stiffness and ductility values that can be intentionally controlled by adjusting these two design parameters. A response modification factor approach reduces this variability, enhancing performance. This benefit is significant because of a wealth of knowledge developed over several decades on using friction devices as supplemental damping devices for optimizing seismic performance [5–7,18–20]. There is a broad consensus that a well-defined optimal performance point exists for friction-damped structures, and it depends on the structural frame's dynamic properties. Given this context, the benefits of the proposed R-factor design approach, which seems to rely on the bilinearization that eliminates (or at the very least discourages) design optimization, should be more clearly highlighted. Therefore, this paper addresses these issues by evaluating the SFRS reduction factors, including the overstrength, ductility, and response modification factors of four-, eight-, and fourteen-story ductile concrete frames with inline seismic friction dampers. The evaluation involves detailed nonlinear static analysis procedures. Additionally, the effects of building height and span length are considered. Moreover, the seismic performance of the buildings is assessed using nonlinear response history analysis.

2. Design of Structural Models

In this study, the buildings are assumed to be in San Bernardino, California, with a latitude and longitude of 34.108 and −117.289, corresponding to a high seismic zone and site class "D." Diagonal Braces with various span lengths of 6 m and 8 m are considered in each building. All three models have five bays in each direction, as demonstrated in Figure 1.

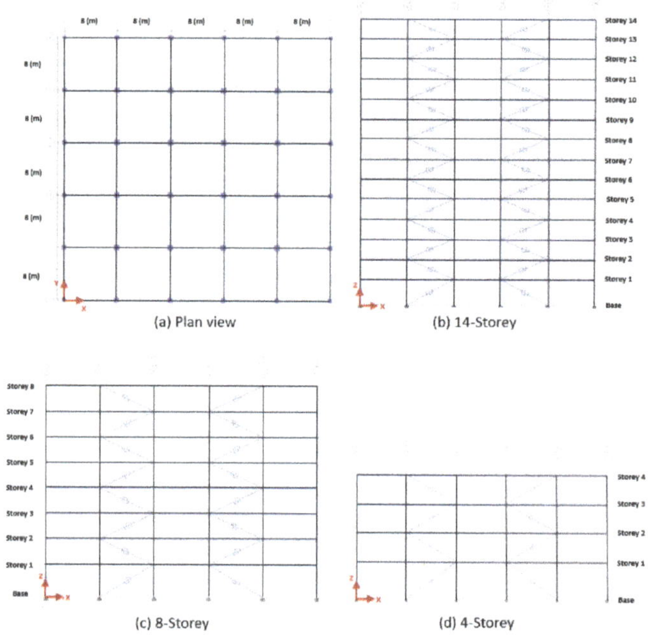

Figure 1. Structural models with Single Diagonal Braces (SBD).

The 2015 NBCC and ETABS software (Version 16.2.1) [14,21] were utilized to design four-, eight-, and fourteen-story ductile concrete frames with inline seismic friction dampers to evaluate the overstrength, ductility, and response modification factors. When it was established that the dampers activate before yielding in the primary members of the frame subjected to the design earthquake, CSA A23.3-19 [22] was employed for detailing the concrete buildings, following the Ultimate Limit States (ULS) or strength design method [22]. For the cylindrical specimen, the assumed compressive strength at 28 days f_c' is 30 MPa, with a modulus of elasticity Ec of 24,500 MPa, and the unit weight of reinforced concrete is 24 KN/m^3. The concrete cover for members exposed to the weather is 40 mm, while for non-exposed members, it is 30 mm. The section properties of the columns are considered

70% of the moment inertia, and for beams, it is 35% of the moment inertia. To prevent column plastic hinges, all beam-to-column joints in structural frames must satisfy the weak beam–strong column criteria, ensuring adequate shear strength in the concrete moment-resisting frame joints to withstand the maximum expected force in the adjoining brace(s). This design procedure includes determining the effective area of the joint, determining the panel-zone design shear force, and controlling the panel-zone shear stress. The assumed design live and dead loads for all models are 1.5 KN/m^2 and 2.4 KN/m^2, respectively, while the snow load acting on the roof is 1.64 KN/m^2. The design details and the brace sections are presented in Table 1.

Table 1. Design details for different models.

		Columns				Beams		Brace Section	
		Interior (Cm)		Exterior (Cm)		Interior (Cm)			
Story	Level	6 m Span	8 m Span	6 m Span	8 m Span	6 m Span	8 m Span	6 m Span	8 m Span
4	4	45 × 45	50 × 50	40 × 40	45 × 45	35 × 35	40 × 40	W14 × 34	W14 × 38
	3	45 × 45	50 × 50	40 × 40	45 × 45	35 × 35	40 × 40	W14 × 48	W14 × 53
	2	50 × 50	55 × 55	45 × 45	50 × 50	40 × 40	45 × 45	W14 × 61	W14 × 68
	1	55 × 55	60 × 60	50 × 50	55 × 55	40 × 40	45 × 45	W14 × 90	W14 × 99
8	8	45 × 45	50 × 50	40 × 40	45 × 45	40 × 40	45 × 45	W14 × 38	W14 × 43
	7	45 × 45	50 × 50	40 × 40	45 × 45	40 × 40	45 × 45	W14 × 53	W14 × 53
	6	60 × 60	65 × 65	55 × 55	60 × 60	55 × 45	55 × 50	W14 × 61	W14 × 68
	5	60 × 60	65 × 65	55 × 55	60 × 60	55 × 45	55 × 50	W14 × 68	W14 × 74
	4	65 × 65	70 × 70	60 × 60	65 × 65	55 × 45	55 × 50	W14 × 74	W14 × 82
	3	65 × 65	70 × 70	60 × 60	65 × 65	60 × 50	65 × 55	W14 × 90	W14 × 99
	2	70 × 70	75 × 75	65 × 65	70 × 70	60 × 50	65 × 55	W14 × 90	W14 × 120
	1	70 × 70	75 × 75	65 × 65	70 × 70	60 × 50	65 × 55	W14 × 159	W14 × 193
14	14	50 × 50	55 × 55	45 × 45	50 × 50	45 × 45	45 × 45	W14 × 34	W14 × 38
	13	50 × 50	55 × 55	45 × 45	50 × 50	45 × 45	55 × 45	W14 × 43	W14 × 53
	12	55 × 55	60 × 60	50 × 50	55 × 55	45 × 45	55 × 45	W14 × 53	W14 × 68
	11	55 × 55	60 × 60	50 × 50	55 × 55	55 × 45	60 × 50	W14 × 68	W14 × 82
	10	55 × 55	60 × 60	50 × 50	55 × 55	55 × 45	60 × 50	W14 × 74	W14 × 90
	9	60 × 60	70 × 70	55 × 55	60 × 60	60 × 50	65 × 55	W14 × 99	W14 × 120
	8	60 × 60	70 × 70	55 × 55	60 × 60	60 × 50	65 × 55	W14 × 109	W14 × 120
	7	60 × 60	70 × 70	60 × 60	65 × 65	60 × 50	65 × 55	W14 × 109	W14 × 132
	6	70 × 70	75 × 75	60 × 60	65 × 65	60 × 50	65 × 55	W14 × 132	W14 × 159
	5	70 × 70	75 × 75	60 × 60	65 × 65	65 × 55	65 × 55	W14 × 145	W14 × 176
	4	70 × 70	75 × 75	60 × 60	65 × 65	65 × 55	70 × 60	W14 × 145	W14 × 176
	3	75 × 75	80 × 80	65 × 65	70 × 70	65 × 55	70 × 60	W14 × 159	W14 × 176
	2	75 × 75	80 × 80	65 × 65	70 × 70	65 × 55	70 × 60	W14 × 193	W14 × 211
	1	75 × 75	85 × 85	65 × 65	70 × 70	65 × 55	70 × 60	W14 × 193	W14 × 233

3. Design of the Inline Seismic Friction Dampers

Buildings designed and constructed following earlier codes and standards often need to meet the life safety criteria based on the current seismic criteria objectives as earthquake requirements evolved. Therefore, in high seismicity zones, and sometimes even in low to moderate seismicity regions, supplemental energy dissipation devices can be employed as part of structural design concepts for new buildings, upgrades, or retrofits of existing structures. In seismically isolated structures, these devices can provide additional dampening and are broadly categorized into two groups, displacement dependent, such as frictional sliding or metallic yielding, and velocity dependent, including viscoelastic and viscous fluid dampers. The 2015 NBCC [14] recommends supplemental energy dissipation but does not explicitly address friction dampers. It specifies that a supplemental energy dissipation system should complement the seismic force-resisting system (SFRS). The modeling of the system should consider greater ductility than that of the SFRS nonlinear hysteretic behavior, and the inherent equivalent viscous damping of the system should not exceed 2.5 percent of the critical damping, excluding the damping provided by energy dissipation devices and other structural elements. The SFRS and components of the supplemental energy dissipation system should be modeled elastically. According to ASCE 7-22, a structure with a damping system must have an SFRS that can resist 100 percent of the load path, while a structure without a damping system must have at least 75 percent of the required design strength [2,23].

Various supplemental energy dissipation devices exist, employing different mechanisms to dissipate energy. These mechanisms include yielding mild steel, viscoelastic behavior in rubber-like materials, shearing of viscous fluids, orifice flow, and sliding friction. Inline Seismic Friction Dampers are utilized in structural engineering to mitigate the impact of seismic forces on buildings and structures. They aim to absorb and dissipate the energy generated during earthquakes, reducing the structural response and minimizing damage. These dampers dissipate energy via the frictional interaction between two sliding solid components. This principle of solid friction is also observed in controlling tectonic movement and earthquake generation and in smaller-scale applications such as automotive brakes, which dissipate kinetic energy. By slipping at a predetermined load before structural yielding, friction dampers effectively dissipate a significant portion of the energy during a major earthquake event. This capability not only safeguards against extensive damage but also offers cost savings compared to the expense of new construction or retrofitting existing buildings, as it provides high levels of energy dissipation.

A clear example of attached damping, specifically Coulomb friction, can be illustrated by Equation (1). This example involves a block moving horizontally on a rough surface, where the equation states that the frictional force is directly proportional to the normal force.

$$F_{friction} = \mu_k N \tag{1}$$

where μ_k is the dynamic friction coefficient or kinetic energy, and N is the normal force [24]. When subjected to cyclic loads, frictional damping devices waste energy via a non-elliptical hysteretic loop. A single-degree-of-freedom system consisting of a mass, spring, and friction can simulate this behavior. Figure 2 depicts an ideal Coulomb damper as a simplified depiction of this notion.

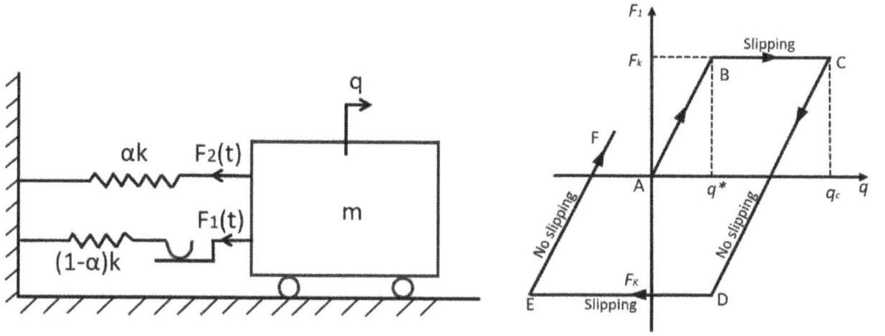

Figure 2. Mass-spring Coulomb damping and hysteresis loop [24].

When the mass, denoted as q, experiences a displacement, and the applied force (F_t) is less than the Coulomb friction force (F_k), the system does not slip. In this scenario, the force–displacement relationship, represented by F_1-q, demonstrates a linear characteristic along the AB line with a slope of $(1 - \alpha)$k. However, sliding occurs when the applied force (F_1) equals F_k, the displacement exceeds the critical value of q, and the force–displacement relationship follows the BC line, establishing a loop. Slipping stops when the mass reverses its motion, the applied force (F_t) falls below F_k, and the force drops down the CD line with a slope of $(1 - \alpha)$k. At point D, where the displacement is $q_c - 2q$, the compressive force equals F_k, resulting in slipping along the DE line. As the mass reaches point E, moving in the opposite direction, the force follows the EF line, and the cycle continues [24].

The dynamic friction behavior of the Inline Seismic Friction Damper (ISFD) is responsible for its characteristic hysteresis pattern observed during the slipping stage. To accurately represent the elastoplastic behavior of the damper, it is recommended to utilize the Wen model, as illustrated in Figure 3. This model is ideal for representing the dynamic friction properties of ISFDs [2,3,25].

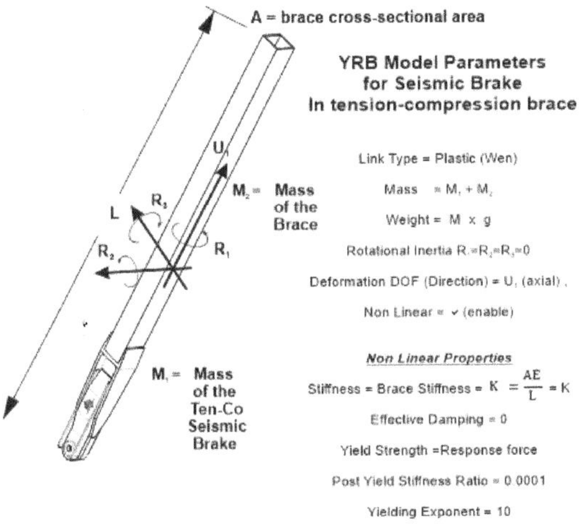

Figure 3. Parameters of Wen model for SFD, courtesy of manufacturer (2023) [26].

Inline Seismic Friction Dampers (ISFDs) should be balanced to ensure structural integrity and prevent excessive design force. Excessive ISFDs deployment leads to rigidity while reducing their numbers eliminates their impact on the system. Therefore, it is crucial to have optimal shear forces on the ISFDs to minimize the demand on the frame and maximize energy dissipation. For this calibration, nonlinear response history analysis should be employed. The central concept is to conduct static analysis and estimate the approximate forces acting on an inline seismic friction damper [16,17].

The equivalent lateral static analysis was performed to calculate the base shear and stiffness of each floor and the design parameters for each damper. These values are presented in Table 2.

Table 2. ISFDs design parameters for different types of structures.

		K_f (KN/mm)		K_d (KN/mm)		K_{ed} (KN/mm)		Post Yield Stiffness Ratio	Yielding Exponent
Story	Level	6 m Span	8 m Span	6 m Span	8 m Span	6 m Span	8 m Span		
4	4	124	134	136	146	148	165	0.0001	10
	3	182	192	198	210	208	231	0.0001	10
	2	238	262	260	286	265	296	0.0001	10
	1	308	331	353	380	373	409	0.0001	10
8	8	139	143	152	156	165	186	0.0001	10
	7	196	198	214	216	231	231	0.0001	10
	6	233	243	254	265	265	296	0.0001	10
	5	266	275	290	300	296	322	0.0001	10
	4	289	305	315	333	322	355	0.0001	10
	3	335	361	366	394	392	430	0.0001	10
	2	359	461	392	503	392	522	0.0001	10
	1	567	625	650	717	656	798	0.0001	10

Table 2. *Cont.*

Story	Level	K_f (KN/mm)		K_d (KN/mm)		K_{ed} (KN/mm)		Post Yield Stiffness Ratio	Yielding Exponent
		6 m Span	8 m Span	6 m Span	8 m Span	6 m Span	8 m Span		
14	14	132	147	144	161	148	165	0.0001	10
	13	157	211	171	230	186	231	0.0001	10
	12	209	269	228	293	231	296	0.0001	10
	11	255	314	278	343	296	355	0.0001	10
	10	282	346	307	377	322	392	0.0001	10
	9	390	458	426	499	430	522	0.0001	10
	8	404	472	441	515	473	522	0.0001	10
	7	429	499	468	544	473	573	0.0001	10
	6	510	582	557	635	573	690	0.0001	10
	5	571	647	623	706	631	765	0.0001	10
	4	567	639	619	698	631	765	0.0001	10
	3	622	685	679	748	690	765	0.0001	10
	2	724	801	791	874	839	916	0.0001	10
	1	690	813	792	933	798	963	0.0001	10

The slip loads are calculated using a method proposed for an equivalent single degree of freedom with idealized lateral load deformation. The ratio of the total braces' stiffness of each floor to the total braces' stiffness plus the story stiffness of each floor is calculated using an iterative procedure, assumed to be 0.8. Additionally, nonlinear time history analysis can be used as an iterative procedure to design the sections of braces and friction dampers [15]. Table 3 represents the equivalent static force analysis results and the slip force for all three models.

Table 3. Seismic analysis results, slip loads, and calculated mass per damper.

Story	Bracing	Level	Seismic Shear Forces (KN)		Seismic Brake Slip Forces (KN)		Seismic Weight (KN)		Mass per Damper (Kg)	
			6 m Span	8 m Span	6 m Span	8 m Span	6 m Span	8 m Span	6 m Span	8 m Span
4	SBD	4	1314	1364	140	150	12,877	16,523	80	80
		3	901	949	240	250	12,877	16,523	115	115
		2	638	676	310	330	12,877	16,523	115	115
		1	369	387	370	390	12,877	16,523	115	115
8	SBD	8	1652	1685	180	180	28,174	33,923	80	80
		7	996	1009	290	290	28,174	33,923	80	115
		6	857	873	380	390	28,174	33,923	115	115
		5	753	761	460	470	28,174	33,923	160	115
		4	615	623	530	540	28,174	33,923	160	160
		3	472	480	580	590	28,174	33,923	160	160
		2	332	339	620	630	28,174	33,923	160	160
		1	190	195	670	680	28,174	33,923	160	195
14	SBD	14	1579	1773	170	190	49,723	65,167	80	80
		13	670	781	250	280	49,723	65,167	115	115
		12	622	727	310	360	49,723	65,167	115	115
		11	600	706	380	440	49,723	65,167	115	115
		10	547	644	440	510	49,723	65,167	115	115
		9	521	616	500	570	49,723	65,167	160	160
		8	470	556	550	630	49,723	65,167	160	160
		7	414	490	590	690	49,723	65,167	160	195
		6	360	425	630	730	49,723	65,167	160	195
		5	318	378	670	770	49,723	65,167	160	195
		4	258	306	690	810	49,723	65,167	160	240
		3	199	236	720	830	49,723	65,167	195	240
		2	139	166	730	850	49,723	65,167	195	240
		1	80	95	780	910	49,723	65,167	195	240

Note: Single Diagonal Braces (SBD).

4. Modal Analysis

The Eigenvalue analysis calculates the natural modes of the system, providing the free-vibration mode shapes and frequencies. The Ritz value analysis determines modes based on specific loading conditions. Since Ritz value yields a better basis than eigenvalue analysis, especially for analyses involving superposition such as response spectrum or time-history analysis, it was utilized to determine the natural periods of the YRB framed system. Table 4 displays the fundamental period of the structure based on the modal analysis, and the results for braced frames obtained from the 2015 NBCC [14] can be determined with empirical Equation (2).

$$T_a = 0.025h_n \quad (2)$$

where $T_a(s)$ represents the fundamental lateral period, and h_n denotes the height of the structure in the meter. Including ISFDs in the frame reduced approximately 60% to 70% compared to the bare frames in the fundamental period. Moreover, the empirical equation results are about 15 to 60% lower than those obtained by the Ritz analysis.

Table 4. The fundamental period of the structures.

Story	T (sec)-Analytical ISFD		T(sec)-Analytical Bare Frame		T_a (s)
	6 m-Span	8 m-Span	6 m-Span	8 m-Span	
4	0.56	0.56	0.81	0.78	0.37
8	0.89	0.85	1.36	1.33	0.73
14	1.46	1.41	1.85	1.82	1.25

5. Response Spectrum

The buildings in this study are assumed to be in San Bernardino, California, with a trend design response spectrum like Tofino on Vancouver Island in British Columbia. A single target response spectrum was developed for the maximum considered earthquake (MCE_R) with 5% damping, obtained by multiplying the design response spectrum by a factor of 1.5, as presented in Figure 4. By selecting a relative location in the United States, the MCE_R was incorporated into the Canadian code for performance assessment of the inline friction dampers.

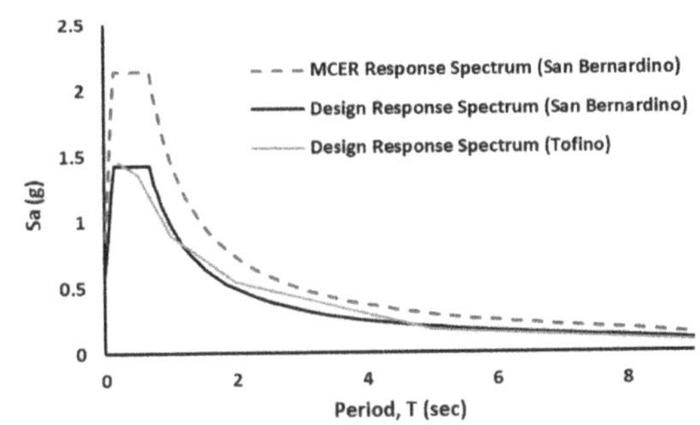

Figure 4. MCER and DER.

The spectral acceleration for a site in San Bernardino, USA, was calculated for different hazard levels, considering both the basic safety earthquakes (BSE) for new and existing building standards. An idealized relationship between base shear and displacement was established using a similar displacement approach for all the models. The analysis was continued until the frame's maximum interstudy drift met the 2.5 percent design limit. The elastic lateral stiffness (K_i), effective lateral stiffness (K_e), effective yield strength (V_y), and target displacements were determined accordingly.

6. Nonlinear Static Analysis

Nonlinear static pushover analysis is carried out to calculate the structural strength capacities and displacement demand. This procedure involves pushing the structure under a lateral load pattern to the level of displacement expected in the design earthquake. The main goal of this analysis is to assess displacement demands in critical elements that exhibit the recommended value by the NBCC 2015 [14] (R = 4.8). These values decrease with the increasing height of the structure and the span length. Moreover, the average values of response modification factors for all three models are 6.87 and 7.29 for 8 m and 6 m span lengths, respectively.

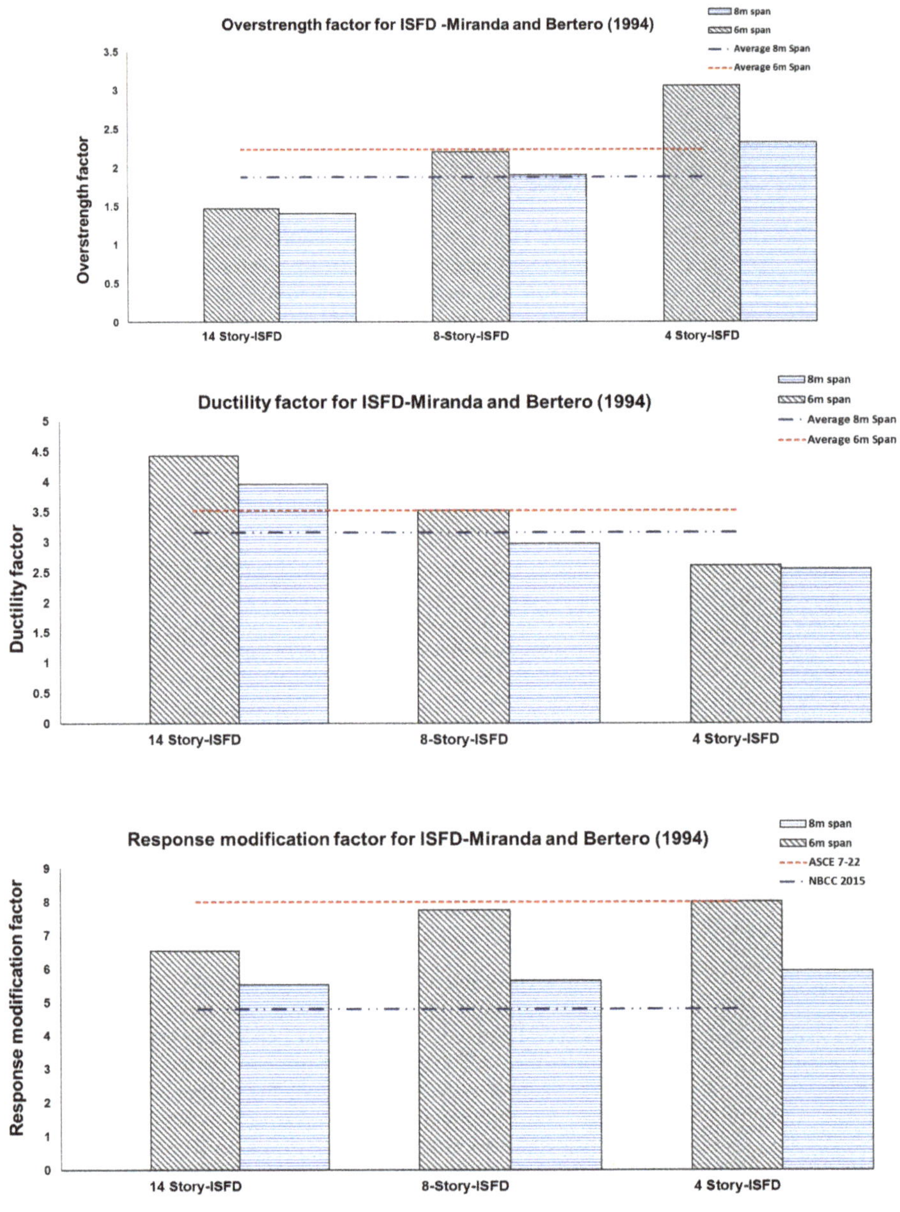

Figure 6. Overstrength, ductility, and response modification factors for 14, 8, and 4 stories.

7. Inelastic Response History Analysis

The relationship between the response of the structure and ground-motion parameters has been explored through a different set of strategies [31,33,34]. Scaling and spectral matching are two approaches for adjusting time series to be consistent with the design response spectrum. Ground motions modification includes multiplying the initial time series by the scaling factor. Then, the matched spectrum equals or exceeds the design spectrum over a specified period range. Matching the time series frequency content to be consistent with the design spectrum is Spectral matching [30,35].

This study obtained 25 different ground motion records from the Pacific Earthquake Engineering Research Center (PEER) database [36], as presented in Table 7. Including a diverse range of ground motions leads to a more comprehensive assessment of seismic performance results. The SeismoMatch 2018 [37] software was used to match the accelerograms based on the design spectrum spectrally. The results of this matching process are illustrated in Figure 7.

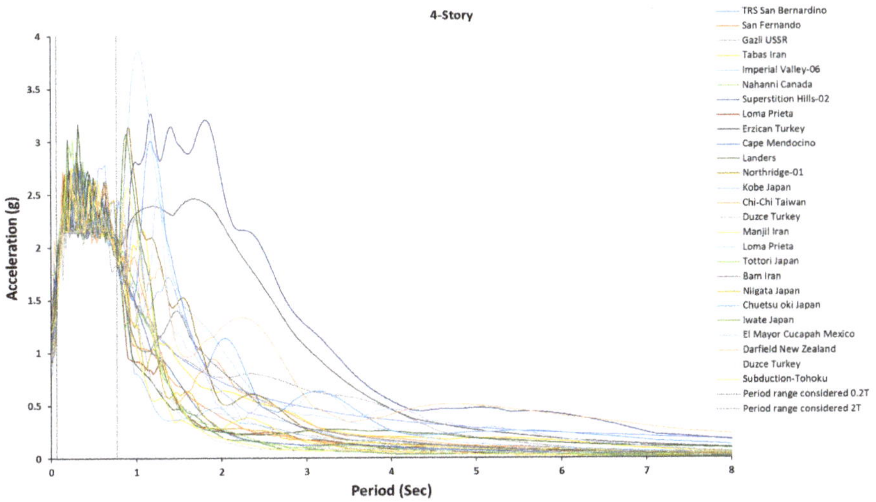

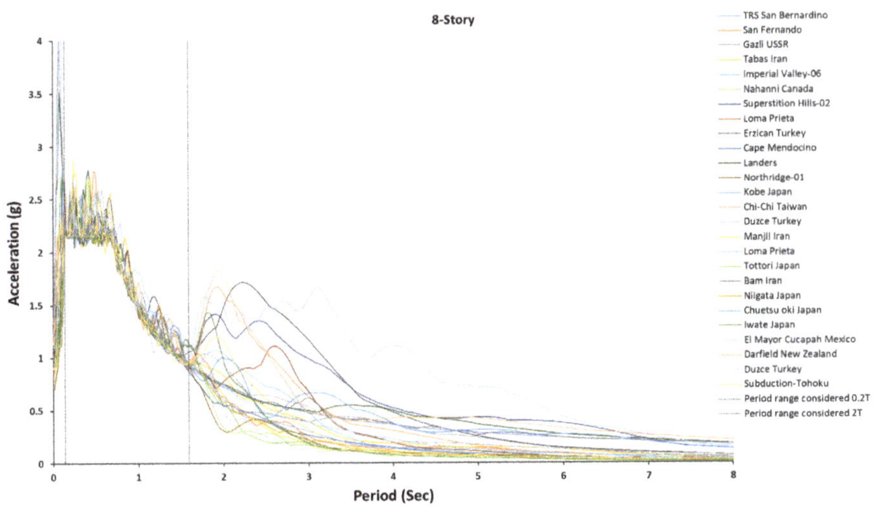

Figure 7. *Cont.*

The results of the nonlinear response history analysis for the base shear demands are presented in Figure 10. These values represent the maximum base shear for each ground motion. It can be observed that all these values are below the capacity of each system. Furthermore, the base shear demands are higher with longer span lengths and in taller buildings.

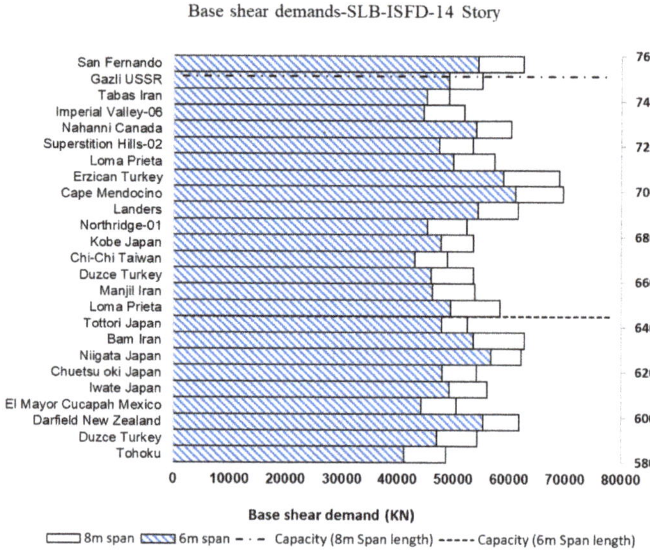

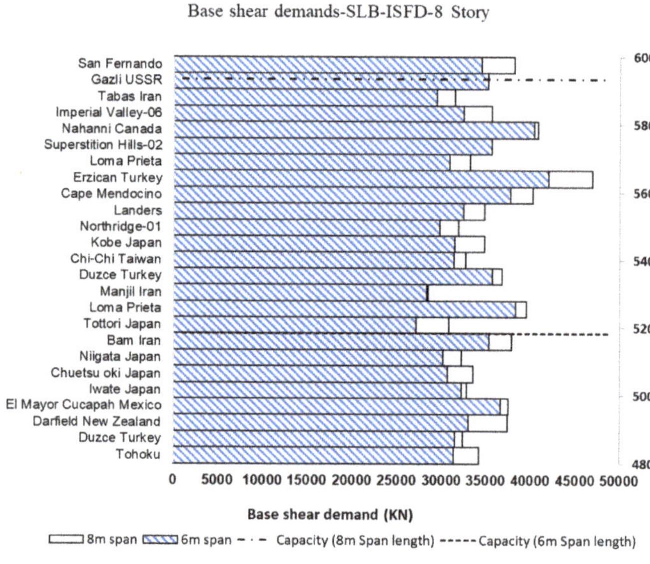

Figure 10. *Cont.*

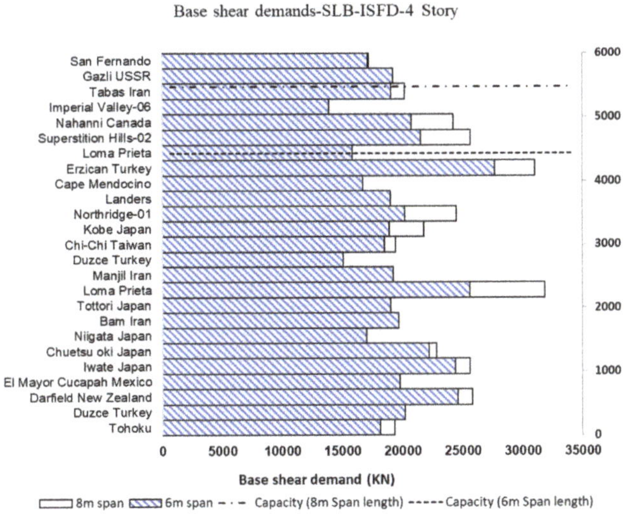

Figure 10. Base shear demands for 4, 8, and 14 stories.

9. Results and Discussions

Friction dampers, like other displacement-dependent devices, exhibit amplitude-dependent damping. While the slip force remains constant, displacements might alter. However, the slip force can be adjusted depending on the required displacement. In this instance, the force–displacement ratio is not proportional, and decoupling the stiffness from the yielding point is one of its advantages in high seismic zones. This article outlines the methodology for determining overstrength, response modification, and ductility factors for various Inline Seismic Friction Dampers (ISFD) systems. Slip loads are calculated based on the lateral shear forces, shear deflection at each floor, the lateral stiffness of braces, and the lateral stiffness of existing frames and braces. The response modification factors of four- eight- and fourteen-story ductile concrete structures equipped with Inline Seismic Friction Dampers (ISFDs) of different span lengths and heights are determined using the method by Miranda and Bertero's (1994) method [32]. For this purpose, normalized moment rotation component models are calculated for each beam and column based on ASCE 41 [38]. These component modeling parameters are applied to each structural member. The SLE, DLE, and MSE target displacement is computed according to ASCE 41 [38]. Nonlinear static analysis is then performed to assess the seismic characteristics of each system. The performance of each system is evaluated through nonlinear response history analysis, considering twenty-five different ground motions matched to the target response spectrum and the structure's period within the range of 0.2T to 1.5 T. The key observations and conclusions are summarized as follows:

In the NBCC 2015 [14], the closest ductility and overstrength factors for an equivalent system are provided for a ductile buckling restrained braced frame. The overstrength factors ranging from 1.41 to 3.06, surpass the prescribed value ($R_0 = 1.2$) specified in the NBCC 2015 [14]. These factors decrease with increasing height and are higher in shorter span lengths. The ductility reduction factors, however, exhibited an upward trend when the height increased. The ductility reduction factors range from 2.55 to 4.43 compared to the NBCC 2015 [14] value 4.

The response modification factor exhibits a decreasing trend as the height increases, and it is more significant for smaller span lengths. The response modification factor ranges between 5.53 and 8.01, with an average of 6.56 and an average minus standard deviation of 5.57, whereas the recommended value in the NBCC 2015 [14] for BRBF is 4.8. These results suggest that structures with a height equal to or less than 50 m adopt a response modification factor of 5.5. This factor covers approximately 95% of the cases and falls within the safe range. The slip loads are calculated based on various factors, including the brace angle, lateral shear forces, shear deflection at each floor, the lateral stiffness of the bare frame, as well as the lateral stiffness of braces and their combination. Therefore, for simplicity and a practical approach used by engineers, it is recommended that the system be modeled as a braced frame, and the axial loads in the braces should be equal to or greater than the slip load of the friction dampers. This approach ensures accurate representation and sufficient capacity for the system. The results acquired from nonlinear response history analysis, explicitly the STDEVA +Mean values, indicated a notable reduction in the story drift ratio ranging from 60% to 70% for all three models incorporating ISFDs, compared to the bare frames.

Moreover, the concentration of story drift demands was higher in taller buildings while remaining relatively consistent or decreasing as the span length decreased. The results from base shear demand indicated that increased structural height led to an elevated base shear demand. Conversely, a decrease in the span length resulted in a reduction of about 10% in the base shear demand.

The analysis of hysteresis curves demonstrated that longer span lengths exhibited higher dissipated energy than shorter span lengths. Implementing the ISFD system directly affects the reduction in the formation of the plastic hinges, resulting in an approximate 45% improvement in structural performance, which aligns with the objective of life safety or better.

There was an accompanying increase in the base shear demand as the height of the structure increased. On the other hand, a smaller span length resulted in a decrease. Notably, implementing inline friction dampers resulted in a substantial reduction in the dissipated energy, particularly in structures with larger span lengths.

10. Conclusions

This paper focuses on applying seismic reduction factors in line with the recommended R values specified in the NBCC 2015 [14]. These reduction factors can support minimizing the number of variable design parameters, enhancing structural performance. Because of this, additional work is warranted to capture such performance for friction dampers. Observations indicate that increasing the height of a structure reduces the overstrength factor while increasing the ductility factor. Shorter span lengths result in greater response modification factors; a suggested response modification factor for structures 50 m in height is 5.5. Base shear demand increases with height and decreases with span length (10% reduction). Hysteresis curve analysis demonstrates greater dissipated energy for longer spans. Nonlinear response history analysis shows a substantial reduction (60–70%) in story drift ratio when Inline Seismic Friction Dampers (ISFDs) are employed, compared to bare frames. The concentration of story drift demands is higher in taller buildings but remains consistent or decreases with shorter spans.

Further investigations are needed to comprehend and capture the performance of friction dampers fully. Additional investigations can be conducted to assess the seismic characteristics of buildings using different bracing configurations, floor heights, and bracing angles. More detailed modeling approaches for structures equipped with ISFDs shall be considered.

Author Contributions: Conceptualization, A.N., A.B. and F.M.T.; methodology, A.N., A.B. and F.M.T.; software, A.N.; validation, A.N., A.B. and F.M.T.; formal analysis, A.N.; investigation, A.N., A.B. and F.M.T.; resources, A.B. and F.M.T.; data curation, A.N.; writing—original draft preparation, A.N.; writing—review and editing, A.B. and F.M.T.; visualization, A.N., A.B. and F.M.T.; supervision, A.B. and F.M.T.; project administration, A.B. and F.M.T. All authors have read and agreed to the published version of the manuscript.

Funding: This research received no external funding.

Institutional Review Board Statement: Not applicable.

Informed Consent Statement: Not applicable.

Data Availability Statement: Not applicable.

Acknowledgments: The support of the Natural Sciences and Engineering Research Council (NSERC) and the IC-IMPACTS Research Network is gratefully acknowledged. The authors would also thank O. Galindo, Quaketech Inc., for his valuable comments and suggestions.

Conflicts of Interest: The authors declare no conflict of interest.

Glossary

R_0	Overstrength factor
R_d	Ductility-related force modification factor
Ta(s)	Fundamental lateral period
h_n	Height of the structure
K_i	Elastic lateral stiffness
K_e	Effective lateral stiffness
V_y	Effective yield strength
V_d	Design Strength
ϕ	Function of ductility
R_μ	Ductility reduction factor
R	Response modification factor
μ	Ductility
Δ_{max}	Maximum displacement
Δ_Y	Yield displacement
SLE	Service Level Event
DLE	Design Level Event
MSE	Maximum Seismic Event
ISFD	Inline Seismic Friction Damper
NBCC	National Building Code of Canada
MRF	Moment Resisting Frame
SFRS	Seismic Force Resisting System
MC_E	Maximum Considered Earthquake
SBD	Single Diagonal Braces

References

1. Akiyama, H. Evaluation of Fractural Mode of Failure in Steel Structures Following Kobe Lessons. *J. Constr. Steel Res.* **2000**, *55*, 211–227. [CrossRef]
2. Naghshineh, A. Seismic Resilience and Performance Design Approach for Concrete Moment Resisting Frame Buildings Equipped with Yielding Restrained Braces. Ph.D. Thesis, Concordia University, Montreal, QC, Canada, 2021.
3. Constantinou, M.; Soong, T.; Dargush, G. *Passive Energy Dissipation Systems for Structural Design and Retrofit*; MCEER Monograph No. 1; University at Buffalo: Buffalo, NY, USA, 1998; ISBN 0-9656682-1-5.
4. Soong, T.T.; Dargush, G.F. *Passive Energy Dissipation and Active Control, Boca Raton: Structural Engineering Handbook*; CRC Press LLC: Boca Raton, FL, USA, 1999; pp. 1–28.
5. Ciampi, V.; De Angelis, M.; Paolacci, F. Design of yielding or friction-based dissipative bracings for seismic protection of buildings. *Eng. Struct.* **1995**, *17*, 381–391. [CrossRef]
6. Fu, Y.; Sherry, S. Design of friction damped structures using lateral force procedure. *Earthq. Eng. Struct. Dyn.* **2000**, *29*, 989–1010. [CrossRef]
7. Levy, R.; Lavan, O.; Rutenberg, A. Seismic design of friction braced frames based on historical records. *Earthq. Spectra* **2005**, *21*, 761–778. [CrossRef]
8. Nabid, N.; Hajirasouliha, I.; Petkovski, M. A Practical Method for Optimum Seismic Design of Friction Wall Dampers. *Earthq. Spectra* **2017**, *33*, 1033–1052. [CrossRef]
9. Couch, L.; Tehrani, F.M.; Naghshineh, A.; Frazao, R. Shake Table Response of a Dual System with Inline Friction Damper. *Eng. Struct.* **2023**, *281*, 115776. [CrossRef]
10. Montuori, R.; Nastri, E.; Piluso, V. Theory of plastic mechanism control for the seismic design of braced frames equipped with friction dampers. *Mech. Res. Commun.* **2014**, *58*, 112–123. [CrossRef]
11. Nabid, N.; Hajirasouliha, I.; Escolano Margarit, D.; Petkovski, M. Optimum Energy Based Seismic Design of Friction Dampers in RC Structures. *Structures* **2020**, *27*, 2550–2562. [CrossRef]

12. Alam, M.S.; Moni, M.; Tesfamariam, S. Seismic Overstrength and Ductility of Concrete Buildings Reinforced with Superelastic Shape Memory Alloy Rebar. *Eng. Struct.* **2012**, *34*, 8–20. [CrossRef]
13. Asgarian, B.; Shokrgozar, H.R. BRBF Response Modification Factor. *J. Constr. Steel Res.* **2009**, *65*, 290–298. [CrossRef]
14. NBCC. *National Building Code of Canada*; NBCC: Ottawa, ON, USA, 2015.
15. Tirca, L.; Serban, O.; Tremblay, R.; Jiang, Y.; Chen, L. Seismic Design, Analysis and Testing of a Friction Steel Braced Frame System for Multi-Storey Buildings in Vancouver. *Key Eng. Mater.* **2018**, *763*, 1077–1086. [CrossRef]
16. Naghshineh, A.; Kassem, A.; Pilorge, A.G.; Galindo, O.R.; Bagchi, A. *Seismic Performance of Reinforced Concrete Frame Buildings Equipped with Friction Dampers. Structures Congress 2018*; American Society of Civil Engineers: Fort Worth, TX, USA, 2018; pp. 94–101. [CrossRef]
17. Galindo, O.; Frazao, R.; Pastor, C.; Coronado, G.; Gonzales, D. Supplemental Damping for the seismic retrofit of 8-storey RC Hotel building in the Mexican Pacific using Yielding Restrained Braces—Part A, Comparison of alternatives. In Proceedings of the 12th Canadian Conference on Earthquake Engineering, Quebec City, QC, Canada, 17–20 June 2019.
18. Cherry, S.; Filiatrault, A. Seismic Response Control of Buildings Using Friction Dampers. *Earthq. Spectra* **1993**, *9*, 447–466. [CrossRef]
19. Christopoulos, C.; Mansour, N. *Performance Based Seismic Design of Structures Equipped with Hysteretic Dampers*; Structural Engineering Research Report No. UTCE-05/01; University of Toronto: Toronto, ON, Canada, 2005.
20. Guo, W.W.J.; Christopoulos, C. Performance spectra based method for the seismic design of structures equipped with passive supplemental damping systems. *Earthq. Eng. Struct. Dyn.* **2012**, *42*, 935–952. [CrossRef]
21. CSI. *Integrated Building Design Software, ETABS, CSI Analysis Reference Manual*; Computers and Structures, Inc.: Berkeley, CA, USA, 2016; Available online: https://www.csiamerica.com/products/etabs (accessed on 18 December 2022).
22. *CSA A23.3-19*; Design of Concrete Structures. Canadian Standards Association: Toronto, ON, Canada, 2019.
23. *ASCE/SEI 7-22*; Minimum Design Loads and Associated Criteria for Buildings and Other Structures. American Society of Civil Engineers, ASCE Standard: Reston, VA, USA, 2022.
24. Roberts, J.B.; Spanos, P.D. *Random Vibration and Statistical Linearization*; Courier Corporation: Chelmsford, MA, USA, 2003.
25. Tehrani, F.M.; Nazari, M.; Naghshineh, A. Role of Seismic Isolation and Protection Devices in Enhancing Structural Resilience. In *Objective Resilience: Technology*; EMI ORC MOP 2020, Part III; American Society of Civil Engineers: Reston, VA, USA, 2022; Chapter 21; pp. 161–209.
26. Quaketek—Earthquake Protection—Seismic Dampers. Available online: https://quaketek.com/ (accessed on 11 May 2023).
27. NBC. *Structural Commentaries User's Guide 2019*; National Research Council of Canada: Ottawa, ON, Canada.
28. Lawsson, R.S.; Vance, V.; Krawinkler, H. Nonlinear static pushover analysis—Why, When, and How. *Earthq. Eng.* **1994**, *1*, 283–292.
29. Krawinkler, H.; Seneviratna, G.D.P.K. Pros and cons of a pushover analysis of seismic performance evaluation. *Eng. Struct.* **1998**, *20*, 452–464. [CrossRef]
30. Newmark, N.M.; Hall, W.J. *Earthquake Spectra and Design*; Earthquake Engineering Research Institute: Berkeley, CA, USA, 1982.
31. Gupta, A.K. *Response Spectrum Method in Seismic Analysis and Design of Structures*; Blackwell Scientific Publications: Boston, MA, USA, 1990.
32. Miranda, E.; Bertero, V.V. Evaluation of Strength Reduction Factors for Earthquake-Resistant Design. *Earthq. Spectra* **1994**, *10*, 357–379. [CrossRef]
33. Baker, J.W. Conditional Mean Spectrum: Tool for Ground-Motion Selection. *J. Struct. Eng.* **2011**, *137*, 322–331. [CrossRef]
34. Cordova, P.P.; Deierlein, G.G.; Mehanny, S.S.F.; Cornell, C.A. Development of a Two-Parameter Seismic Intensity Measure and Probabilistic Assessment Procedure. In Proceedings of the 2nd U.S.-Japan Workshop on Performance-Based Earthquake Engineering Methodology for Reinforced Concrete Building Structures 2000, Sapporo, Japan, 11–13 September 2000; pp. 187–206. Available online: https://apps.peer.berkeley.edu/research/peertestbeds/Cct/Cordova%20et%20al%202001.pdf (accessed on 18 September 2018).
35. Gavin, H.P.; Dickinson, B.W. Generation of Uniform-Hazard Earthquake Ground Motions. *J. Struct. Eng.* **2011**, *137*, 423–432. [CrossRef]
36. PEER Ground Motion Database—PEER Center. Available online: https://ngawest2.berkeley.edu/ (accessed on 5 May 2021).
37. SeismoSoft. SeismoStruct-User Manual for Version 7.0, Pavia, Italy. 2014. Available online: www.seismosoft.com (accessed on 5 October 2020).
38. *ASCE/SEI 41-17*; Seismic Evaluation and Retrofit of Existing Buildings. American Society of Civil Engineers: Reston, VA, USA, 2017.

eng

MDPI

Review

Surface Waterproofing Techniques: A Case Study in Nova Lima, Brazil

Lorena Carias de Freitas Gomes, Henrique Comba Gomes * and Elvys Dias Reis *

Federal Center for Technological Education of Minas Gerais, Belo Horizonte 30421-169, MG, Brazil; lorena.carias@gmail.com
* Correspondence: henriquecombagomes@gmail.com (H.C.G.); elvysreis@yahoo.com.br (E.D.R.)

Abstract: Considering the various problems caused by infiltration in civil construction, this study aimed to identify the most appropriate waterproofing methods for different types of surfaces. A study was conducted on the mechanisms of water infiltration on surfaces and the waterproofing methods available on the market, focusing on asphalt blankets, in addition to a literature review highlighting state-of-the-art methods on this topic. A case study was also conducted in a residence in Nova Lima, Brazil, analyzing different waterproofing techniques, including their characteristics and stages. Among the conclusions, it is highlighted that the implementation of adequate project, installation, inspection, and maintenance techniques can significantly reduce the waterproofing failure rate and repair costs, and that the excellent choice of materials, along with the skill of the labor force in the application, is fundamental to guarantee the adequate performance of these materials in buildings.

Keywords: durability; waterproofing; infiltration; asphalt blankets; pathologies

Citation: Gomes, L.C.d.F.; Gomes, H.C.; Reis, E.D. Surface Waterproofing Techniques: A Case Study in Nova Lima, Brazil. *Eng* **2023**, *4*, 1871–1890. https://doi.org/10.3390/eng4030106

Academic Editor: Alessio Cascardi

Received: 31 May 2023
Revised: 28 June 2023
Accepted: 2 July 2023
Published: 4 July 2023

1. Introduction

Building infiltration is a common problem when water penetrates the walls, floors, or ceilings. This problem can be caused by several factors, such as lack of maintenance, poor waterproofing, and exposure to rain or wind, among others, and has been a point of extreme concern in all buildings since it is the leading cause of several pathologies in buildings [1]. Among the various damages to the building structure, one can mention the weakening of the walls, the appearance of mold and mildew, rotting materials, corrosion of metals, and the proliferation of insects and other animals [2]. In addition, the moisture generated by infiltration contributes to an unhealthy environment due to the proliferation of fungi and bacteria, which can cause allergies, respiratory irritation, and other health problems for residents [3].

Investing in good waterproofing during the construction or renovation of the building is crucial to avoid infiltration in buildings. In this sense, construction companies seek to isolate moisture and take necessary measures at each stage of construction, considering which material is the most suitable for the place to be waterproofed and aiming to avoid as much water action as possible. The safest and most cost-effective way to ensure the water-tightness of a building is to conduct waterproofing planning from the project's conception [4]. This is important because, if infiltration has already occurred, it is necessary to solve the problem as soon as possible, which may involve identifying the source of the infiltration, repairing the affected areas, and eliminating the conditions that allowed the infiltration. In more severe cases, hiring professionals who specialize in waterproofing and building restoration may be necessary to solve the problem correctly and safely. All this makes the project more costly [5].

In this context, waterproofing is a fundamental technique to protect structures against the action of water and moisture. One of the primary forms of waterproofing is the application of blankets, which consists of placing a layer of resistant and waterproof material on the surfaces to be protected. Blankets can be used in various construction areas, such as slabs, pools, bathrooms, kitchens, reservoirs, and basements [6]. Thus, it is essential to know the proper techniques for applying blankets and the different types of materials available on the market to ensure effective and lasting waterproofing.

With this perspective, the general objective of this paper is to present the surface-waterproofing methods available in the construction market and their main applications, focusing on asphalt blankets. It is worth mentioning that, in this paper, the term "asphalt" refers to the asphalt binder, also known as "asphalt cement" or "bitumen," and not to a mixture of binder and aggregate. Specifically, it aims to: (i) explain the parameters involved in the mechanism of water infiltration on surfaces; (ii) identify the techniques used for surface waterproofing, focusing on asphalt membrane; (iii) describe the current state of knowledge on the use of surface-waterproofing membranes; and (iv) present the waterproofing techniques used in a real case that involved the application of membranes.

Research Significance

Concerns about waterproofing surfaces arose a long time ago. However, it was only in the 20th century that it gained greater importance in civil construction. With the advancement of technology and the development of new materials, applying more efficient and durable techniques in different structures became possible. Today, waterproofing is considered an essential step in the construction or renovation of buildings because it prevents problems caused by moisture and ensures the durability and safety of buildings [4].

Infiltration is a problem that can compromise the life of a structure, causing pathologies in buildings and affecting the aesthetics of buildings. When uncontrolled, it can cause severe damage, such as metal corrosion, the proliferation of fungi and bacteria, and rotting materials, among others. These problems can decrease the structure's life and affect its stability, putting the safety of users at risk [7,8].

Therefore, proper waterproofing is a crucial element in planning construction projects, as it prevents the occurrence of infiltrations and pathologies in buildings. It is essential that waterproofing be planned from the project's conception, considering the characteristics of the region and the terrain, the purpose of the structure, and the materials used. Using good-quality techniques and materials in waterproofing increases the structure's durability, maintains aesthetics, and avoids constant repairs [9].

With this perspective, studying the forms of waterproofing for each type of surface becomes relevant because each surface presents different characteristics that can influence the choice of the most appropriate technique. This study should involve, but not be limited to, factors such as durability, resistance, ease of application, costs, and the necessary maintenance.

2. Theoretical Background

2.1. Surface Infiltration

Surface infiltration is a process by which water penetrates porous surfaces, such as walls, roofs, slabs, floors, and sidewalks, or by flaws in the structure, such as cracks, fissures, or leaks. It can occur in various ways, depending on the surface type and environmental conditions, and cause damage and problems, such as humidity, mold, and structural deterioration, among others [10].

The main mechanisms by which surface infiltration can occur include percolation, capillarity, leaks, drainage-system failures, and condensation, described below.

2.1.1. Water Percolation

Water percolation on surfaces occurs when water penetrates a structure through cracks, fissures, or other defects. It is a continuous process that can cause damage to the structure over time [11].

The process begins when water from rain or another source comes into contact with the structure's surface. If the surface is in good condition and impermeable, the water will run off it until it reaches the drainage system. However, if the surface has cracks or fissures, the water will penetrate the structure.

Cracks and fissures in the structure can be caused by several factors, such as natural wear and tear, aging of the structure, and water damage, among others. When water penetrates the structure, it follows the easiest path through the cracks and fissures, infiltrating the porous structure [12].

As water penetrates the structure, it can cause significant damage. For example, water penetrating a concrete wall can corrode the steel bars and weaken the structure. If water penetrates a brick wall, it can damage the plaster and weaken the structure [13].

It is crucial to perform regular inspections and maintenance, repairing cracks, fissures, and other defects immediately to prevent water from percolating onto building surfaces. It is also essential to ensure the drainage system works correctly and uses waterproof and water-resistant materials whenever possible.

2.1.2. Capillarity

Capillarity of water on surfaces occurs when water is absorbed by porous materials, such as bricks, concrete, stones, and mortar. These materials have microscopic pores, also known as capillaries, which allow water to rise through the surface through the action of surface tension [4].

Surface tension is the physical property that causes water molecules to attract each other, creating a force that holds water together on a surface. When water comes into contact with a porous material, it spreads across surfaces. It penetrates capillaries, attracted by the surface-tension force, which propels it and causes it to rise through the pores. As the process occurs, water can cause damage to the structure, such as staining, discoloration, deterioration, and corrosion of materials [10].

Water capillarity on surfaces can be minimized through waterproofing and surface-treatment techniques. Waterproof coatings, such as paints, varnishes, and sealants, can help prevent water penetration into porous materials. In addition, the use of water-resistant materials, such as stone and tile, can also minimize capillarity. However, this natural process can be difficult to prevent entirely. Therefore, regular maintenance and monitoring of the structure are essential to ensure its integrity and to prevent water damage [14].

2.1.3. Condensation

Water condensation on surfaces occurs when water vapor in the air comes into contact with a cold surface and condenses into liquid form. This process can occur on various surfaces in buildings, such as walls, windows, and roofs [13].

Naturally, the amount of condensed water will depend on the surface temperature and the relative humidity of the air. Thus, condensation may be more common in environments with high humidity, such as bathrooms, kitchens, and laundry rooms [4].

Water infiltration caused by condensation can occur when condensed water penetrates surfaces and seeps into the structure. This can cause damage, such as staining, discoloration, detachment of coatings, and, in more severe cases, damage to the structure [15].

To prevent condensation and the infiltration caused by it, it is important to maintain an adequate temperature on the surfaces and a balance in the relative humidity of the air, which can be achieved through thermal insulation and adequate ventilation, among other solutions. It is also essential to ensure that surfaces are adequately waterproofed to prevent water infiltration and that the use and maintenance of waterproofing materials, such as paints, varnishes, sealants, and membranes, is practical [6].

2.1.4. Leaks

Water leaks on surfaces occur when there is a rupture or failure in hydraulic systems or structures, allowing the passage of water to unwanted areas. This can occur in different parts of buildings, such as slabs, roofs, walls, floors, and windows [9].

Among the leading causes of leaks are the aging of hydraulic systems, problems in connections and welds, failures in construction materials, and changes in water pressure. The infiltration caused by leaks can cause damage such as humidity, stains, mold, discoloration, corrosion of structures, damage to coatings, and even structural problems [2].

To prevent infiltration caused by leaks, it is essential to perform regular maintenance on hydraulic systems and building structures, ensuring that they are in good working order. In addition, it is essential to ensure that areas vulnerable to leaks, such as slabs and roofs, are adequately waterproofed. Suitable materials, such as asphalt blankets and liquid waterproofing, can help prevent the problem [16].

2.1.5. Drainage-System Failures

Drainage-system failures can cause infiltration on surfaces, such as slabs and walls, when water accumulates on the structure's surface that needs to be adequately drained. This accumulation can occur due to several factors, such as heavy rains, clogged gutters and drains, and problems in rainfall drainage systems [17].

Water accumulating on the structure's surface can infiltrate pores, cracks, and fissures in the surface, which can be aggravated by water and weather. Water infiltration in slabs can generate several problems, such as the formation of moisture, stains, bubbles, and detachment of coatings, in addition to contributing to the corrosion process of structures and aggravating structural problems [18].

To prevent infiltration caused by failures in the drainage system, it is vital to ensure that the system is appropriately sized and sized for local rainfall conditions, in addition to performing regular maintenance to ensure that gutters and drains are unobstructed [6]. In addition, one should ensure that the structure's surface, such as the slab, is adequately waterproofed, utilizing appropriate materials and conducting regular maintenance to ensure its effectiveness.

2.2. *Surface-Waterproofing Techniques*

Surface-waterproofing techniques prevent water from entering structures and prevent problems caused by infiltration, such as corrosion, humidity, and detachment of coatings. Several waterproofing techniques can be chosen according to the characteristics of the structure, exposure conditions, and level of demand regarding effectiveness and durability [19].

In the construction market, waterproofing techniques are classified into flexible and rigid. Rigid waterproofing does not resist the movement of the structure and, therefore, should be used in stable locations that do not suffer much temperature variation. The primary example of this technique is polymeric mortar, commonly used in base beams, floors, walls, and basements. Scheidegger [20] states that flexible waterproofing, with asphalt and polymers in its composition, better supports the contraction and dilation of the substrate. Asphalt blankets applied to slabs, roofs, pools, and water tanks are mainly used in the rigid system.

Furthermore, it is important to consider the water-resistance characteristics of building materials, especially in areas exposed to water. Concrete, a widely used material in construction, has specific test methods and requirements to assess its water resistance [21]. These tests evaluate water absorption, permeability, and durability under moisture conditions. Adhering to the appropriate standards and conducting thorough testing can contribute to the waterproofing strategies' overall effectiveness.

Among the most common surfaces, waterproofing techniques are the application of asphalt blankets, polymers, polymeric mortars, and special paints, as described below.

2.2.1. Waterproofing with Asphalt Membrane

Waterproofing surfaces using asphalt membrane is one of the most widely used in construction due to its high efficiency and durability. The process involves applying asphalt blankets over the surface to be waterproofed, forming a continuous and water-resistant layer [4,6]. Figure 1 shows the main steps of this technique [22].

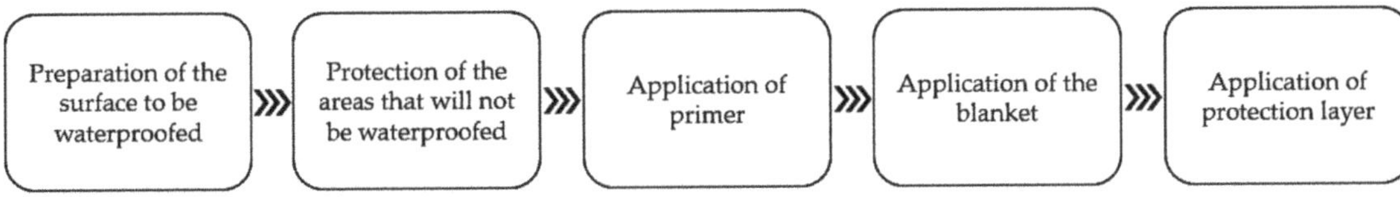

Figure 1. General stages of waterproofing with asphalt blankets.

The first stage consists of the adequate preparation of the surface to ensure the adherence of the asphalt membrane. This includes cleaning the surface, removing loose or poorly adhered parts, leveling, and smoothing imperfections.

The second occurs before the application of the mat, protecting the areas that will not be waterproofed with adhesive tape or other materials to prevent the asphalt membrane from sticking to these surfaces.

The third involves applying primer, a layer of diluted asphalt product, which is applied over the surface to be waterproofed. The primer improves the adhesion of the asphalt membrane and ensures its fixation.

The fourth consists of applying the mat over the primer, using adhesives and heat for fixation. The asphalt membrane comprises layers of polymer-modified asphalt and polyethylene or aluminum coatings, ensuring high mechanical resistance and durability.

The fifth and last stage, in turn, comprises applying a protective layer, such as mortar or a ceramic coating, on some types of asphalt membrane to protect it against the action of ultraviolet rays, mechanical aggression, and other external agents.

In Brazil, there are technical standards that regulate the use of blankets for waterproofing surfaces, among which NBR 9574: Waterproofing execution [23], NBR 9575: Waterproofing—Selection and design [24], NBR 9952: Asphalt blanket for waterproofing [25], and NBR 9685—Asphalt blankets with reinforcement for waterproofing—Specification [26] stand out.

There are different types of asphalt blankets on the market, and the choice is based on the characteristics of each one. The main types used are aluminized blankets, slate blankets, and polyethylene blankets, which vary in additives, finish, and thickness. These variations establish the best technique to be used according to the place to be waterproofed.

The aluminized blanket comprises modified asphalt and polymers, is structured with polyester, and is covered with an aluminized layer. One of the advantages of using this type of blanket is its ability to reflect the sun's rays and reduce indoor temperatures. The disadvantage of its use is its low mechanical resistance, so it should be used in places with no traffic. The aluminized asphalt blanket is used mainly on slabs without traffic, roofs, ruffles, and gutters.

The slate blanket is composed of modified asphalt and polymers, and is structured with polyester, ensuring excellent stability and flexibility. Its surface is coated with granules of mineral aggregate (slate) available in green, gray, and red, providing an esthetic finish on roofs. This type of asphalt blanket should be used in places with no traffic, as it has low resistance to mechanical stress. Some examples of its application are in slabs with no traffic, marquees, and eaves.

In this context, it is worth mentioning that a protective layer does not cover the slate and aluminized mats, although Figure 1 suggests this in a generalized way.

The polyethylene blanket comprises modified asphalt and polymers and has a polyethylene finish, providing greater flexibility, adherence, and durability. Unlike other types of asphalt membranes, the polyethylene membrane cannot be exposed to the sun for more than 72 h and must be covered with mechanical protection, such as ceramic tiles. The waterproofing with a polyethylene blanket can be conducted on slabs, pools, and reservoirs, all with mechanical protection.

2.2.2. Waterproofing with Polymers

The method of waterproofing surfaces using polymers is a process that involves applying a polymer layer over the surface to be protected in order to prevent the penetration of water or other liquids that could cause damage. Sometimes these technologies are also called "liquid membranes". The most common waterproof polymers are polyurethane, acrylic, epoxy, and silicone [27].

The process is similar to using an asphalt blanket and begins with surface preparation. The surface must be clean, dry, and free of dust, oil, or other contaminants that can impede the adhesion of the polymer. Next, a primer is applied to help the polymer adhere to the surface. After the primer is dry, a coat of the chosen polymer is applied, usually using a roller or a sprayer. Depending on the thickness and desired properties, the final layer can be wrapped in several thin layers [28].

Polymer curing, in turn, can vary depending on the type chosen but, generally, occurs through solvent evaporation or a chemical reaction process [29]. It is worth noting that the manufacturer's instructions regarding curing time and the waiting time between layers must be respected. Once the polymer layer is fully cured, it forms a liquid-resistant barrier. This technique is commonly used in bathrooms, kitchens, slabs, roofs, and foundations to protect against infiltrations and leaks.

Not least, it should be noted that the waterproofing of surfaces using polymers must be carried out by qualified professionals, following the applicable safety standards and regulations. In addition, it is essential to choose a high-quality polymer that is compatible with the surface to be waterproofed to ensure the protection's effectiveness and avoid future problems [30].

2.2.3. Waterproofing with Polymeric Mortar

The technique of waterproofing surfaces using polymeric mortar is a process that consists of applying a mixture of cement, aggregates, and polymers on the surface to be protected to create a layer resistant to water and other liquids. Polymeric mortars are composed of an acrylic polymer, which gives the mixture waterproofing properties [11].

As with the techniques presented previously, the process begins with preparing the surface, which must be dry and clean. Then, a primer is applied to help the mortar adhere to the surface. When the primer dries, the polymeric mortar is applied using a straightener or a spatula. The thickness of the layer depends on the area to be waterproofed and the desired properties [31].

The curing of polymeric mortar can vary according to the type chosen but generally occurs through water evaporation or a chemical reaction process. Following the manufacturer's instructions regarding mixing, curing time, and the waiting time between coats is essential. Once the mortar layer is cured, it will waterproof the surface [32].

The use of polymeric mortars in the waterproofing of surfaces, such as that of polymers, requires skilled labor and care with safety during application. In addition, the material chosen must be compatible with the surface to be staked.

2.2.4. Waterproofing with Special Paints

Waterproofing surfaces with special paints involves applying a layer of paint or special coating over the surface to be staked. There are several special paints for waterproofing options, including epoxy, polyurethane, acrylic, and liquid rubber [11].

Waterproofing begins by preparing the surface to be clean, dry, and free of dust, oil, or other contaminants that can impede the adhesion of the special paint or coating. Next, a primer is applied, and when it is dry, the paint is applied to the surface, usually using a brush, roller, or sprayer. As with polymers, several thin layers can be applied to comprise the final coat, depending on the expected properties [33].

The curing of the paint can vary depending on the type chosen but also occurs by chemical reactions and/or solvent evaporation. After the special coating cures, a liquid-resistant barrier is formed, waterproofing the surface [34].

Once again, it is emphasized that trained professionals should perform this type of service, using high-quality materials compatible with the surface to be waterproofed.

2.2.5. Waterproofing with Cold-Applied and Hot-Applied Bituminous Mastic Compounds

Cold-applied mastic compounds are bitumen-based materials typically supplied in a ready-to-use form. They consist of bitumen, fillers, and additives that enhance their performance. These mastics are applied directly onto the surface to be waterproofed, providing a flexible and durable barrier against water ingress. The cold application eliminates the need for heat during installation, making it a convenient and time-saving solution for various waterproofing applications [35,36].

On the other hand, hot-applied mastic compounds involve heating bitumen to a suitable temperature, typically using specialized equipment. The molten bitumen is then applied to the prepared surface, forming a thick, cohesive layer upon cooling. This technique offers excellent adhesion, ensuring a solid bond with the substrate and providing long-lasting waterproofing protection. Hot-applied mastics are commonly used in high-performance waterproofing applications that require superior durability and resistance to environmental factors [37].

Both cold-applied and hot-applied bituminous mastic compounds offer distinct advantages in surface waterproofing. They possess excellent water-resistance properties, ensuring the integrity of the protected surfaces. Additionally, their flexibility accommodates structural movements and temperature variations without compromising the effectiveness of waterproofing [36,38].

These bituminous mastics are compatible with various substrates, including concrete, masonry, and metal. They can be applied to horizontal and vertical surfaces, such as roofs, terraces, foundations, and walls. The choice between cold-applied and hot-applied mastic compounds depends on specific project requirements, including the expected durability, application method feasibility, and environmental conditions.

3. Materials and Methods

3.1. *Selecting the State-of-the-Art Bibliography*

This step aims to select the bibliographic portfolio to present state-of-the-art information on applying surface-waterproofing blankets. In other words, it aims to provide a general and critical overview of the most relevant research and studies published so far to situate the work concerning the advances already achieved in the study area, identify gaps, and propose directions for future research.

The ProKnow-C (Knowledge Development Process—Constructivist) method [39], commonly used in systematic literature reviews [40–43], was used to select the bibliographic reference. Table 1 details all the steps of ProKnow-C.

Table 1. Stages of bibliographic selection by ProKnow-C.

Stage	Description	Details	Articles
1	Preliminary research	Definition of databases: Science Direct, Scopus, and Google Acadêmico. Time frame: 2015 to 2022. Keywords: (1) Waterproofing, sealing, or impermeabilization; (2) Blanket, coating, or covering; (3) Slab or floor; (4) Roof; (5) Infiltration or seepage; (6) Pathology or condition. Keyword combinations: (a) 1 + 2 + 3; (b) 1 + 2 + 4; (c)1 + 2 + 5; (d) 1 + 4 + 5; (e) 3 + 5 + 6; (f) 4 + 5 + 6; (g) 2 + 3 + 5. Database filters: title, abstract and keywords.	447
2	Extraction of articles from the databases	Type of work: full papers published in journals or conferences	250

Table 1. *Cont.*

Stage	Description	Details	Articles
4	Elimination of duplicates	Exclusion of articles published in more than one database	232
5	Alignment by title	Articles with a title aligned with the research topic	45
6	Alignment by abstract	Articles with the abstract fully aligned with the research topic	26
7	Full-text retrieval	Manuscripts made available in full by the databases	26
8	Full alignment	Articles fully aligned with the research theme	17

Table 2 presents the articles that comprise the selected bibliographic portfolio and their primary information: authors, title, journal/congress, and year of publication.

Table 2. Details of the selected bibliographic portfolio.

Reference	Title	Journal/Conference	Year
Alev et al. [44]	Air leakage of concrete floor and foundation junctions	Energy Procedia	2015
Figueiredo et al. [45]	Impermeabilização com manta asfáltica de uma laje plana de cobertura	Construindo	2017
Pinto and Aguiar [46]	Sistema de impermeabilização com manta asfáltica e manta líquida em lajes de coberturas	Projectus	2017
Sriravindrarajah and Tran [47]	Waterproofing practices in Australia for building construction	4th International Conference on Rehabilitation and Maintenance in Civil Engineering	2018
Coppola et al. [48]	Innovative carboxylic acid waterproofing admixture for self-sealing water-tight concretes	Construction and Building Materials	2018
Gonçalves et al. [49]	Environmental and economic comparison of the life cycle of waterproofing solutions for flat roofs	Journal of Building Engineering	2019
Ribeiro et al. [50]	Impermeabilização a base de polímero acrílico: estudo de caso no tratamento de infiltração da laje de cobertura	Revista Eletrônica da Estácio Recife	2019
Ksit and Plich [51]	Liquid plastic films as a solution in terms of tightness problems and roofs aesthetics improvement–review	Budownictwo i Architektura	2019
Scheidegger [20]	Impermeabilização de edificações: mantas asfálticas e argamassas poliméricas	Revista Científica Multidisciplinar Núcleo do Conhecimento	2019
Pinto et al. [52]	Infiltration: The disease of the structure	ITEGAM-JETIA	2020
Pettersson et al. [53]	On the impact of porous media microstructure on rainfall infiltration of thin homogeneous green roof growth substrates	Journal of Hydrology	2020
Van Linden and Van Den Bossche [54]	Airtightness of sealed building joints: Comparison of performance before and after artificial ageing	Building and Environment	2020

Table 2. *Cont.*

Reference	Title	Journal/Conference	Year
Neves et al. [55]	Desenvolvimento de um guia de reparos para diferentes tipos de patologia relativos à umidade nas edificações	XVI Congreso Latinoamericano de Patología de la Construcción	2021
Silva and Coelho [56]	Pathological manifestations due to infiltration in garage floors–Case study in Salvador-BA	XVI International Conference on Pathology and Constructions Rehabilitation	2021
Reis et al. [57]	Pathologies caused by infiltration in a long-stay institution for the elderly: A case study	RECIMA21-Revista Científica Multidisciplinar	2021
Heinlein et al. [58]	Pre-applied bonded waterproofing membranes: A review of the history and the state of the art in Europe and North America	Construction and Building Materials	2021
Maj and Ubysz [59]	The reasons for the loss of polyurea coatings adhesion to the concrete substrate in chemically aggressive water tanks	Engineering Failure Analysis	2022

3.2. Case Study

A case study was conducted to analyze the application of surface-waterproofing techniques and provide insights that can be useful for future research.

The study object was a high-standard residential building located in Nova Lima, in Minas Gerais State, Brazil, built on a 526.23 m^2 lot and with a constructed area of 245.77 m^2, distributed in four suites, two living rooms, two bathrooms, one kitchen, one leisure area with a swimming pool, one fireplace, one laundry room, and two garages. For confidentiality reasons, this work did not disclose the residence's address and the company name responsible for the construction.

The materials used and the execution of the waterproofing of the following environments or structures were evaluated: (i) floors and external areas; (ii) slab and water tank; (iii) bathrooms; and (iv) swimming pool, fireplace, and deck. The technical standards NBR 9574 [23], NBR 9575 [24], NBR 9952 [25], and NBR 9685 [26] were taken as references.

4. Results and Discussions

4.1. Review and Discussion of the Selected Papers

The studies presented below, in chronological order, refer to those shown in Table 2 and address different aspects related to waterproofing, including sealing methods, slab waterproofing procedures, concrete admixtures, and environmental and economic impacts of different waterproofing options.

Alev et al. [44] investigated different sealing methods to evaluate their effect on air leakage in construction joints. The research considered the air-leakage rate of several commonly used foundation and slab-on-ground joints, such as limestone, external, and separation walls constructed of lightweight blocks. Different sealing materials were applied to these constructions, and it was found that the air-leakage rate decreased between 1.3 and 86 times, depending on the method used. In addition, the researchers found that the quality of the constructions, such as the materials, method, and labor used, also significantly affected the air-leakage rate.

Figueredo et al. [45] analyzed the waterproofing procedure of a roof slab in development intending to achieve 100% water-tightness. The study emphasized the importance of care in the preparation of the substrate, the knowledge of professionals, and the end user's attention regarding the use of the environment, maintenance, and periodic inspections necessary to maintain efficient waterproofing.

The research by Pinto and Aguiar [46], in turn, aimed to clarify the importance of waterproofing and the relevance of well-executed projects. The researchers concluded that the correct execution of waterproofing, with adequate materials and qualified professionals, can avoid future inconveniences and high maintenance costs. Furthermore, it is important to point out that the cost of this measure is reasonable if executed during construction.

Sriravindrarajah and Tran [47] point out that the successful waterproofing of a building depends on considering several factors, such as product selection, membrane detailing, drainage design, substrate preparation, design, installation, quality assurance, and maintenance. The authors conducted a study in Australia intending to discuss the standards and codes, membrane systems and performance, waterproofing practices, design and installation techniques, inspection and testing, and quality assurance adopted by the waterproofing industry. They stated that implementing proper design and installation techniques and performing maintenance procedures can significantly reduce the waterproofing failure rate and repair costs. Therefore, it is essential to consider all these factors to ensure the effectiveness of a building's waterproofing system.

In addition to waterproofing methods applied to surfaces, it is also possible to improve the water-tightness of construction from an admixed concrete. In this sense, the study by Coppola et al. [48] aimed to evaluate the performance of a carboxylic acid-based mixture added to concrete. The results obtained experimentally showed that adding the additive did not negatively affect the mechanical properties of concrete. It was observed that by adding 1% of the additive per mass of cement, there was a halving in water penetration under pressure compared to concrete without the waterproofing additive at 28 days. In addition, the mixture proved capable of sealing cracks resulting from plastic shrinkage, showing that the additive is a promising solution to ensure the water-tightness of a construction.

The study by Gonçalves et al. [49] compared different waterproofing options for various slabs, considering environmental and economic impacts. The authors analyzed the risk of global warming, consumption of non-renewable primary-energy resources, and life-cycle costs. After the analysis, it was concluded that bituminous membranes were the best option for all slabs, while synthetic EPDM (ethylene–propylene–diene) membranes showed the highest values.

Ribeiro et al. [50], from another perspective, pointed out the lack of waterproofing as a significant flaw in daily construction, neglected by the lack of prioritization for the prevention of construction problems in the country. They pointed out that this practice can cause structural problems, resulting in much higher repair costs than prevention. In their case study, they used acrylic and flexible polymer-based products to waterproof the roof of a residential building, concluding that the roof infiltration was resolved and the products used ensured the desired service life of the building.

Meanwhile, Ksit and Plich [51] presented liquid-plastic films as a waterproofing solution composed of a mixture of polymers or polymers and bitumen. These authors' studies demonstrated that using these plastic films could provide a smooth and uniform surface, resulting in an excellent aesthetic result and even a "cool roof," capable of reducing the greenhouse effect in large cities.

Scheidegger [20] conducted a literature review on the two main waterproofing products: asphalt blankets and polymeric mortar. It was concluded that polymeric mortar is easier to apply, but it is not recommended for structures subjected to thermal or settlement movements, which can compromise its water-tightness. Asphalt blankets, on the other hand, are more resistant to thermal movements or settlements, but their application may be more complicated, generating possible complications later on.

In the work of Pinto et al. [52], the infiltration process and its consequences were analyzed, concluding that the lack of adequate waterproofing is generally caused by the lack of specialization of professionals and the lack of standards for project preparation. The prioritization of cost over waterproofing can result in maintenance costs of up to 15% of the total value of the work. In contrast, if proper waterproofing is not performed during construction, the repair cost can reach up to 50% of the value of the work.

Considering green roofs, Pettersson et al. [53] warn about the need for more data on their ability to retain and reduce stormwater under different climatic conditions. However, their study examined green roof water infiltration at the pore scale and evaluated the relationship between porous microstructure in thin substrate layers and water infiltration. Studies evaluating different microstructures and different heights of standing water on the porous medium surface were conducted, which demonstrated the relationship between microstructure and flow homogeneity by analyzing saturation, liquid–gas interfacial area, pore-size distribution, porosity, and infiltration depth. Thus, the authors concluded that the knowledge of these relationships allows for optimizing the performance of the green roof concerning rainfall duration time from the material used in the substrate treatment.

Van Linden and Van Den Bossche [54] conducted an experimental study in Belgium to evaluate the performance of different joint-sealing materials for the water-tightness of constructed buildings. The materials evaluated were tapes, silicone, base materials, coatings, foam sealing tapes, and polyurethane foams, which were subjected to different parameters. The study's results indicated that the choice of waterproofing materials should be reached carefully, as all tested materials performed well in waterproofing. However, it was observed that the labor skill used in applying the materials had a significantly more significant impact on the materials' performance than the life cycle. Therefore, it was concluded that skilled labor is critical to ensure the proper performance of joint-sealing materials in buildings.

Neves et al. [55] developed a guide to detect and solve the main pathological manifestations in Brazilian residences related to moisture. The guide was created from a literature review of the leading causes of pathology and included the types, manifestations, origins, risks, and a repair guide. The material also contains information about the leading suppliers of products for these repairs and a study of the processes to be carried out. Although the valuable life of a repaired construction does not compare with a well-developed, rendered compatible, and well-executed construction, the authors indicated, among other aspects, that the Brazilian market has good-quality brands that offer good services and assist their consumers.

Silva et al. [56] conducted a case study evaluating pathological manifestations due to infiltrations in a building in Salvador, Bahia, Brazil. Diagnostic techniques were used to identify the primary damage and determine the appropriate repair methodologies. It was concluded that the investigation methods were essential to identify the risk in each pathology found in the building and find the necessary repair method to prevent degradation.

Reis et al. [57], in turn, investigated the leading causes of pathology and moisture damage to enable better health and mobility conditions for the elderly at the Carlos Romeiro Asylum, located in Conselheiro Lafaiete, Minas Gerais State, Brazil. On-site inspections were conducted to identify and analyze the damage and define the appropriate repairs in each case. The authors concluded that, in addition to repairs, repairs to leaking pipes, improving the water-tightness of the reservoirs, and cleaning and unblocking surface drainage devices would be necessary. Installation of roofing and/or application of waterproofing blankets were raised as ways to treat the damage.

According to Heinlein et al. [58], the use of waterproofing membranes has grown in recent years, evidenced by the increase in the number of manufacturers, patents, and membranes on the market. However, it is essential to note that these membranes are only efficient if installed by qualified professionals due to possible adhesion problems that can arise in the application and preparation of the surfaces to be waterproofed.

The study by Maj and Ubysz [59] finally analyzed the use of polyurea coating to protect concrete surfaces, intending to reinforce waterproofing in water and other liquid tanks. To ensure water-tightness, it is essential to pay attention to the phenomena of diffusion and osmosis, and to take prudent measures in preparing the substrate, using primers, adequate coating thickness, and avoiding micropores and holes. In addition, the authors state that it is essential to schedule periodic coating inspections.

The analysis of the articles mentioned above reveals several relevant types of research on the application of surface waterproofing blankets, covering a broad theme. These studies address water infiltration in constructions, the importance of waterproofing forecasting, analysis of existing techniques, and the feasibility of new approaches. The diversity of studies on the subject highlights the importance of waterproofing in civil construction, with the growing need to improve construction quality and increase structures' life span.

4.2. *Case Study*

A case study was conducted in a residence located in Nova Lima, Minas Gerais State, Brazil. The construction in question comprises structural masonry, two floors, and a waterproofed slab roof. The project included solar and photovoltaic heating, air-conditioning infrastructure, a sound system, an electric car-charging point, a swimming pool, and a fireplace.

Next, the waterproofing methods used in this residence are presented, divided into different areas: floors and external areas; slab and water tank; bathrooms; swimming pool, fireplace, and deck. An analysis of the chosen waterproofing method and its application techniques was performed based on the NBR 9574 standard [23].

4.2.1. Floors and External Areas

The floors were waterproofed from the beginning of the project in different stages and methods, according to the infiltration risks and use of the area.

When the first slab was made, supported on the ground at the beginning of the project, it was necessary to waterproof the contact between the concrete and the soil to avoid the capillarity phenomenon. The first waterproofing was conducted using a 150-micron tarp (Figure 2), thicker than an ordinary tarp, usually with 50 to 100 microns, and, therefore, offering a better performance and useful life. After applying the tarp, EPS (expanded polystyrene) sheets were placed over it (Figure 3) to increase the distance between the ground and the concrete, and to take advantage of the characteristics of EPS, which is highly resistant to mold and moisture. The tarp method was also used under the concrete ramp for vehicle access to the upper garage (Figure 4) to avoid capillarity between soil and concrete.

Figure 2. Application of tarp on the floor before the concreting.

Figure 3. Positioning of EPS plates.

Figure 4. Application of tarp on the ramp floor before concreting.

4.2.2. Slab and Water Tank

Using waterproof slabs for the conventional roof has become more common in high-end residences. The conventional roof demands more time to build and requires frequent maintenance. On the other hand, the waterproofed slab requires an advanced waterproofing system because it is directly exposed to rain and humidity, and concrete is not a waterproof material.

The method chosen to waterproof the slab in this study was aluminized asphalt blanket. Despite the high cost of the material and labor, the material was chosen due to its high waterproofing capacity. In addition, as the slab is a non-traffic area, no additional coating (mechanical protection) was required, which avoids extra costs. The same material was used to waterproof the water tank, an area with a high risk of infiltration due to possible leaks.

A specialized team should apply asphalt blanket on the slab due to the complexity of the service, handling, and adherence of the blanket. The professional must be careful when joining blanket rolls, dealing with corners, and, especially, when making the necessary holes in the slab, such as drains for drainage and air-conditioning pipes.

The application method consists of preparing the slab by applying a layer of mortar to render the surface smooth and free of imperfections to receive the blanket. This mortar layer is also essential to ensure water drainage, foreseen at the moment of the slab's concreting, directing it to the drains and avoiding the accumulation that could damage the slab. Next, general cleaning is conducted on the surface to be waterproofed, using a broom and a powerful vacuum cleaner.

After this process, a primer composed of polycondensed asphalt cement diluted in solvents is applied (Figure 5a,b), which increases the adhesion between the asphalt-based waterproofing system and the surface. With the primer applied, modified asphalt sheets are heated to dissolve and apply them, thus, beginning the process of adherence of the asphalt mats to the surface (Figure 5c,d)—although not detailed above, applying a hot layer is usually required for the blanket application. The professional distributes the roll of the mat on the slab and uses a blowtorch to ensure the adherence of the mat and the modified asphalt to the surface (Figure 5e).

After the blanket application is completed on the entire slab (Figure 5f), it is necessary to wait for drying time and perform the water-tightness test. This test aims to verify the effectiveness of the blanket applied and correct any details, if necessary. In the water-tightness test, the slab is completely flooded with water to a certain height, usually above the finished floor level. It is then observed whether leaks or seepage occur over a certain time, usually 24 h. If any leakage is found by the inspection in the lower floor below the slab, it will be evident by moisture stains under the slab, and the responsible team must correct this failure and perform the test again, ensuring no more leaks. It is essential to mention that no standards exist for performing this test. It is conducted based on the experience of the professionals when dealing with these waterproofing materials.

Figure 5. (**a**) Primer application on the slab; (**b**) primer application in water tank room; (**c**) use of a blowtorch to melt the modified asphalt; (**d**) application of modified asphalt to receive the blanket; (**e**) positioning of the blanket roll on the slab; and (**f**) aluminized asphalt blanket applied on the slab.

4.2.3. Bathrooms

As this is an environment with much contact with humidity, it is also necessary to worry about infiltration in this area. In the building under study, the method used for waterproofing the bathrooms was the application of polymeric mortar. This material was chosen mainly due to its easy application, not requiring specialized labor, and low cost. The product used, the waterproofing material "tec plus top" by Quartzolit, is intended for use in humid areas such as bathrooms, kitchens, service areas, walls, baseboards, and planters.

Applying the waterproofing material followed the manufacturer's recommendations and the NBR 9574 standard [21]. First, the liquid component was mixed with the powder component using a metal rod attached to a drill. Then, using a brush, the first layer of the mixture was applied on the floors and walls (up to a height of approximately 1.80 m) of the bathrooms (Figure 6) and then reapplied after drying in the opposite direction to the application of the first coat. This way, the bathrooms were released for the finishing stage, covering floors and walls with ceramic tiles and protecting them from humidity.

Figure 6. Bathroom after applying waterproofing.

4.2.4. Pool, Fireplace, and Deck

The pool, fireplace, and deck waterproofing were conducted using a polyethylene asphalt blanket. This type of blanket is chosen for these areas due to its greater flexibility, resistance, and durability. Unlike other surfaces, the pool, the fireplace, and the deck are subject to a much higher volume and time of exposure to humidity, besides being places with foot traffic, which require mechanical protection on the asphalt membrane.

The waterproofing process of the deck, pool, and fireplace surfaces started with verifying the drainage system and testing the water flow toward the area's drains. After this verification, these areas were cleaned using brooms and vacuum cleaners. Then, the primer was applied to the pool, fireplace, and deck surfaces (Figure 7a), preparing them to receive the asphalt. Then, a layer of modified asphalt was applied and melted with a blowtorch to adhere the mat to the surface (Figure 7b)—while it was not explicitly mentioned earlier, it is typically necessary to apply a layer of hot material when using the blanket application method.

(**a**) (**b**)

Figure 7. (**a**) Pool with primer applied; and (**b**) use of modified asphalt on the pool surface.

Then, applying the polyethylene asphalt blanket on the surfaces was started, using a blowtorch to melt the modified asphalt and ensure the adherence of the blanket. After finishing the application of the blanket on the pool surfaces (Figure 8), deck, and fireplace, the water-tightness test was performed, as was conducted on the slab, to verify the waterproofing guarantee. Next, the pool, fireplace, and deck were filled with water up to the limit of the edges (Figure 9).

Figure 8. Swimming pool with polyethylene asphalt blanket applied.

(**a**) (**b**)

Figure 9. (**a**) Pool, fireplace, and deck in the water-tightness test; and (**b**) pool in the water-tightness test.

As proven to be water-tight after the test, the next step would be plastering the pool and fireplace and the counter flooring of the deck to apply the lining later. For this type of blanket used, before receiving the lining, chicken wire was applied to the entire surface of the pool and fireplace (Figure 10a) using mortar to facilitate the adherence of the plaster to the blanket and prevent future detachment of the ceramic linings. After the preparation with the chicken wire and the mortar on the polyurethane asphalt blanket, the pool and the fireplace were plastered, and the counter-flooring was added to the deck. Then, these surfaces were covered with porcelain tiles or ceramic pieces (Figure 10b).

(**a**) (**b**)

Figure 10. (**a**) Application of chicken wire over pool blanket; and (**b**) pool, fireplace, and deck.

5. Conclusions

This paper aims to present the surface-waterproofing methods available on the Brazilian construction market and their main applications, focusing on asphalt blankets. The study analyzed how infiltration occurs and its pathologies, thus, showing the importance of surface waterproofing for construction. Specifically, the following conclusions were reached:

(i) The main mechanisms of water infiltration into surfaces in construction are percolation, capillarity, condensation, leaks, and drainage-system failures. These mechanisms can cause damage to structures, such as humidity, stains, mold, deterioration, and corrosion, compromising the integrity and durability of buildings;
(ii) Surface waterproofing techniques are widely used in civil construction to avoid problems caused by water infiltration, such as corrosion, humidity, and detachment of coatings. Among the available techniques are the application of asphalt blankets, polymers, polymeric mortars, special paints, and cold or hot mastic compounds. Waterproofing with asphalt blankets is one of the most used due to its high efficiency and durability;
(iii) State-of-the-art information has shown, among several aspects, that (a) different sealing methods can significantly reduce the rate of air leakage in construction joints, depending on the sealing material used; (b) the correct execution of waterproofing, with appropriate materials and qualified professionals, can avoid future inconveniences and high maintenance costs; (c) the implementation of proper design, installation, inspection, and maintenance techniques can significantly reduce waterproofing failure rates and repair costs; (d) concrete admixtures can be used to improve the water-tightness of construction without compromising the mechanical properties of the material; and (e) the proper choice of waterproofing materials, along with the skill of labor in the application, is critical to ensure the proper performance of these materials in buildings;
(iv) The case study described the methods and materials used to waterproof different areas of the residence. Specifically, it was shown that (a) for waterproofing the floors and external areas, a 150-micron canvas, thicker than common canvas, was used, along with EPS sheets to avoid the capillarity phenomenon; (b) the slab and the water tank were waterproofed with the use of aluminized asphalt blanket due to its high waterproofing capacity; (c) in the bathrooms, a polymeric mortar was used as the waterproofing method due to its ease of application and low cost, with the subsequent laying of ceramic tiles; (d) the waterproofing of the pool, fireplace, and deck was conducted with the use of polyethylene asphalt blanket, due to its flexibility, resistance, and durability, and later these surfaces were coated with porcelain tiles or ceramic tiles. Thus, the importance of choosing the appropriate method for each environment was highlighted, aiming to prevent damage caused by water infiltration.

Future Research Directions

This work presented the primary forms of waterproofing available on the market, focusing on asphalt blankets. However, it was limited to a case study in a single residence. The following topics are suggested for future research:

(i) Investigate the importance of labor skills and training in adequately applying sealing materials and their relationship to the overall performance of building waterproofing;
(ii) Conduct experimental studies to evaluate the performance of different admixtures that can be added to concrete to improve its waterproofing, considering mechanical properties, resistance to water penetration, and crack-sealing ability;
(iii) Investigate the environmental, economic, and performance impacts of different waterproofing options available on the market, considering factors such as global warming, consumption of energy resources, life-cycle cost, and durability.

Author Contributions: Conceptualization, L.C.d.F.G.; methodology, L.C.d.F.G., H.C.G. and E.D.R.; formal analysis, E.D.R.; investigation, L.C.d.F.G.; resources, E.D.R.; data curation, H.C.G. and E.D.R.; writing—original draft preparation, L.C.d.F.G. and E.D.R.; writing—review and editing, H.C.G.; visualization, E.D.R.; supervision, E.D.R.; project administration, E.D.R.; and funding acquisition, E.D.R. All authors have read and agreed to the published version of the manuscript.

Funding: This research received no external funding.

Institutional Review Board Statement: Not applicable.

Informed Consent Statement: Not applicable.

Data Availability Statement: All the data in the analyses of this study have been listed in the paper.

Conflicts of Interest: The authors declare no conflict of interest.

References

1. Belous, D.A.C.; Souza, L.C. Estudo de Caso de Manifestação Patológica do Tipo Fissuras e Trincas com a Impermeabilização como Forma de Tratamento na laje de Uma Escola, na Região de Joinville/SC. Bachelor's Thesis, Sociedade Educacional de Santa Catarina, Joinville, Brazil, 2022.
2. Yazigi, W.A. *Técnica de Edificar. Revisada e Atualizada*, 10th ed.; Pini, SIDUSCON-SP: São Paulo, Brazil, 2009; p. 517.
3. Lopes. Como o Bolor e a Umidade em sua Casa Afetam seu Corpo e Podem Prejudicar sua Saúde. Redação Lopes. 2018. Available online: https://www.lopes.com.br/blog/na-minha-vida/estilo-de-vida/como-o-bolor-e-a-umidade-em-sua-casa-afetam-seu-corpo-e-podem-prejudicar-sua-saude (accessed on 6 May 2023).
4. Pereira, F.S. Impermeabilização de Lajes de Cobertura Com Manta Asfáltica. *Rev. Científica Sem. Acadêmica* **2018**, *1*, 000148.
5. Apolinário, M.S. *Danos Causados por Falhas na Impermeabilização da Infraestrutura de Edificações Térreas Residenciais Privativas Unifamiliares com Área até Oitenta Metros Quadrados*; Revista Especialize Online: São Paulo, Brazil, 2013; pp. 1–14.
6. Costa, E.U.; Pinto, J.R.; Miranda, M.L. Procedimento Visando a Diminuição de Manifestações Patológicas no Sistema de Impermeabilização em Lajes de Cobertura Expostas com Mantas Asfálticas. Bachelor's Thesis, IDD Institute, São Paulo, Brazil, 2012.
7. Machado, K.M.; Alencar, E.A.B. Levantamento de Patologías Causadas por Umidade nas Edificações da Cidade de Manaus—AM. Bachelor's Thesis, Centro Universitário do Norte, Manaus, Brazil, 2019.
8. Rigui, G.V. Estudos dos Sistemas de Impermeabilização: Patologias, Prevenções e Correções—Análise de Casos. Master's Thesis, Federal University of Santa Maria, Santa Maria, Brazil, 2009.
9. Valera, G.T.V.; Kawakami, T.; Payot, B.D. Mixing, fluid infiltration, leaching, and deformation (MILD) processes on the slab-mantle wedge interface at high T and P conditions: Records from the Dalrymple Amphibolite, Philippines. *Chem. Geol.* **2022**, *604*, 120941. [CrossRef]
10. Queruz, F. Contribuição Para Identificação dos Principais Agentes e Mecanismos de Degradação em Edificações em Vila Belga. Master's Thesis, Federal University of Santa Maria, Santa Maria, Brazil, 2007.
11. Campos, L.S.; Soares, T.A. Processo de Escolha de um Sistema de Impermeabilização Para a laje de um Estacionamento. Bachelor's Thesis, Technological Institute of Caratinga, Caratinga, Brazil, 2015.
12. Thomaz, E. *Trincas em Edifícios: Causas, Prevenção e Recuperação*, 2nd ed.; Oficina de Textos: São Paulo, Brazil, 2020; pp. 1–34.
13. Grachev, A. Waterproofing Methods Comparison in Russia and Finland. Bachelor's Thesis, Hämeenlinna University Centre, Keinusaari, Finland, 2021.
14. Santana, B.V.; Aleixo, I.V. *Avaliação da Absorção por Capilaridade de Argamassas para Revestimento com Diferentes Aditivos Impermeabilizantes*; Final Report of Undergraduate Scientific Research; Relatório Final de Pesquisa de Iniciação Científica, Faculdade de Tecnologia e Ciências Sociais Aplicadas: Brasília, Brazil, 2017.
15. Peixoto, V.F.; Torres, M.I.; Guimarães, A.S. *Humidade Ascensional*, 1st ed.; FEUP: Porto, Portugal, 2008.
16. Siqueira, V. Impermeabilização em Obras de Construção Civil. Bachelor's Thesis, University of Southern Santa Catarina, Palhoça, Brazil, 2018.
17. Martins, F.S. Estudo das Manifestações Patológicas Pós-Ocupação Relacionadas à Impermeabilização da Cobertura do Centro Pedagógico Paulo Freire—Cidade Universitária Dom Delgado—UFMA. Bachelor's Thesis, Federal University of Maranhão, São Luís, Brazil, 2018.
18. Salomão, M.F. Estudo da Umidade Ascendente em Painéis de Alvenaria de Blocos Cerâmicos. Master's Thesis, Federal University of Uberlândia, Uberlândia, Brazil, 2012.
19. Silva, F.L.; Oliveira, M.P.S.L. Manifestações patológicas causadas pela ausência ou falha de impermeabilização. *Rev. Científica Multidiscip. Núcleo Conhecimento* **2018**, *1*, 76–95.
20. Scheidegger, G.M. Impermeabilização de edificações: Mantas asfálticas e argamassas poliméricas. *Rev. Científica Multidiscip. Núcleo Conhecimento* **2019**, *4*, 126–151.
21. Lauch, K.S.; Desmettre, C.; Charron, J.P. Self-healing of concrete containing different admixtures under laboratory and long-term real outdoor expositions based on water permeability test. *Constr. Build. Mater.* **2022**, *324*, 126700. [CrossRef]

22. Aecweb. Como Impermeabilizar com Mantas Asfálticas? Assista ao Passo a Passo. 2019. Available online: https://www.aecweb.com.br/revista/materias/como-impermeabilizar-com-mantas-asfalticas-assista-ao-passo-a-passo/18482 (accessed on 27 April 2023).
23. Associação Brasileira De Normas Técnicas. *NBR 9574: Execução de Impermeabilização*; ABNT: Rio de Janeiro, Brazil, 2010.
24. Associação Brasileira De Normas Técnicas. *NBR 9575: Impermeabilização—Seleção e Projeto*; ABNT: Rio de Janeiro, Brazil, 2010.
25. Associação Brasileira De Normas Técnicas. *NBR 9952: Manta Asfáltica para Impermeabilização*; ABNT: Rio de Janeiro, Brazil, 2014.
26. Associação Brasileira De Normas Técnicas. *NBR 9685—Mantas Asfálticas com Armadura para Impermeabilização—Especificação*; ABNT: Rio de Janeiro, Brazil, 2011.
27. Aecweb. Conheça os Principais Tipos de Selantes Para Impermeabilização. 2020. Available online: https://www.aecweb.com.br/revista/materias/conheca-os-principais-tipos-de-selantes-para-impermeabilizacao/20264 (accessed on 28 April 2023).
28. Fibersals. Tudo Sobre Impermeabilização com Poliuretano. 2021. Available online: https://fibersals.com.br/blog/impermeabilizacao-com-poliuretano/ (accessed on 26 April 2023).
29. Lee, H.L. *The Handbook of Dielectric Analysis and Cure Monitoring*; Lambient Technology LLC: Boston, MA, USA, 2014; Volume 156.
30. Woodson, R.D. *Concrete Structures: Protection, Repair and Rehabilitation*; Butterworth-Heinemann: Oxford, UK, 2009.
31. Fibersals. Tudo Sobre Impermeabilização com Argamassa Polimérica. 2020. Available online: https://fibersals.com.br/blog/impermeabilizacao-com-argamassa-polimerica/ (accessed on 26 April 2023).
32. Moreira, A.A.A.; Vermelho, L.C.; Zani, M.C. Estudo da Argamassa Polimérica de Assentamento de Blocos e Tijolos Segundo Aspectos Técnicos, Econômicos, Mercadológicos e de Clima Organizacional. *Espacios* **2017**, *38*, 14–29.
33. Fibersals. Impermeabilização com Resina Epóxi. 2020. Available online: https://fibersals.com.br/blog/impermeabilizacao-com-resina-epoxi/ (accessed on 26 April 2023).
34. Superfície, E. Tinta Epóxi, o Que é e Qual Sua Aplicação? 2020. Available online: https://engenhariadesuperficie.com.br/piso-epoxi/tinta-epoxi-o-que-e-e-qual-sua-aplicacao/ (accessed on 28 April 2023).
35. Dutton, E.M. *Asphalt Coatings and Mastics for Roofing and Waterproofing*; Marcel Dekker: New York, NY, USA, 1997.
36. Dickson, W.J. Cold applied asphalt coatings. *Ind. Eng. Chem.* **1966**, *58*, 28–32. [CrossRef]
37. Biggins, T. *Waterproofing Underground Concrete Structures*; University of Florida: Gainesville, FL, USA, 1990.
38. Brown, A. Cold-Applied Roofing Systems and Waterproofing and Dampproofing. In *Roofing Workbook and Tests*; ERIC: Washington, DC, USA, 1982.
39. Ensslin, L.; Ensslin, S.R.; Lacerda, R.D.O.; Tasca, J.E. ProKnow-C, knowledge development process-constructivist. Processo técnico com patente de registro pendente junto ao INPI. *Brasil* **2010**, *10*, 2015.
40. Resende, H.F.; Reis, E.D.; Fernandes, F.M.; Rodrigues, L.A.; Ângelo, F.A. Uso de resíduos de construção e demolição como agregado reciclado no concreto: Uma breve revisão de literatura. *Rev. Principia-Divulg. Científica Tecnológica IFPB* **2022**. *Early View*. [CrossRef]
41. Reis, E.D.; Resende, H.F.; Ludvig, P.; De Azevedo, R.C.; Poggiali, F.S.J.; Bezerra, A.C.D.S. Bonding of Steel Bars in Concrete with the Addition of Carbon Nanotubes: A Systematic Review of the Literature. *Buildings* **2022**, *12*, 1626. [CrossRef]
42. Gomes, H.C.; Reis, E.D.; Azevedo, R.C.D.; Rodrigues, C.D.S.; Poggiali, F.S.J. Carbonation of Aggregates from Construction and Demolition Waste Applied to Concrete: A Review. *Buildings* **2023**, *13*, 1097. [CrossRef]
43. Reis, E.D.; Borges, L.A.; Camargos, J.S.F.; Gatuingt, F.; Poggiali, F.S.J.; Bezerra, A.C.S. A systematic review on the engineering properties of concrete with carbon nanotubes. *J. Braz. Soc. Mech. Sci. Eng.* **2023**, *45*, 205. [CrossRef]
44. Alev, Ü.; Antsov, P.; Kalamees, T. Air leakage of concrete floor and foundation junctions. *Energy Procedia* **2015**, *78*, 2046–2051. [CrossRef]
45. Figueredo, V.S.; Rinaldi, V.S.; Abi-Ackel, E. Impermeabilização com manta asfáltica de uma laje plana de cobertura. *Construindo* **2017**, *9*, 62–72.
46. Pinto, J.B.; Aguiar, L.E.A. Sistema de impermeabilização com manta asfáltica e manta líquida em lajes de coberturas. *Projectus* **2017**, *1*, 141–151. [CrossRef]
47. Sriravindrarajah, R.; Tran, E. Waterproofing practices in Australia for building construction. In Proceedings of the 4th International Conference on Rehabilitation and Maintenance in Civil Engineering, Solo Baru, Indonesia, 11–12 July 2018.
48. Coppola, L.; Coffetti, D.; Crotti, E. Innovative carboxylic acid waterproofing admixture for self-sealing watertight concretes. *Constr. Build. Mater.* **2018**, *171*, 817–824. [CrossRef]
49. Gonçalves, M.; Silvestre, J.D.; de Brito, J.; Gomes, R. Environmental and economic comparison of the life cycle of waterproofing solutions for flat roofs. *J. Build. Eng.* **2019**, *24*, 100710. [CrossRef]
50. Ribeiro, A.C.R.G.; Azevedo, F.G.; Ferreira, F.G.D.; De Azevedo Filho, R.D. Impermeabilização a base de polímero acrílico: Estudo de caso no tratamento de infiltração da laje de cobertura. *Rev. Eletrônica Estácio Recife* **2019**, *5*, 1–13.
51. Ksit, B.; Plich, B. Liquid plastic films as a solution in terms of tightness problems and roofs aesthetics improvement–review. *Bud. Archit.* **2019**, *18*, 51–58. [CrossRef]
52. Pinto, D.E.B.; Firmino, B.C.C.; Dos Santos, M.C.; Carvalho, W.B.D.R.C.; De Oliveira, L.C.S.; Farias, R.E. Infiltration: The disease of the structure. *ITEGAM-JETIA* **2020**, *6*, 53–57. [CrossRef]
53. Pettersson, K.; Maggiolo, D.; Sasic, S.; Johansson, P.; Sasic-Kalagasidis, A. On the impact of porous media microstructure on rainfall infiltration of thin homogeneous green roof growth substrates. *J. Hydrol.* **2020**, *582*, 124286. [CrossRef]

54. Van Linden, S.; Van Den Bossche, N. Airtightness of sealed building joints: Comparison of performance before and after artificial ageing. *Build. Environ.* **2020**, *180*, 107010. [CrossRef]
55. Neves, J.; Garcia, D.; Santana, V.; Lopes, C.; Mendes, J. Desenvolvimento de um guia de reparos para diferentes tipos de patologia relativos à umidade nas edificações. In Proceedings of the XVI Congreso Latinoamericano de Patología de la Construcción, Online, 19–21 October 2021.
56. Silva, F.G.S.; Coelho, V.A. Pathological manifestations due to infiltration in garage floors–Case study in Salvador-BA. In Proceedings of the XVI International Conference on Pathology and Constructions Rehabilitation, Online, 13–15 July 2020.
57. Reis, E.D.; Leite, T.A.; Magalhães, J.C.; Barbosa, L.D.A.; Leandro, W.S.; Da Costa, A.C.; Sima, B.R.; Lisbôa, E.L.M.; de Oliveira, W.J.; de Souza Ferreira, M.B.; et al. Pathologies caused by infiltration in a long-stay institution for the elderly: A case study. *RECIMA21-Rev. Científica Multidiscip.* **2021**, *2*, e27532.
58. Heinlein, U.; Thienel, K.; Freimann, T. Pre-applied bonded waterproofing membranes: A review of the history and state-of-the-art in Europe and North America. *Constr. Build. Mater.* **2021**, *296*, 123751. [CrossRef]
59. Maj, M.; Ubysz, A. The reasons for the loss of polyurea coatings adhesion to the concrete substrate in chemically aggressive water tanks. *Eng. Fail. Anal.* **2022**, *142*, 106774. [CrossRef]

eng

MDPI

Article

Improved Structural Health Monitoring Using Mode Shapes: An Enhanced Framework for Damage Detection in 2D and 3D Structures

Marzieh Zamani Kouhpangi [1], Shaghayegh Yaghoubi [1] and Ahmadreza Torabipour [2,*]

[1] Department of Marine Engineering, Amirkabir University of Technology (Tehran Polytechnic), Tehran 15916-34311, Iran
[2] Department of Civil and Water Engineering, University of Laval, Quebec, QC G1V 0A6, Canada
* Correspondence: ahmadreza.torabipour.1@ulaval.ca

Abstract: Structural health monitoring (SHM) is crucial for ensuring the safety and performance of offshore platforms. SHM uses advanced sensor systems to detect and respond to negative changes in structures, improving their reliability and extending their life cycle. Model updating methods are also useful for sensitivity analysis. It is feasible to discuss and introduce established techniques for detecting damage in structures by utilizing their mode shapes. In this research, by considering reducing the stiffness of elements in the damage scenarios, we conducted simulations of the models in MATLAB, including both two-dimensional and three-dimensional structures, to update the method suggested by Wang. Wang's method was improved to produce a sensitivity equation for the damaged structures. The sensitivity equation solution using a subset of mode shapes data was found to evaluate structural parameter changes. Comparing the updated results for Wang's method and the suggested method in the two- and three-dimensional frames showed a noticeable modification in damage recognition. Furthermore, the suggested method can update a model containing measurement errors. Since Wang's damage detection formulation is suitable only for 2D structures, this modified framework provides a more accurate decision-making tool for damage detection of structures, regardless of whether a 2D or 3D formulation is used.

Keywords: structural health monitoring; damage detection; model updating; modal parameters; offshore platform

Citation: Zamani Kouhpangi, M.; Yaghoubi, S.; Torabipour, A. Improved Structural Health Monitoring Using Mode Shapes: An Enhanced Framework for Damage Detection in 2D and 3D Structures. *Eng* **2023**, *4*, 1742–1760. https://doi.org/10.3390/eng4020099

Academic Editors: Antonio Gil Bravo and Alessio Cascardi

Received: 11 April 2023
Revised: 12 June 2023
Accepted: 15 June 2023
Published: 19 June 2023

1. Introduction

Due to distinct reasons, such as earthquakes, waves crashing, fatigue, and erosion, offshore platforms are faced with some changes in their structural properties, which can cause structural failure, people's deaths, and economic damage [1]. Thus, structural safety and performance require early detection, monitoring, and analysis of the damaged structure. Different methods have been developed to identify the damage location and severity using non-destructive tests and detection algorithm responses [2]. According to the damaging effect, the damage detection algorithms can lead to a detectable change in a structure's response [3].

There are generally two methods used to update the trends in structural health monitoring: applying forces or stresses and measuring the structural response. Damage detection using dynamic data has more advantages in comparison with static techniques, which can be more sensitive to changes in structural parameters and structural damage [4]. The dynamic characteristics of a structure, such as its mass, stiffness, and damping, can vary because of damage. As a result, damage detection techniques often rely on changes in these characteristics, as measured by modal data, such as natural frequencies, mode shapes, and damping ratios. In practice, natural frequencies can be measured more accurately than mode shapes and are, therefore, often used to update structural models [5]. However, changes in natural frequencies alone do not provide detailed information about the location or extent of damage. Therefore, utilizing mode shape data is considered to be a more valuable approach for model updating, despite the mathematical and practical challenges that it poses [6]. Modal data covering the natural frequencies and the mode shapes have been considered by researchers in recent decades.

Regarding the response, updating techniques can be classified into methods based on the mode and response. Updating a technical model based on the mode involves obtaining modal properties, which are given by indirectly measured frequency response function (FRF) data. Structural parameters are updating techniques that are applied using direct methods or iterative approaches [7,8]. A straightforward method for calibrating structural models involves adjusting the mass and stiffness parameters to match the numerical data to the measured modal data [9]. However, these changes may not be easily interpreted in terms of physical changes to the structure. An iterative updating approach that uses experimental data or frequency response function (FRF) data and techniques for updating finite element (FE) models to account for sensitivity may be required to better understand the effects of damage on a structure [10–13].

Detecting damage using experimental modal data [14] can be difficult due to noise and limited nodal points. There are different methods to address this issue, such as using the imperialist competitive optimization algorithm with damage functions and incorporating variable widths and weights. Another approach [15] involves combining incomplete modal data with sparse regularization techniques to improve the accuracy and robustness of damage identification. This method does not require sensitivity analysis or complete modal data and uses a new goal function. Structural damage is detected by using mode shape changes or the mode shape itself. Wang et al. [16] undertook a sensitivity analysis of the structural parameters. They also achieved dynamic modification directly and indirectly through modal data analysis. Kim and Stubbs [17] presented an algorithm for identifying structural damage in jacket offshore structures using few-mode shapes. Ren et al. [18] used mode shape data to locate and quantify structural damage. They proposed using sensitivity equations that involve all mode shapes of the undamaged and damaged structures. However, the need for complete mode shape data can be a limitation in structures with a high degree of freedom, making it difficult to apply the method in real-world scenarios. Lee et al. [19] used the cross-modal method as a reliable proposed method to recognize the damage in jacket offshore platforms when incomplete modal data is available.

Shahsavari et al. [20] undertook a mode shape analysis with the likelihood ratio (LR) test to specify damage. Wang et al. [21] introduced the kriging model as a secondary model to optimize by frequency response function instead of focusing just on the one FE model repeatedly. Umar et al. [22] remarked on a new response surface methodology (RSM), which used both the natural frequencies and mode shapes to upgrade the damage recognition.

Hariri-Ardebili et al. [23] coupled a polynomial chaos expansion metamodel and modal analysis to find the most vulnerable areas of arch dams in an efficient way. Dahak et al. [24] declared an approach based on normalized natural frequencies using experimental data in order to recognize damage in a cantilever beam. Khatir et al. [25] achieved a new method for the determination of cracks, which simultaneously investigates changes from measured frequencies experimentally and by using particle swarm optimization.

Model updating methods use the modal parameters elicited from vibration data in the structures, which were measured before and after damage. In order to distinguish the damage intensity in structures, Park et al. [26] presented a damage intensity function to clarify the relationship between damage intensity and multiple-per-revolution (MPR) changes for the baseline model. Schacht et al. [27] produced a modified method for damage evaluation in which the modified indexes were found using a given function combined with a genetic algorithm (GA). The identification of damage in truss and beam-like structures which was achieved using FRF data combined with optimization techniques [28], presenting the fastest solution compared with other approaches that used a GA and bat algorithm (BA) to find the location and severity of the damage.

Pérez and López [29] developed a fully functional SHM system based on spectral domain indices. They characterized a comparison between experimental data and numerical data, which led to a careful assessment of proposed spectral correlation indices (SCIs) for structural evaluation and damage identification. Khatir et al. [30] indicated one specific approach that has two steps in order to assess damage in beam-like structures using two dimensions, namely, isogeometric analysis (IGA) and finite element method besides optimization techniques. Abasi et al. [31] presented a damage identification method using the nearest neighbor search method to evaluate structures. Consequently, the nearest neighbor search method was rather strong for noisy data in comparison to the artificial neural network.

Sensitivity-based model updating can be done based on mode shape changes [32]. The accurate assessment of mode shape changes by considering unknown structural parameters is a crucial issue for updating a model. A sensitivity equation can be linear or of a higher degree. Linear sensitivity needs measurement in all freedom degrees, which is a challenging issue for real structures. Furthermore, the first-order Taylor series expansion enables the calculation of a linear formula for updating the model.

Iannelli et al. [33] researched using deep learning to identify damage in large space antennas. They explored the use of a long short-term memory neural network for the classification of various types of damage, and the results indicate successful detection. Wang ad Wu [34] suggested an enhanced version of explicit connectivity Bayesian networks for analyzing system reliability. Their approach considers multiple failure modes and the probabilistic relationships between different components.

To understand the relationship between changes in the mode shapes and structural parameters, a sensitivity equation was employed to revise the model. This equation utilizes eigenvector derivatives, and various techniques were proposed by researchers to calculate these derivatives. One method is the modal method, which is used to estimate the derivative of mode shapes based on design variables. Another method, which is called the Nelson method, is a reliable way to evaluate eigenvalue derivatives. Additionally, the Wang method, which improves upon the modal method by incorporating a quasi-static term into the eigenvector derivative equation, was shown to accelerate the convergence of the modal method [35–37]. Damage detection by using both modal and Wang methods leads to similar results, which were confirmed using the equation. The superiority of the Wang method is because of its better convergence [38].

In this article, a brief consideration of both standard damage recognition methods (modal and Wang's methods) is presented. An improved model updating method based on natural frequencies combination was suggested. To enhance Wang's method [37] for accurate damage recognition in both two-dimensional and three-dimensional frames, an updated model incorporating measurement errors and precise damage identification was employed. This update involved utilizing the natural frequency of the damaged structure to establish a more precise correlation between changes in mode shapes and structural parameters. Wang's damage detection formula is limited to 2D structures, but this adapted framework provides a precise tool for identifying damage in structures, irrespective of their 2D or 3D formulation.

According to our detailed review in Section 1, the need for proposing a modified damage detection framework based on Wang's model formed the backbone of the current study. Since Wang's damage detection formulation is suitable only for 2D structures, this study was centered on extending the applicability of Wang's model to 3D structures. Furthermore, one may note that it can be appealing to increase the accuracy of Wang's model by adding an error metric during updating the model. To meet these two major contributions of this study, some ingredients are required, which are comprehensively described in the following.

In Section 2, the background theory of Wang's model is introduced, along with the basis of the updating procedure. Next, assumptions for numerical simulation and the configuration of strain gauges are described briefly in Sections 3 and 4, respectively. The application of the proposed formulation for 2D and 3D structures is demonstrated in Section 5. Finally, the major contributions and outputs of this study, in conjunction with some recommendations for future works, are described in Section 6. This research developed an improved SHM method that extends the capability of existing techniques, enabling accurate damage detection in both 2D and 3D structures, and accommodating model updates and measurement errors. This advancement can significantly enhance the reliability and effectiveness of structural health monitoring systems for offshore platforms and other structures.

2. Description of Theory and Analysis Method

2.1. Theory

This section provides a brief overview of the methods presented in Wang's model [37] and the modal method. It also mentions the characteristic equation of eigenvalues for an undamaged structure (without damping), which is introduced by Equation (1).

$$K\varphi_i = (\lambda_i M)\varphi_i \; for \; i = 1, \ldots, n \tag{1}$$

In the equation provided, K_{nxn} and M_{nxn} represent the structural stiffness matrix and mass matrix, respectively. The variable φ_i is the eigenvector and λ_i is the ith eigenvalue. The square of the ith natural frequency is represented by ω_i and the number of freedom degrees is represented by n. By taking the derivative of both sides of the equation, the equation can be rewritten.

$$(K' - \lambda_i M')\varphi_i + (K - \lambda_i M)\varphi_i' - \lambda_i' M\varphi_i = 0 \tag{2}$$

The derivative of an eigenvalue can be found by multiplying Equation (2) through by φ_i^T, as shown in Equation (3).

$$\lambda_i' = \varphi_i^T (K' - \lambda_i M')\varphi_i \tag{3}$$

where λ_i', K', and M' give the derivatives of eigenvalues, stiffness matrix, and mass matrix, respectively. Based on the modal method, the derivative of the ith eigenvector is expressed as a series expansion of all eigenvectors:

$$\varphi_i' = \sum_{j=1}^{n} c_{ij}\varphi_j \tag{4}$$

Moreover, the factor c_{ij} is defined as follows:

$$c_{ij} = \left\{ \begin{array}{ll} \frac{1}{\lambda_i - \lambda_j}\varphi'_j(K' - \lambda_i M')\varphi_i & if \; i \neq j \\ -0.5\varphi'_j M\varphi_j & if \; i = j \end{array} \right\} \tag{5}$$

The modal method or Fax formula is a common way to appraise the derivative of a mode shape. Wang [37] presented the modified modal method. Using Wang's method leads to expressing the derivative of the ith mode shape as a linear combination of mode shapes, as indicated by Equation (6).

$$\varphi_i' = \varphi_{i0}' + \sum_{j=1}^{n} a_{ij}\varphi_j \tag{6}$$

$$\varphi_{i0}' = -K^{-1}(K' - \lambda_i' M - \lambda_i M')\varphi_i \tag{7}$$

In Equation (6), the combinatory coefficients demonstrate the share of the jth mode shape, as expressed by Equation (8).

$$a_{ij} = \left\{ \begin{array}{ll} \frac{\lambda_i}{\lambda_j}\frac{1}{\lambda_i-\lambda_j}\varphi_j{}^T(K' - \lambda_i M')\varphi_i & if\ i \neq j \\ -0.5\varphi_j{}^T M'\varphi_j & if\ i = j \end{array} \right\} \quad (8)$$

Wang's method [37] is employed for determining the derivative of the modal shapes. In this method, the derivatives of the eigenvalues must be taken into consideration. Equation (3) serves as the sole means of evaluating the derivatives of natural frequencies in response to design variables. Accordingly, the initial evaluation of alterations in modal shapes is presented by Equation (9), wherein p constitutes a diagonal matrix of stiffness parameters.

$$\delta\varphi_i = \frac{\partial\varphi_i}{\partial p}\delta p = -K^{-1}\left(\frac{\partial k}{\partial p}\delta p - \left(\frac{\partial\lambda_i}{\partial p}\delta p\right)M - \lambda_i\frac{\partial M}{\partial p}\delta p\right)\varphi_i + \sum_{j=1}^{n}\varphi_j\varphi_j^T\frac{\lambda_i}{\lambda_j}\frac{1}{\lambda_i-\lambda_j}\left(\frac{\partial K}{\partial p}\delta p - \lambda_i\frac{\partial M}{\partial p}\delta p\right)\varphi_i \quad (9)$$

The matrices of mass and stiffness of linear function are unknown matrices from structural parameters:

$$\frac{\partial K}{\partial p}\delta p = \delta K \quad (10)$$

$$\frac{\partial M}{\partial p}\delta p = \delta M \quad (11)$$

As a result, eigenvalue changes can be found as follows:

$$\frac{\partial\lambda_i}{\partial p}\delta p = \delta\lambda_i \quad (12)$$

A method was developed for determining structural damage by analyzing changes in the natural frequencies. $\delta\lambda_i$ can be calculated using the following equation:

$$\delta\lambda_i = \lambda_{id} - \lambda_{iI} \quad (13)$$

Equation (13) can be used to account the accuracy and logical value for $\delta\lambda_i$ in comparison to Equation (3). Consequently, the mode shape changes can be revealed by replacing Equation (13) in Equation (9), and thus, rewriting it as Equation (14).

$$\delta\varphi_i = \delta\lambda_i K^{-1}M\varphi_i - K^{-1}(\delta K - \lambda_i\delta M)\varphi_i + \sum_{j=1}^{n}\varphi_j\varphi_j^T\frac{\lambda_i}{\lambda_j}\frac{1}{\lambda_i-\lambda_j}(\delta K - \lambda_i\delta M)\varphi_i \quad (14)$$

The first term of Equation (14) is calculated by utilizing the damaged structure, which measures the natural frequencies, mass, stiffness, and mode shapes. Therefore, it can be utilized instead of the parametric formula of this term for updating the model.

To define the term $\delta\varphi_{ia}$, Equation (9) is rewritten as follows:

$$\delta\overline{\varphi_i} = \delta\varphi_i - \delta\varphi_{ia} = -K^{-1}(\delta K - \lambda_i\delta M)\varphi_i + \sum_{j=1}^{n}\varphi_j\varphi_j^T\frac{\lambda_i}{\lambda_j}\frac{1}{\lambda_i-\lambda_j}(\delta K - \lambda_i\delta M)\varphi_i \quad (15)$$

$$\delta\varphi_{ia} = \delta\lambda_i K^{-1}M\varphi_i \quad (16)$$

Since the changes in the structure's mass matrix are minimal, even after destruction, the difference in mass can be discarded. The stiffness matrix as a function of structural parameters is depicted by Equation (17).

$$[K] = [A][P][A]^T \quad (17)$$

Matrix A, with dimensions of npxnp, is referred to as the stiffness connectivity matrix. It is a diagonal matrix, with the stiffness parameters of elements, which are represented by npx1 P-forms and P-vectors, serving as its input.

$$diag[P] = \{P\} \tag{18}$$

The total stiffness matrix, as stated in Equation (19), is a linear function of the stiffness parameters. Furthermore, the matrix [A] is not dependent on the matrix [P].

$$[K + \delta K] = [A][P + \delta P][A]^T \tag{19}$$

Expanding Equation (19) and subtracting it from Equation (17) leads to obtaining a parametric form of the total stiffness matrix, as revealed in Equation (20).

$$[\delta K] = [A][\delta P][A]^T \tag{20}$$

If Equation (20) is accommodated in Equation (19), the changes in the *i*th mode shape can depend on changes in the stiffness parameters:

$$\delta \overline{\varphi_i} = S^K_{\overline{\varphi} i} \delta P_k \tag{21}$$

Equation (22) illustrates the relationship between the strain and the displacement.

$$\delta \varepsilon_i = S^K_{\varepsilon i} \delta P_k \tag{22}$$

$$S^K_{\overline{\varepsilon} i} = -K^{-1} A_S diag(A_S^T \varphi_i) + \sum_{j=1}^{n} \varphi_j \varphi_j^T \frac{\lambda_i}{\lambda_j} \frac{1}{\lambda_i - \lambda_j} A_S diag(A_S^T \varphi_i) \tag{23}$$

As an overview, the steps for updating the model are provided in the following flowchart (Figure 1). The purpose of updating the model is to align the predicted results of the model with actual measurements by adjusting certain key parameters. The distance between the theoretical and experimental outcomes is often measured using the mean of the least squares, and thus, updating the model can be seen as a problem of structural optimization. Since the variation of the *i*th mode shape is related to changes in the stiffness parameters, by altering the natural frequency and mode shape of healthy modes and measuring the strain, we can obtain changes in the diagonal matrix, including the eigenvalues of the combined stiffness matrix of the components. Finally, the structural update process is compared with the individual components. If the results are not in agreement, the process is repeated with variations in stiffness. The steps in the flowchart are as follows:
$\varnothing_i^I$: *i*th eigenvector in the healthy state;
λ_i^I: *i*th eigenvalue in the healthy state;
K_{new}: updated stiffness;
δk: stiffness variation;
$\delta \varepsilon_i$: strain variation.

2.2. Analysis Method

In practical applications, the successful utilization of vibration information models is limited due to measurement errors. These errors can arise from factors such as noise during frequency and mode measurements of the structure being tested. Such errors have undesirable consequences, as they can lead to failures going undetected or result in excessive predictions of failure. To simulate the presence of errors and noise in the measurement process, random noise of 1% was generated using MATLAB software. This noise was then multiplied by the mode shape of the damaged structure. The noise application loop, which operated outside the failure detection loop, generated a noise matrix for each iteration. The loop ran fifty times, creating fifty noise matrices. Each time, failure detection was performed, and the average of the detected failures over fifty iterations was calculated. Additionally, the standard deviation was computed and plotted to indicate the dispersion of failure detections across the fifty noise applications. By simulating fifty measurement errors, the goal was to mimic a scenario where the structure's response was measured fifty times in a laboratory, with failure detection performed each time. This method aimed to assess the accuracy of the approach in the presence of measurement errors.

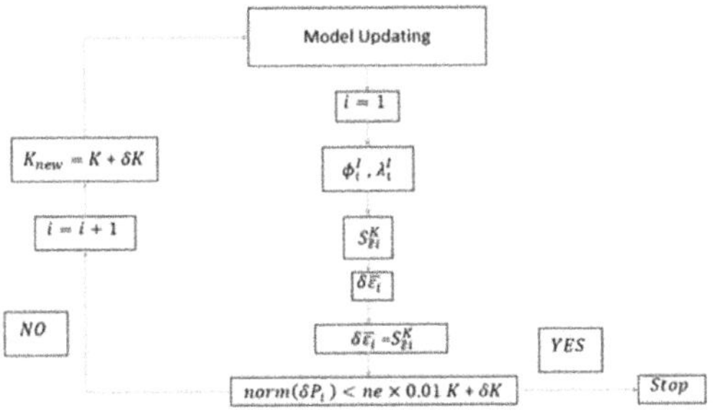

Figure 1. Flowchart of the steps for model updating.

3. Assumptions and Constraints of Numerical Modeling

In this research, we assumed that damage causes a change in the stiffness of structural elements, and this change, along with reduced flexural stiffness, applies to the structure. The proposed method links the alterations in the valid parameters to the mode shape of a structurally damaged object, as informed by the data. The stiffness matrix can be computed by adding together the stiffness matrices of the parts, as outlined in Equation (24).

$$K = \sum_{r}^{n_E} K_r \tag{24}$$

Here, K_r represents the contribution of a specific part to the overall stiffness matrices, and n_E stands for the number of elements. Hence, the behavior of the stiffness matrix can be rewritten as follows:

$$\delta K = \sum_{r=1}^{n_E} \delta K_r = \sum_{r=1}^{n_E} (K_{rd} - K_r) = \sum_{r=1}^{n_E} \delta K_r K_r \tag{25}$$

K_{rd} symbolizes the stiffness of the rth element when it is damaged, and δK_r is a scalar multiplier that demonstrates the proportional changes in the stiffness of the rth element in both damaged and undamaged states. The fluctuation of axial rigidity is indicated by the value of δK_r, which is obtained from the truss elements. Equation (25) is utilized to modify both the bending and axial rigidities in frame elements through decomposition.

To use methods of damage identification, the response of the structure should continue to display linear behavior after a seismic load; this method supposed which analytical model of the structure should be obtained before applying this method. The linear finite element modeling for a healthy structure was monitored as a basic model update to detect parameters of structural model stiffness [26]. For verification, the simulated structure was updated with the finite element method of the structural model and the behavior of the existing model was consistent with the realistic structural model.

The proposed method's effectiveness was examined in two models in this study. The approach was tested on a two-dimensional frame with 32 elements that underwent multiple damage scenarios. Additionally, the same method was tested on a three-dimensional frame that had 48 elements.

In the study, the mode shapes of the damaged structure were subjected to noise [14] about fifty times. Furthermore, by applying the same noise level (around 1%), the measurement error and standard deviation value were calculated. To ensure a reliable solution, several factors needed to be considered, including the type and location of the sensor, the type and location of the excitation, the quality of the measurement data (measurement error), the accuracy of the mathematical model (modeling error), and the numerical methods used.

4. Installation of the Strain Gauge

The placement of strain gauges on 2D and 3D frames depends on the specific requirements of the structure and the type of information needed for the analysis. Typically, strain gauges are placed at critical locations on the structure, such as near joints, supports, or areas susceptible to damage. For 2D frames, strain gauges are typically installed at the top and bottom of the frame, while for 3D frames, they can be placed on different faces of the structure to capture the deformations in different directions. The specific positioning of strain gauges can be determined through careful analysis and modeling of the structure to ensure the accuracy and reliability of the collected data. In this study, natural frequencies and mode shapes were acquired from sensors and applied to the damaged structure. The data was used to detect damage by using optimized equations, which improved Wang's method. The performance and validity of this approach were verified and investigated for both 2D and 3D frames. Due to various factors that affect the detection of the structure, a limited number of sensors must be used to measure strains, including determining the sensor location for access to different sides of the structure and measuring capabilities. Hence, the sensor location was determined based on the element's strain energy density, which is calculated using Equation (26).

$$U_i^e = \frac{1}{2}\{\varphi_i^e\}^T k_e \{\varphi_i^e\} \quad for\ e = 1, \ldots, ne \quad i = 1, \ldots, n \tag{26}$$

The strain energy density is close to the same strain energy density of the whole structure assigned as a target element, and it can generally be used to find the behavior of the structure. According to the circumscription of the use of strain gauges, the number of gauges in the 2D frame and 3D frames were limited to 18 and 20 units, respectively. Moreover, by using trial and error as a fundamental method for solving the problem, the strain gauges were placed in various positions along the length of the elements, where these distances were 3/4, 1/2, and 1/4. The positions of the strain gauges in the 2D and 3D frames are depicted in Figure 2.

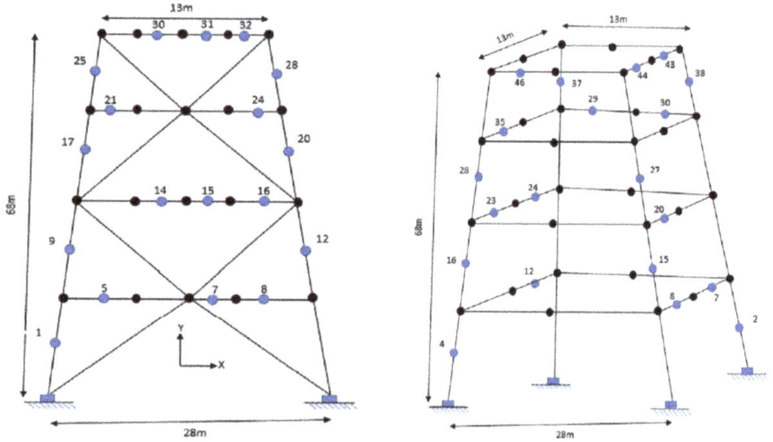

(a) (b)

Figure 2. Installing the strain gauges in (**a**) two-dimensional and (**b**) three-dimensional frames.

5. Damage Detection of Offshore Structures Using the Proposed Formulation

5.1. Two-Dimensional Frame Analysis

Based on Figure 3, a one-bay four-story frame was modeled to investigate the sufficiency of the proposed method in order to complete the model and identification of the structural damages. All 32 elements of the frame were constructed of steel material. The specification of the steel structure, including the elasticity modulus, density, circular cross-section with a thickness of 2 cm, and moment of inertia were 2.1 GPa, 7850 kg/m^3, 307 cm^2, and 91,430 kg/m^2 respectively. During the simulation of a damaged two-dimensional frame structure, measurements were taken at specific locations. These locations include elements at positions 32, 31, 30, 28, 25, 24, 21, 20, 17, 16, 15, 14, 12, 9, 8, 7, 5, and 1.

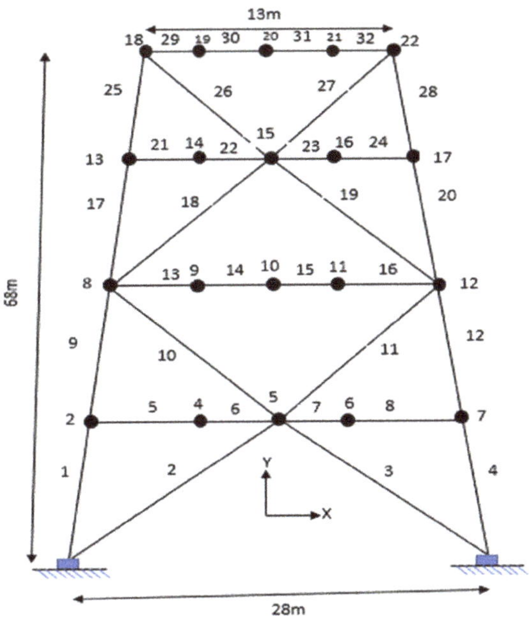

Figure 3. Two-dimensional frame structure equipped with gauges.

The unknown parameters in the two-dimensional frame were flexural rigidity (EI) and axial rigidity (EA), where E and A are used to define the elasticity modulus and cross-section of the element, respectively. The structural response of the two-dimensional frame is influenced by both the flexural and axial behavior, by which the rigidities should be allocated to update the model.

However, the excitation of a structure in axial mode differs from reality. The measurable mode shapes are often low mode shapes, which are dominated by flexural behavior. The axial rigidity as an unknown parameter has a minimum contribution rate on the sensitivity equations, which can lead to obtaining incompatible sensitivity equations and a lack of convergence. Hence, this article places significant emphasis on the consideration of flexural rigidity.

The geometry of the frame was entered into the numerical modeling and the damage was modeled. To estimate and investigate the effectiveness of the parameters, the location and severity of several instances of damage were classified, as shown in Table 1.

Table 1. Percentage of stiffness reduction of elements for the considered damage scenarios(2D).

Damage Case	1	2	3	4	5	6
Element no.	11	25	22	7	21	15
Damage rate	20%	20%	30%	20%	30%	20%
Element no.	13	26	27	10	23	17
Damage rate	40%	30%	20%	30%	20%	20%
Element no.	16	32	31	20	25	24
Damage rate	20%	40%	40%	30%	20%	30%
Element no.	...	...	...	30	29	25
Damage rate	...	...	...	40%	40%	30%
Element no.	...	...	...	...	...	29
Damage rate	...	...	...	...	...	40%

For instance, the damage for case 3 was modified by three modes, where an increasing trend in element number from 22 to 31 correlated with downward and upward percentage changes in damage of approximately 10% and 20%, respectively.

Measurement errors are inevitable, where this issue is due to the preparation process of the experimental and real data; determining the damage using Wang's method [37] and the proposed method was influenced by measurement errors. In this study, we assumed the measurement spread between the lowest and highest measured values. To resemble the measurement errors, some errors with uniform distribution were added to the correct information. To correctly simulate the measured data, one percentage of error was added to strain data with a uniform distribution.

Monte Carlo simulation (MCS) is the most robust and accurate method for estimating the probability of failure [39]. MCS [40] is a statistical method that uses random variables with a probability distribution to simulate measurement errors for unknown inputs in a problem. The damage detection algorithm repeatedly uses a set of data errors that are randomly simulated, and similar conditions are applied in the damage recognition process based on the absolute difference between the methods. Numerical simulations were performed using 50 observations, and the average of all results was used as the predicted damage for each damage scenario. The results of the parameter estimation, including the predicted and actual damage ratio of the 2D frame model in six cases, are displayed in Figures 4 and 5.

Figure 4 illustrates that the proposed method was accurate in diagnosing the location and severity of the damage while the Wang method [37] did not have the ability to recognize the damage. The given information and the input data were mostly the same for both methods. Consequently, the sensitivity equation was accurate and reliable, and it improved the results of the model updating.

Scattering of the obtained damage detection results using 50 Monte Carlo simulations [40] around the average value could be measured by using a standard statistical measurement known as the coefficient of variation (COV) or relative standard deviation (RSD), which gives the ratio of the standard deviation to the average value. The COV index is used for robustness evaluation against measurement errors, and a low level of the COV denotes the power of the method. In more detail, a higher COV value indicates that small changes in the monitored variable can have a larger impact on the structure, making it more sensitive to variability. In terms of health monitoring, higher COV values can affect reliability and safety, introducing uncertainty and variability that may impact the overall performance. When assessing risks and making decisions related to health monitoring, higher COV values suggest a higher level of risk and uncertainty, calling for additional measures to mitigate potential failures or accommodate variability in the design. Therefore, COV values in health monitoring provide important insights into the variability and its implications for reliability, safety, and risk assessment.

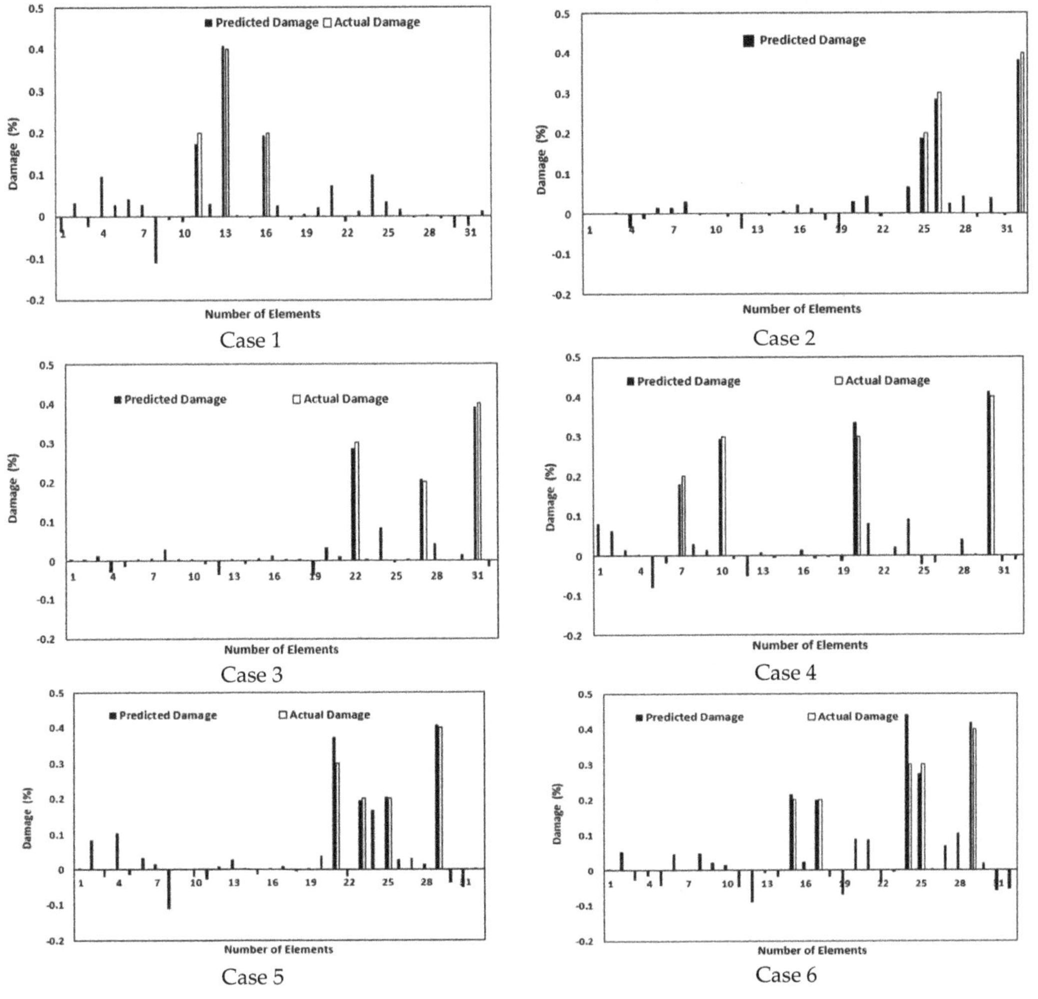

Figure 4. Predicted and actual damage ratio of the 2D frame model using the present method.

The COV does not have a limited number or an upper bound. The COV is calculated as the ratio of the standard deviation to the mean of a variable, expressed as a percentage. Since both the standard deviation and mean can take any positive value, the COV can theoretically range from 0% to positive infinity. However, it is important to note that the COV is not a suitable measure for variables with a mean close to zero or approaching zero, as it can lead to undefined or misleading results. The coefficients of variation of the predicted damage are shown in Figure 5.

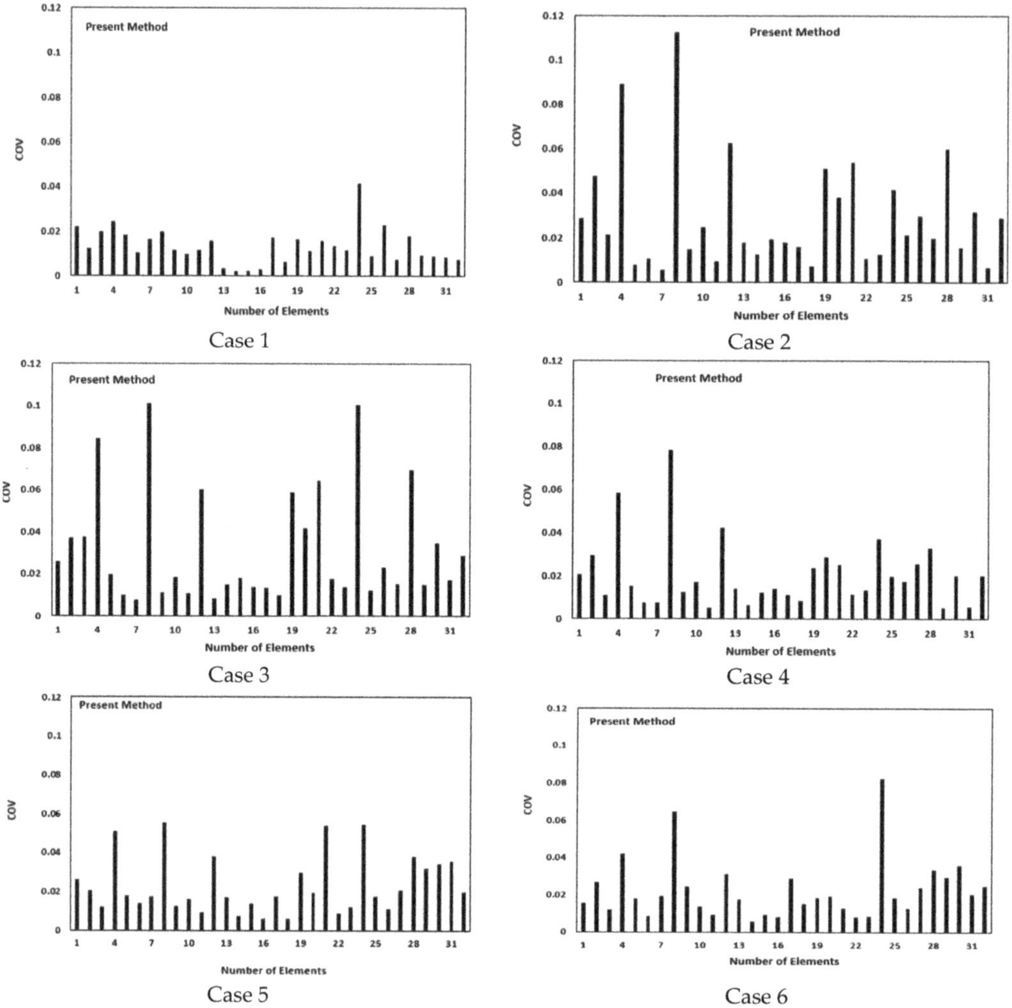

Figure 5. The coefficient of variation of the predicted damage ratio for the 2D frame model using the current method.

The COVs indicated less variability in the results of the evaluated parameters in the proposed method. Certain metrics are employed to quantitatively compare the accuracy of the results. This accuracy can be assessed through the closeness index (CI), which measures the proximity between the actual and predicted damage vectors.

$$CI = 1 - \frac{|\delta P_P - \delta P_a|}{|\delta p_p|} \tag{27}$$

The closeness index (CI) [41] is a measure of the accuracy of the predicted damage parameters. It is calculated as the distance between the actual damage vector (δp_p) and the predicted damage vector (δP_a), with dimensions of npxp, as shown in Equations (17) and (27).

The CI quantitatively evaluates how similar the predicted damage vector is to the actual damage vector. Higher values of the CI indicate greater similarity between the predicted and actual damage vectors, leading to more accurate damage detection results.

Using the CI as a performance metric allows researchers to compare the effectiveness of different damage detection methods, which can result in the development of more precise and dependable techniques for structural systems. The use of the CI can also enhance the overall safety and performance of structural systems by enabling the early detection of damage, which can be repaired or reinforced before a catastrophic failure occurs, thus increasing the system's durability and resilience.

The accuracy of the proposed method based on the results obtained is presented in Table 2. The CI index is defined in this table to determine the accuracy of the failure detection. The closer the CI value is to one, the more accurately the failure value is predicted, and the closer it is to zero, the less accurately the failure value is calculated [41] The proposed method in this study identified the damaged elements (Table 1) with high accuracy. The accuracy of the method was evaluated by calculating the proximity index; for example, in the first case, 0.64768 was obtained, indicating the accuracy of the proposed method. Additionally, this table demonstrates that the results were reliable and had low variability.

Table 2. Closeness indexes of the estimated parameters of the 2D frame model using the present method.

Damage Case	Closeness Index (CI)
1	0.64768
2	0.74341
3	0.76866
4	0.66808
5	0.62217
6	0.64667

The results of the CI for different damage cases are presented in Table 2 and indicate the effectiveness of the proposed method in improving the sensitivity equation.

5.2. The Assessment of the 3D Frame

Referring to Figure 6, a one-bay four-story frame was modeled to investigate the usefulness of the proposed method in order to produce the model and identify the damage. Finite element analysis was performed for a three-dimensional frame element to simulate and compare the experimental data.

In the 3D frame, the steel element properties were as follows: elasticity modulus of 2.1 GPa, a density of 7850 kg/m^3, and a moment of inertia limited to 91,430 kg/m^2. Moreover, the cross-section's element was 307 cm^2. To process data of a damaged structure, measurements were made using elements 2, 4, 7, 12, 8, 15, 16, 20, 23, 24, 27, 28, 29, 30, 35, 37, 38, 43, 44, and 46.

The unknown parameters in a three-dimensional frame are the torsional rigidity GJ, flexural rigidity EI, and axial rigidity EA, where E is the elasticity modulus, G is the shear modulus, and A is the cross-section of each element. In three-dimensional mode, torsional and flexural rigidity appeared too. Several instances of damage are explained in Table 3. For instance, for damage case 2, the number of elements changed simultaneously with the damage percent. In more detail, for element number 32, when the damage rate was limited to 50 percent, and for element number 45, the damage rate remained the same.

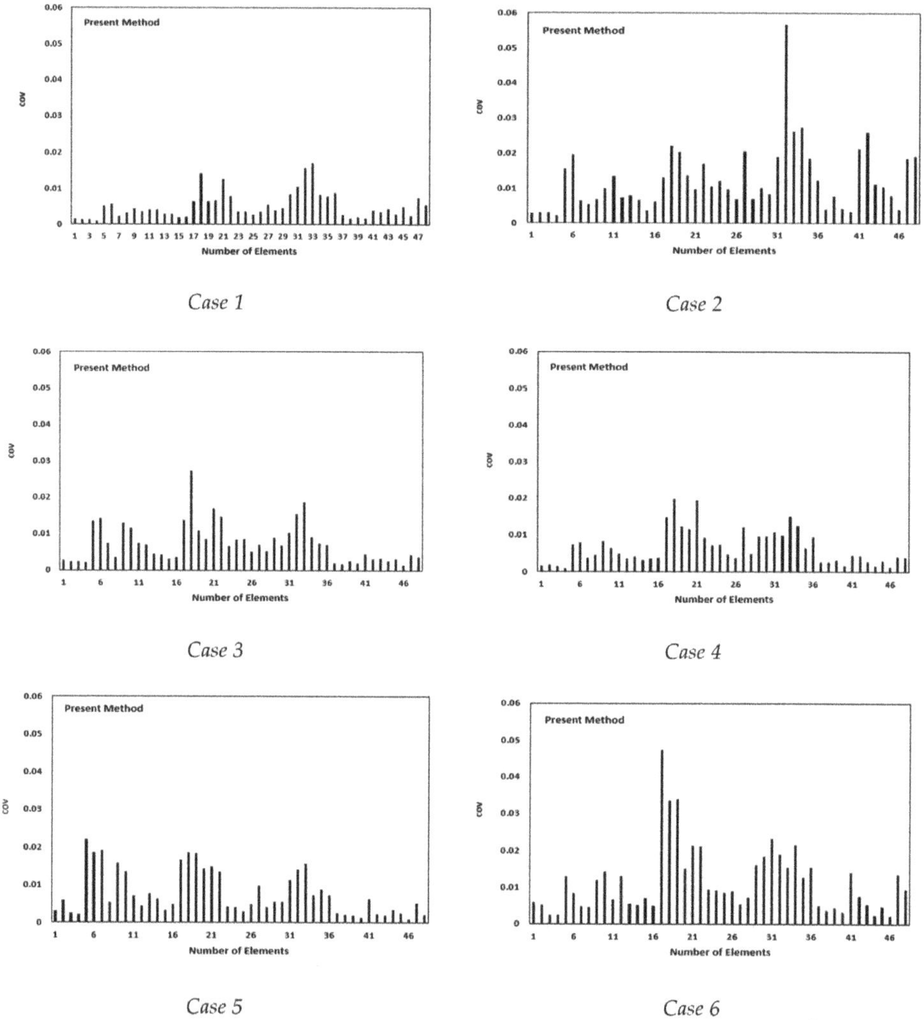

Figure 8. The coefficient of variation of the predicted damage ratio for the 3D frame model using the current method.

6. Conclusions

This research focused on the goals of structural health monitoring (SHM) and the limitations of Wang's method in accurately assessing damage in two-dimensional and three-dimensional structural elements, particularly in offshore structures. This study reviewed Wang's method and proposed a new method to enhance its accuracy and applicability for 2D and 3D structural elements.

In this study, six damage cases were considered for assessing and achieving the CI and COV indexes, and by using the proposed equation, the failure rates in 2D and 3D structural elements were calculated with appropriate accuracies. For example, a COV value of 0.04 suggests that the standard deviation is only 4% of the mean value, indicating that the data points are relatively close to the mean and exhibit limited variability. This level of consistency can be advantageous in applications where precision, accuracy, and reliability are critical, such as in high-precision manufacturing, quality control processes, or safety-critical systems. This method can be used to improve structures with axial and bending behavior. Due to the high sensitivity of the strain to failure and the high accuracy of the sensitivity equations, an enhanced SHM method for accurate damage detection in 2D and 3D structures that accommodates model updates and measurement errors enhances the reliability and effectiveness of structural health monitoring systems.

The following could be the focus of potential future studies: Due to the importance of seismic loads for offshore structures, the impact of the ground motion record [42,43] on the SHM of offshore structures using the proposed damage detection method can be investigated in future works.

Combining the probabilistic risk assessment methodology introduced in [44] and the proposed damage detection formulation for assessing the lifecycle of offshore structures.

A comparison of the proposed method with other well-known damage detection techniques, such as the finite element method, to evaluate the performance and accuracy of the method in different structures.

An evaluation of structural damage through the soil–structure interaction by considering the seismic analysis results of soil deposits, which can obtain a better understanding of the behavior of the soil–structure system and design structures that can withstand the effects of earthquakes [45].

An extension of the proposed method to more complex structures, such as bridges [46] and post-tensioned joints [47,48], to validate the method's adaptability to a wider range of structures. The damage assessment of joints in the frame structure [49] can be a useful way to produce better designs based on performance.

An integration of the proposed method with other structural health monitoring techniques, such as strain and load monitoring, to provide a comprehensive view of a structure's behavior and health.

An extension of the proposed method to include the analysis of non-linear structures, such as those with large deformations and material non-linearities, to evaluate its applicability to a wider range of structures.

Author Contributions: Conceptualization, M.Z.K. and A.T.; methodology, M.Z.K. and S.Y.; software, M.Z.K.; validation, M.Z.K., A.T. and S.Y.; formal analysis, A.T.; investigation, S.Y.; resources, M.Z.K.; data curation, S.Y.; writing—original draft preparation, M.Z.K.; writing—review and editing, A.T.; visualization, S.Y.; supervision, A.T.; project administration, A.T. All authors have read and agreed to the published version of the manuscript.

Funding: This research received no external funding.

Data Availability Statement: Not Applicable.

Conflicts of Interest: The authors declare no conflict of interest.

Nomenclature

A	Structure stiffness joint matrix
Ae	Non-zero eigenvector stiffness matrix
B	Strain–displacement matrix
d	Damage mode index
I	Healthy mode index
K	Structure stiffness matrix
M	Structure mass matrix
ne	Number of elements
P	Diagonal matrix of stiffness parameters
Pe	Stiffness parameters vector
T	Transform matrix
Ue	Displacement vector's eth element
εe	Strain of eth element
∅I	*i*th mode shape
λI	*i*th natural frequency

References

1. Zarrin, M.; Asgarian, B.; Abyani, M. Probabilistic seismic collapse analysis of jacket offshore platforms. *J. Offshore Mech. Arct. Eng.* **2018**, *140*, 031601. [CrossRef]
2. Khedmatgozar Dolati, S.S.; Caluk, N.; Mehrabi, A.; Khedmatgozar Dolati, S.S. Non-Destructive Testing Applications for Steel Bridges. *Appl. Sci.* **2021**, *11*, 9757. [CrossRef]
3. Anders, A. Vibrational Based Inspection of Civil Engineering Structures. Ph.D. Thesis, Aalborg University, Aalborg, Denmark, 1993.
4. Vahedi, M.; Khoshnoudian, F.; Hsu, T.Y.; Partovi Mehr, N. Transfer function-based Bayesian damage detection under seismic excitation. *Struct. Des. Tall Spec. Build.* **2019**, *28*, e1619. [CrossRef]
5. Lee, J.S.; Choi, I.Y.; Cho, H.N. Modeling and detection of damage using smeared crack model. *Eng. Struct.* **2004**, *26*, 267–278. [CrossRef]
6. Modak, S.V. Model updating using uncorrelated modes. *J. Sound Vib.* **2014**, *333*, 2297–2322. [CrossRef]
7. Lim, T.W. Submatrix approach to stiffness matrix correction using modal test data. *AIAA J.* **1990**, *28*, 1123–1130. [CrossRef]
8. Maeck, J.; Wahab, M.A.; Peeters, B.; De Roeck, G.; De Visscher, J.; De Wilde, W.P.; Ndambi, J.-M.; Vantomme, J. Damage identification in reinforced concrete structures by dynamic stiffness determination. *Eng. Struct.* **2000**, *22*, 1339–1349. [CrossRef]
9. Carvalho, J.; Biswa, N.D.; Abhijit, G.; Maitreya, L. A direct method for model updating with incomplete measured data and without spurious modes. *Mech. Syst. Signal Process.* **2007**, *21*, 2715–2731. [CrossRef]
10. Mottershead, J.E.; Link, M.; Friswell, M.I. The sensitivity method in finite element model updating: A tutorial. *Mech. Syst. Signal Process.* **2011**, *25*, 2275–2296. [CrossRef]
11. Charbel, F.; Hemez, F.M. Updating finite element dynamic models using an element-by-element sensitivity methodology. *AIAA J.* **1993**, *31*, 1702–1711. [CrossRef]
12. Brownjohn, J.M.W.; Pin-Qi, X.; Hong, H.; Yong, X. Civil structure condition assessment by FE model updating: Methodology and case studies. *Finite Elem. Anal. Des.* **1995**, *37*, 761–775. [CrossRef]
13. Lin, R.M.; Lim, M.K.; Du, H. Improved inverse eigensensitivity method for structural analytical model updating. *J. Vib. Acoust.* **1995**, *117*, 192–198. [CrossRef]
14. Gerist, S.; Maheri, M.R. Structural damage detection using imperialist competitive algorithm and damage function. *Appl. Soft Comput.* **2019**, *77*, 1–23. [CrossRef]
15. Wang, L.; Lu, Z.R. Sensitivity-free damage identification based on incomplete modal data, sparse regularization and alternating minimization approach. *Mech. Syst. Signal Process.* **2019**, *120*, 43–68. [CrossRef]
16. Wnag, X.; Lian, C.; Zheng, C.; Fen, L. Sensitivity analysis and dynamic modification of modal parameter in mechanical transmission system. *J. Mar. Sci. Appl.* **2005**, *4*, 53–58. [CrossRef]
17. Kim, J.T.; Stubbs, N. Damage detection in offshore jacket structures from limited modal information. *Int. J. Offshore Polar Eng.* **1995**, *5*, 58–66.
18. Ren, W.X.; De Roeck, G. Structural damage identification using modal data. I: Simulation verification. *J. Struct. Eng.* **2002**, *128*, 87–95. [CrossRef]
19. Li, H.; Junrong, W.; Hu, S.-L.H. Using incomplete modal data for damage detection in offshore jacket structures. *Ocean Eng.* **2008**, *35*, 1793–1799. [CrossRef]
20. Shahsavari, V.; Chouinard, L.; Bastien, J. Wavelet-based analysis of mode shapes for statistical detection and localization of damage in beams using likelihood ratio test. *Eng. Struct.* **2017**, *132*, 494–507. [CrossRef]
21. Wang, J.; Wang, C.; Zhao, J. Frequency response function-based model updating using Kriging model. *Mech. Syst. Signal Process.* **2017**, *87*, 218–228. [CrossRef]
22. Umar, S.; Bakhary, N.; Abidin, A.R.Z. Response surface methodology for damage detection using frequency and mode shape. *Measurement* **2018**, *115*, 258–268. [CrossRef]
23. Hariri-Ardebili, M.A.; Mahdavi, G.; Abdollahi, A.; Amini, A. An rf-pce hybrid surrogate model for sensitivity analysis of dams. *Water* **2021**, *13*, 302. [CrossRef]
24. Dahak, M.; Touat, N.; Benseddiq, N. On the classification of normalized natural frequencies for damage detection in cantilever beam. *J. Sound Vib.* **2017**, *402*, 70–84. [CrossRef]
25. Khatir, S.; Dekemele, K.; Loccufier, M.; Khatir, T.; Wahab, M.A. Crack identification method in beam-like structures using changes in experimentally measured frequencies and Particle Swarm Optimization. *Comptes Rendus Mécanique* **2018**, *346*, 110–120. [CrossRef]
26. Park, H.S.; Kim, J.; Oh, B.K. Model updating method for damage detection of building structures under ambient excitation using modal participation ratio. *Measurement* **2019**, *133*, 251–261. [CrossRef]
27. Tiachacht, S.; Bouazzouni, A.; Khatir, S.; Wahab, M.A.; Behtani, A.; Capozucca, R. Damage assessment in structures using combination of a modified Cornwell indicator and genetic algorithm. *Eng. Struct.* **2018**, *177*, 421–430. [CrossRef]
28. Zenzen, R.; Belaidi, I.; Khatir, S.; Wahab, M.A. A damage identification technique for beam-like and truss structures based on FRF and Bat Algorithm. *Comptes Rendus Mécanique* **2018**, *346*, 1253–1266. [CrossRef]
29. Pérez, M.A.; Serra-López, R.A. Frequency domain-based correlation approach for structural assessment and damage identification. *Mech. Syst. Signal Process.* **2019**, *119*, 432–456. [CrossRef]

30. Khatir, S.; Wahab, M.A.; Boutchicha, D.; Khatir, T. Structural health monitoring using modal strain energy damage indicator coupled with teaching-learning-based optimization algorithm and isogoemetric analysis. *J. Sound Vib.* **2019**, *448*, 230–246. [CrossRef]
31. Abasi, A.; Harsij, V.; Soraghi, A. Damage detection of 3D structures using nearest neighbor search method. *Earthq. Eng. Eng. Vib.* **2021**, *20*, 705–725. [CrossRef]
32. Chen, W.; Zheo, W.; Yang, H.; Chen, X. Damage detection based on optimized incomplete mode shape and frequency. *Acta Mech. Solida Sin.* **2015**, *28*, 74–82. [CrossRef]
33. Iannelli, P.; Angeletti, F.; Gasbarri, P.; Panella, M.; Rosato, A. Deep learning-based Structural Health Monitoring for damage detection on a large space antenna. *Acta Astronaut.* **2022**, *193*, 635–643. [CrossRef]
34. Wang, Q.; Wu, Z. Structural system reliability analysis based on improved explicit connectivity BNs. *KSCE J. Civ. Eng.* **2018**, *22*, 916–927. [CrossRef]
35. Chen, H.P.; Maung, T.S. Regularised finite element model updating using measured incomplete modal data. *J. Sound Vib.* **2014**, *333*, 5566–5582. [CrossRef]
36. Nelson, R.B. Simplified calculation of eigenvector derivatives. *AIAA J.* **1976**, *14*, 1201–1205. [CrossRef]
37. Wang, B.P. Improved approximate methods for computing eigenvector derivatives in structural dynamics. *AIAA J.* **1991**, *29*, 1018–1020. [CrossRef]
38. Sanayei, M.; Saletnik, M.J. Parameter estimation of structures from static strain measurements. I: Formulation. *J. Struct. Eng.* **1996**, *122*, 555–562. [CrossRef]
39. Abdollahi, A.; Moghaddam, M.A.; Monfared, S.A.H.; Rashki, M.; Li, Y. A refined subset simulation for the reliability analysis using the subset control variate. *Struct. Saf.* **2020**, *87*, 102002. [CrossRef]
40. Yuen, K.V.; James, L.B.; Siu, K.A. Structural damage detection and assessment by adaptive Markov chain Monte Carlo simulation. *Struct. Control Health Monit.* **2004**, *11*, 327–347. [CrossRef]
41. Esfandiari, A.; Bakhtiari-Nejad, F.; Rahai, A.; Sanayei, M. Structural model updating using frequency response function and quasi-linear sensitivity equation. *J. Sound Vib.* **2009**, *326*, 557–573. [CrossRef]
42. Alkayem, N.F.; Shen, L.; Asteris, P.G.; Sokol, M.; Xin, Z.; Cao, M. A new self-adaptive quasi-oppositional stochastic fractal search for the inverse problem of structural damage assessment. *Alex. Eng. J.* **2022**, *61*, 1922–1936. [CrossRef]
43. Taslimi, A.; Tehranizadeh, M. The effect of vertical near-field ground motions on the collapse risk of high-rise reinforced concrete frame-core wall structures. *Adv. Struct. Eng.* **2022**, *25*, 410–425. [CrossRef]
44. Kia, M.; Amini, A.; Bayat, M.; Ziehl, P. Probabilistic seismic demand analysis of structures using reliability approaches. *J. Earthq. Tsunami* **2021**, *15*, 2150011. [CrossRef]
45. Torabipour, A.; Hamidi, A. Ground Response Analysis of Cemented Alluviums Using Non-Recursive Algorithm. *J. Earthq. Eng.* **2020**, *24*, 1390–1416. [CrossRef]
46. Vahdati, V.J.; Yaghoubi, S.; Torabipour, A.; Correia, J.A.F.O.; Fazeres-Ferradosa, T.; Taveira-Pinto, F. Combined solutions to reduce scour around complex foundations: An experimental study. *Mar. Syst. Ocean Technol.* **2020**, *15*, 81–93. [CrossRef]
47. Shiravand, M.R.; Torabipour, A.; Mahboubi, S. Parametric study on effect of adding stiffener to post-tensioned steel connection. *Int. J. Steel Struct.* **2019**, *19*, 478–494. [CrossRef]
48. Torabipour, A.; Abdollahi, A.; Yaghoubi, S. Assess the Effectiveness of SMA on Response of Steel Connection with PT Strands. *KSCE J. Civ. Eng.* **2019**, *23*, 5133–5142. [CrossRef]
49. Yang, J.N.; Xia, Y.; Loh, C.H. Damage identification of bolt connections in a steel frame. *J. Struct. Eng.* **2014**, *140*, 04013064. [CrossRef]

eng

MDPI

Article

Chemo-Thermo-Mechanical FEA as a Support Tool for Damage Diagnostic of a Cracked Concrete Arch Dam: A Case Study

Noemi Schclar Leitão [1,*] and Eloísa Castilho [2]

[1] Laboratório Nacional de Engenharia Civil (LNEC), Av. do Brasil 101, 1700-066 Lisbon, Portugal
[2] Instituto Superior Técnico, Universidade de Lisboa, Av. Rovisco Pais 1, 1049-001 Lisbon, Portugal; eloisa.castilho@tecnico.ulisboa.pt
* Correspondence: nschclar@lnec.pt

Abstract: Most of the larger hydropower plants in Western Europe, the former Soviet Union, North America and Japan were constructed between the 1940s and 1970s. This implies that the rehabilitation or repair of existing dams is a top priority, which entails new challenges for the dam engineering community. Since no two dams are the same, in cases in which abnormal behavior is suspected, an in-depth diagnosis of the state of the dam to define the causes and consequences of the damage is required. To illustrate the diagnostic process, an old concrete arch dam is presented which showed signs of reservoir water seepage through some construction joints, resulting in a buildup of calcium carbonate on the downstream face. After analyzing the available data, we put forward a hypothesis that the high temperature gradient promoted the opening of some construction joints on the upstream face during the first filling of the reservoir. Over time, water penetration expanded the cracks, reaching the downstream face. To prove our diagnosis, a chemo-thermo-mechanical finite element analysis was carried out in order to simulate the behavior of the dam during its construction and initial impoundment.

Keywords: concrete arch dam; finite element chemo-thermal-mechanical analysis; diagnosis; first filling

1. Introduction

Between the 1940s and 1970s, spurred initially by World War II and followed by strong post-war economic and population growth, state-owned utilities built significant hydropower developments throughout Western Europe, as well as the former Soviet Union, North America and Japan [1]. These water infrastructural projects were critical to economic development of the agriculture and energy sectors. Since the 1980s, however, the number of new dams in developed countries has started to decline. This slowdown in dam construction resulted from several political-ecological factors: (i) the economics of construction (the best and most economical sites were exploited first); and (ii) the increased awareness of the need to consider the environmental impact of large dams [2,3].

Concurrently, water availability and climate change have been topics of increasing concern in recent decades. To face this situation without constructing new dams, developed countries have no other option than continue using the existing dams as much as possible [4]. This implies that the repair, rehabilitation or strengthening of existing dams should be carried out in order to enhance their performance, extend their service life and increase their load-carrying capacity.

Similar to other concrete structures, concrete dams are inherently durable and usually require a minimum amount of repair and maintenance. However, with the passage of time, the exposure of a dam to various external and internal aggressions can lead to the deterioration of the structure.

One of the most harmful distress mechanisms affecting the durability and serviceability of aging structures is the alkali–aggregate reaction (AAR). Due to its impact on the

Citation: Leitão, N.S.; Castilho, E. Chemo-Thermo-Mechanical FEA as a Support Tool for Damage Diagnostic of a Cracked Concrete Arch Dam: A Case Study. *Eng* **2023**, *4*, 1265–1289. https://doi.org/10.3390/eng4020074

Academic Editor: Alessio Cascardi

Received: 21 March 2023
Revised: 15 April 2023
Accepted: 20 April 2023
Published: 22 April 2023

safety and durability of concrete structures, many efforts to harmonize and coordinate the diagnosis, prognosis and assessment of evaluating AAR damage are underway through the International Union of Laboratories and Experts in Construction Materials, Systems and Structures (RILEM) [5,6].

However, there are many causes of concrete damage in dams apart from AARs [7,8]. Freezing–thawing and drying–wetting cycles, structural overload, cracking due to seismic actions or non-uniform foundation movements, thermal and shrinkage volumetric changes, cavitation, abrasion–erosion, and sulfate attack are some examples of causes of damage to concrete. Design and construction defects, poor-quality concrete, poor finishing and poor curing can lead to concrete suffering damage.

In addition, there are a number of other factors that may affect service life, resulting in the necessity of strengthening dams. Changes in the design criteria (hydrology and seismic hazards) based on new information obtained since the initial design of a dam, changes in methods of analysis, and new safety concepts or the results of risk assessments (new risks and changes in risk acceptance criteria) can trigger the need to strengthen actions [8].

The planning, design, implementation and monitoring of a repair and/or strengthening project should always begin with a careful assessment of the existing structure. The purpose of this assessment is to identify all defects and damage, to diagnose their causes and hence to assess the present and likely future adequacy of the structure. The information obtained from the structural assessment can then be used to determine whether corrective work is required. Without prior planning and proper assessment, any program of corrective work is likely to prove ineffective [9].

However, historically, the repair of concrete dams has been based at least as much on art as on science [7]. Determining the cause of damage has often been given very little importance by technicians, who make decisions based on their expertise and intuition [10].

In order to improve practices and knowledge in concrete dam repair, the U.S. Bureau of Reclamation, which operates and maintains hydroelectric and water resource structures in the Western United States, has invested a lot of effort into formulating a consistent and systematic approach to repairing concrete. In this regard, the importance of a correct diagnosis was emphasized:

> The first and very important step of repairing damaged or deteriorated concrete is to correctly determine the cause of damage. Knowing what caused the damage, and reducing or eliminating that cause, will make the repair last longer. If no attempt is made to eliminate the original cause of damage, the repair may fail as the original concrete did, resulting in wasted effort and money (von Fay [7], p. 1–13).

With the purpose of organizing a diagnostic procedure, Pardo-Bosch and Aguado [10] and Blanco et al. [11] outlined a framework to aid in the diagnosis of different pathologies that affect concrete dams based on the theory used in medical diagnosis. In this context, numerical models have been proven to be important tools for the study and validation of hypotheses elaborated during this diagnosis procedure [12–14].

Following the above approach closely, the present article illustrates the diagnostic process of an old concrete arch dam, which shows signs of reservoir water seepage through some construction joints, resulting in a buildup of calcium carbonate on the downstream face. After analyzing the available data, we put forward a hypothesis that the high temperature gradient promoted the opening of some construction joints on the upstream face during the first filling of the reservoir. Over time, water penetration expanded the cracks, reaching the downstream face. To prove this diagnosis, a chemo-thermo-mechanical finite element analysis (FEA) was carried out in order to simulate the behavior of the dam during its construction and initial impoundment.

2. Diagnosis Procedure

According to [10,11], the sequence of activities can be divided into two main stages:

1. The first stage, which usually lasts a few weeks, involves the following:
    a. Clinical history;
    b. Filed (dam) inspection;
    c. Initial cabinet works;
    d. First hypothesis.
2. The second stage, which may take weeks to months, entails the following:
    a. Laboratory tests;
    b. Numerical modeling;
    c. Validation of the hypothesis;
    d. Prediction of future behavior.

3. Dam Description

The dam studied in this article was completed in 1955 and is located in the central region of Portugal. It is a double-curvature, thin concrete arch dam with 17 keyed monoliths, a maximum height of 63 m and a crest length of 175 m at an elevation of 181 m. The width is 2 m at the crest and 7 m at the base of the largest monolith. The dam has a surface uncontrolled spillway along its crest designed for a maximum discharge capacity of 2200 m^3/s. The reservoir's normal level is 175 m. A concrete pad (called socle in Portugal) was added to the foundation to make the site symmetrical and to provide a better distribution of stresses on the foundation, as shown in Figure 1.

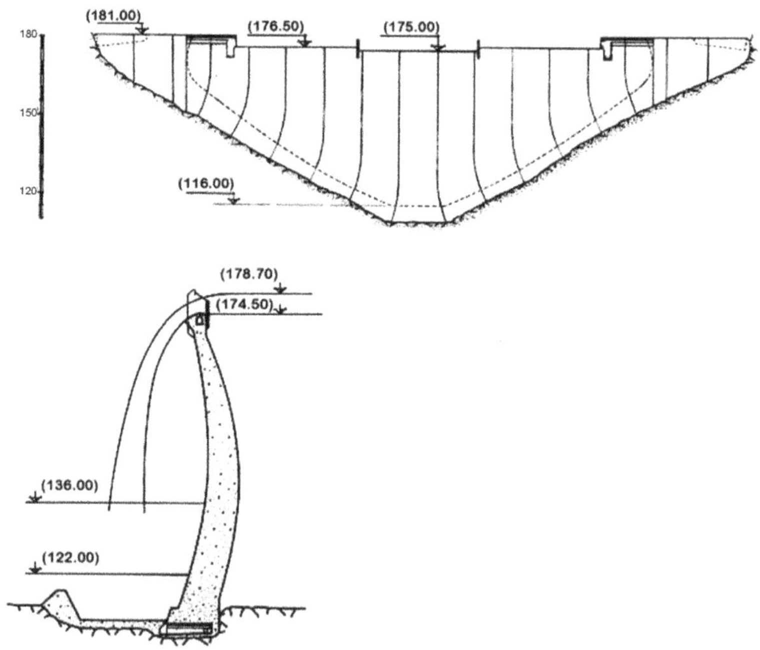

Figure 1. Upstream view and vertical cross section of the dam.

The dam foundation corresponds to a contact zone of Cambrian and Silurian formations, with the special feature being that the left bank consists of granite and the right bank consists of schist. Both rocks are very siliceous and, in general, the rock mass is fractured and presents weathering near the surface [15].

The monitoring system consists of several devices to measure concrete and air temperatures, water level, displacements in the dam and its foundation, joint movements, strains and stresses in the concrete, pressure and discharges in the foundation.

Because of the lack of pendulum, the geodetic surveying method is the exclusive source of horizontal displacements. This planimetric system consists of triangulation networks on the downstream banks, as shown in Figure 2.

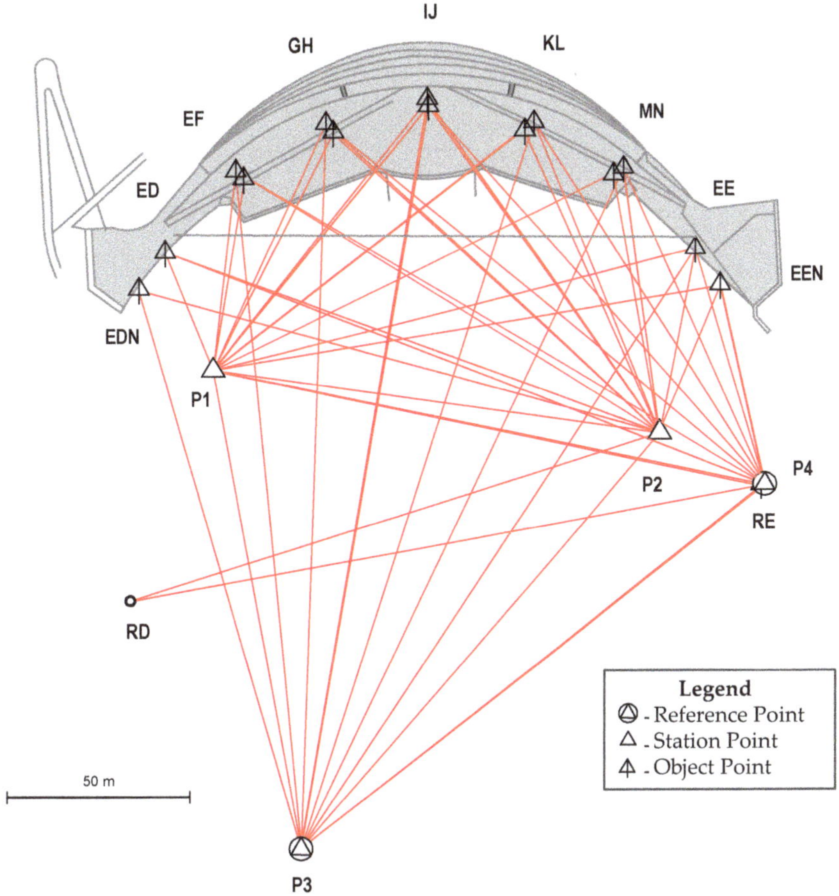

Figure 2. Triangulation network.

4. Clinical History and Field Inspections

Apart from the already cited article [15] published by the former owner of the dam, Hidroeléctrica do Zêzere, which describes the preliminary studies, design, construction and material tests of the dam, most of the information used in the present work comes from internal technical reports by the National Laboratory for Civil Engineering (LNEC). However, it is important to note that the most relevant results of these reports have also been previously published in different conferences and journals [16–19]. Description of the concrete composition used in the dam and other relevant thermal information are also available in Silveira's thesis [20].

4.1. Dam Design

It is important to note that in the 1950s, two different approaches were used to perform stress analyses of dams on both sides of the Atlantic. In the USA, the design of dams was based on an extensive analytical method, the so-called trial-load method, developed by the U.S. Bureau of Reclamation. Meanwhile, European countries, mainly France, Italy and

Portugal, preferred the use of small-scale structural models rather than the time-consuming trial-load method to find the final shape of the dam. As a result, the arch dams constructed in Europe were thinner than those constructed in the USA [21,22].

In the present case, the preliminary design of the dam was based on the "independent arch" theory. According to this theory, the dam was assumed to be divided by horizontal planes into arch rings of units of vertical height. These rings were considered to work as independent arches.

For the final design of the dam, small physical (or scale) model tests were carried out in the LNEC. The tests were performed in homogeneous plaster–diatomite models at a scale of 1/200. Figure 3 illustrates the principal stress path obtained using brittle varnish technique, the average principal stresses obtained in the tests for a reservoir water level of 177.5 m and the composition of these stresses with the stresses computed by analytical method for the self-weight of the dam.

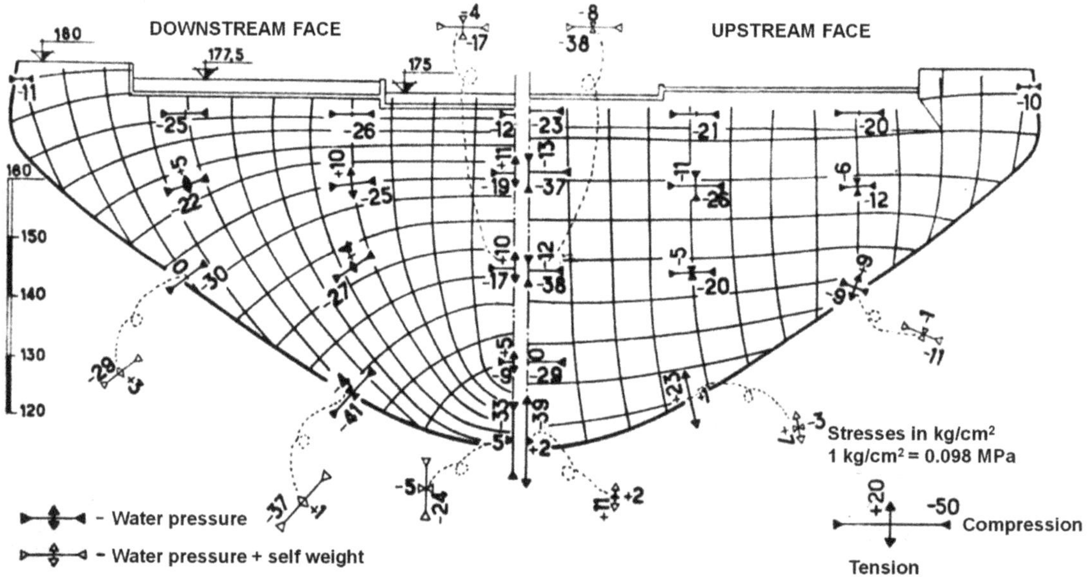

Figure 3. Principal stress on the downstream face (on the left) and on the upstream face (on the right) obtained by model tests (adapted from [16]).

At the same time, in order to reach a more meaningful estimate of the margin of safety, tests up to failure were also performed. To this aim, a technique was employed of loading with jacks that simultaneously increased the dead weight and water load, thus reproducing progressive reductions in the concrete strength, as shown in Figure 4.

4.2. Dam Construction and Initial Impoundment

The construction of the dam took place between October 1954 and September 1955. The first filling of the reservoir began in 1955, with the dam still under construction.

The grouting of the contraction joints was carried out in various stages between May 1955 and March 1956. During May and June 1955, with the dam still under construction, the grouting of the lower part was performed, between the foundation and the level at 135 m. The middle part between the levels at 135 m and 155 m was performed in August 1955. Finally, the grouting of the upper part took place between February and March 1956, in correspondence with the first emptying of the reservoir.

Between August and September 1958, a second emptying of the reservoir took place, however no work was reported at that time.

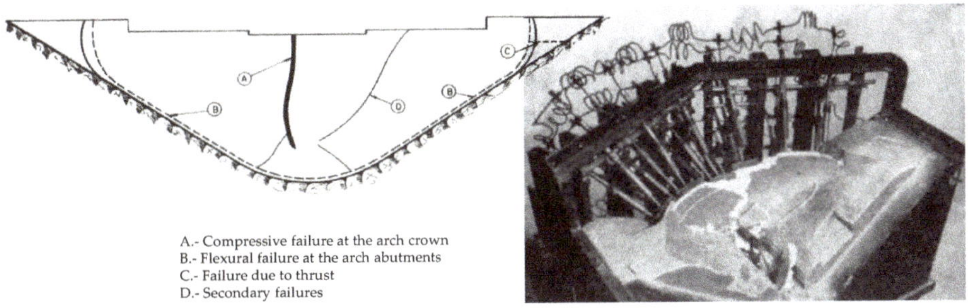

Figure 4. Tests up to failure of the dam (adapted from [18,23]).

4.3. Cracking Evolution

The existence of water seepage through some construction joints has been known since 1976. As time passes, the leaching from the concrete has been forming calcium carbonate deposits on the downstream face of the dam. Besides these water leaking cracks, several construction joint openings also exist at a lower level.

Figure 5 shows the evolution of the downstream face's appearance over nearly 25 years. The oldest photograph of the damage that the authors could find goes back to some time before 1992 and corresponds to the inventory of dams in Portugal published in 1992 [24]. The rest of the photographs correspond to the periodic inspections carried out by LNEC since 2002. The major change observed in this period corresponds to January 2007, when a new seepage path through another construction joint appeared in the monolith G–H.

Figure 5. Downstream surface appearance from early 1990s to January 2015.

As part of the safety inspection, the owner of the dam has performed periodical crack surveys since 1982. These surveys allow us to identify and locate the existent cracks, as well as follow their evolution, as shown in Figure 6.

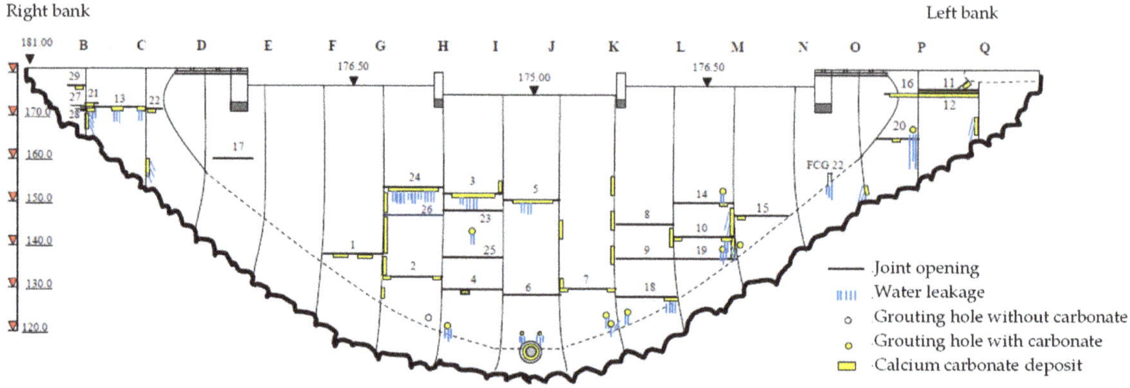

Figure 6. Crack survey carried out in March 2012.

In addition, some cracks were instrumented with joint meters. Figure 7 shows, from top to bottom, in the first graph, the monthly average temperature and the water level, and in the following three graphs, the monitored opening of joint meters installed in crack 3, crack 5 and crack 7, respectively, for the period between 2002 and 2016.

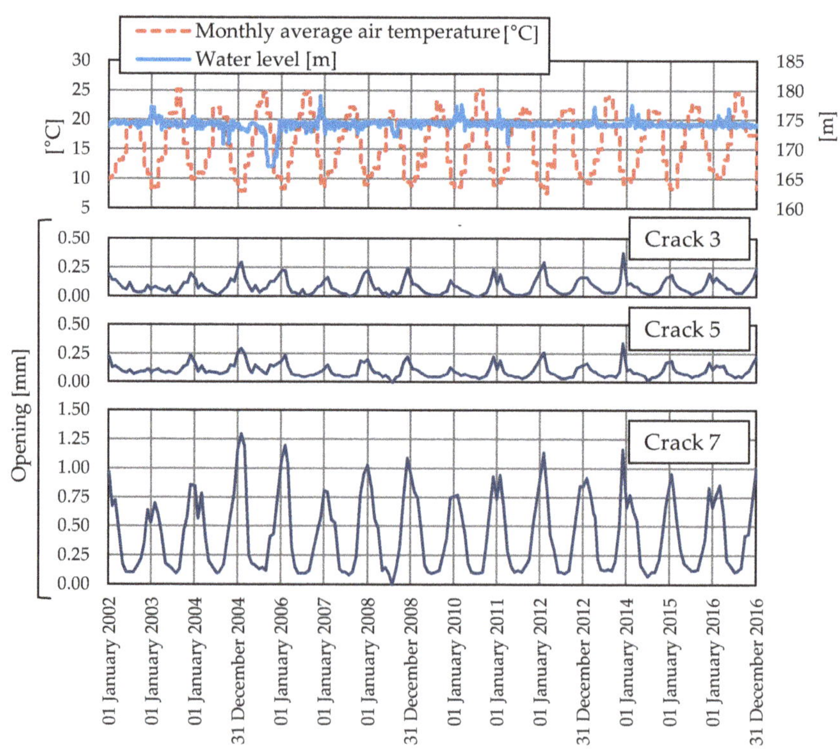

Figure 7. Monthly average air temperature, reservoir levels, and monitored crack openings between 2002 and 2016.

5. Cabinet Works and Initial Diagnosis Hypothesis

Apart from the open construction joints, the dam did not present any other significant signs of concrete deterioration. The classic symptom of map cracking induced by AAR was not observed. Nevertheless, the displacements of the dam were analyzed in order to detect any signs of irreversible displacements. As it has already been established, in arch dams, internal concrete swelling phenomena usually induce upstream drift and crest rises, even before cracking takes place. Since irreversible displacements were not detected, any swelling processes could be excluded from our study.

Additionally, with respect to progressive deformation, it is important to note that, although there is no sign of non-recoverable deformation in the no-stress strain meters located in most of the dam, the no-stress strain meters installed near the foundation have been showing signs of concrete swelling since 2001. However, this phenomenon is limited and does not justify, at this moment, further investigation.

After excluding an ongoing swelling process effect, thermal stresses were identified as the main cause of the opening of the construction joints.

It is worth noting that the properties of construction joints are greatly influenced by how the joints are prepared before pouring the next lift of concrete. Even with good preparation, the strength and fracture energy at a construction joint are much lower than the values for mass concrete, creating horizontal planes of weaknesses [25]. Therefore, it is not surprising that vertical tensile stresses were released by the opening of the nearest construction joints.

As the open construction joints with water seepage (cracks 3, 5 and 24) show a very slight opening seasonal variation compared with the cracks situated under the level of 150 m, as illustrated by Figure 7, the phenomenon associated with each type of cracking was considered as a different type.

For the lower cracks, the high seasonal fluctuation together with the absence of seepage suggest that they are thermal fatigue cracks due to the seasonal thermal variation acting on the downstream face of the dam. These fatigue cracks do not propagate a lot in depth. Thus, they do not compromise the behavior of the structure under static loads and their formation will be not studied in this paper.

On the other hand, the formation of the cracks located over the level of 150 m was associated with the fact that the initial impoundment started with the dam under construction. The upper concrete lifts did not have enough time to release all the heat of hydration before being in contact with the cold water. Once opened, the water pressure within the cracks triggered additional damage and weakened the fracture properties. This effect coupled with the hydrostatic pressure slowly opened a water path to the downstream face, leaching calcium carbonate.

In the following section, the formation of the cracks over the level at 150 m due to thermal stresses generated during the initial impoundment will be investigated using FEA.

6. Finite Element Model

In their guidelines for nonlinear FEA of existing concrete structures and infrastructures, Hendrix et al. remark:

> A finite element model of a structure is an abstraction of the physical structure with a number of assumptions, generalizations, and idealizations. The abstraction process has two distinct steps: first, the abstraction from the structure to the mechanical model, and then the abstraction from the mechanical model to the finite element model. In the first step, assumptions and simplifications have to be made regarding to which extent and to which detail the structure has to be modeled, how the boundaries of the model are described, which loads on the structure are significant and how they are described, et cetera. The second step is to discretize the mechanical model into a finite element model, and attach the necessary attributes such as material models, boundary conditions, and loading to the finite element model (Hendrix et al. [26], p. 9).

Indeed, FEA requires a great number of a priori modeling decisions. These decisions require a certain level of expertise and are quite subjective in nature, leading to considerable differences in the approaches adopted. As pointed out by Saouma and Hariri-Ardebili [27] (p. 243), "the selected finite element analysis is often a compromise between: (a) needs and time constraint, (b) our understanding of the problem and of nonlinear analysis, (c) tools available, and (d) quality of results expected".

Based on previous experience in modeling the behavior of concrete arch dams during initial impounding [28], a chemo-thermal-mechanical model is elaborated to investigate the crack formation during the first filling of the reservoir. First, a chemo-thermal analysis using the chemical affinity concept is carried out to determine the temperature distribution during the construction and the initial impoundment of the dam. Then, a nonlinear viscoelastic analysis is performed in order to obtain the structural response of the dam, with particular focus on the nonlinear behavior caused by the opening and closure of the contraction joints as well as the interface between the dam and the foundation.

6.1. Finite Element Mesh

For the finite element analysis, the double curvature concrete arch dam and an adequate volume of the foundation are represented as shown in Figure 8.

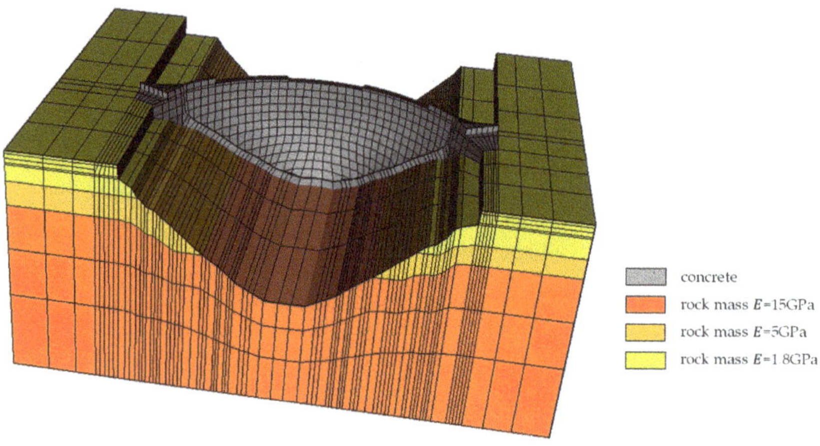

Figure 8. Finite element model of the dam.

The dam model comprises four layers of 20-node solid elements. To facilitate the transmission of data from the chemo-thermal analysis to the mechanical analysis, the same solid elements' mesh is utilized for both the thermal and the mechanical analysis.

In order to take into account the influence of the contraction joints in the mechanical analysis, the arch is divided into 17 monoliths. The contraction joints and the dam–foundation interface are represented by 16-node zero-thickness interface elements. The model consists of 5200 solid elements and 968 interface elements.

6.2. Thermal Analysis

6.2.1. Governing Equations

For a stationary medium, the transient heat conduction equation is given as follows:

$$\frac{\partial}{\partial x}\left[k_x\frac{\partial T}{\partial x}\right]+\frac{\partial}{\partial y}\left[k_y\frac{\partial T}{\partial y}\right]+\frac{\partial}{\partial z}\left[k_z\frac{\partial T}{\partial z}\right]+G=\rho\, c\frac{\partial T}{\partial t} \tag{1}$$

with the following boundary conditions:

$$T=\bar{T}\quad \text{in}\ \Gamma_T \tag{2}$$

$$k_x \frac{\partial T}{\partial x} l + k_y \frac{\partial T}{\partial y} m + k_z \frac{\partial T}{\partial z} n + q_c + q_r + q_q = 0 \quad \text{in } \Gamma_q \tag{3}$$

and the following initial condition:

$$T = T_o \quad \text{in } \Omega \text{ for } t = t_o \tag{4}$$

where t is the time; T is the temperature; k_x, k_y and k_z are the thermal conductivities; G is the internally generated heat per unit of volume and time; ρ is the material density; c is the specific heat; $\overline{T}$ is the temperature at the boundary Γ_T; q_c is the heat flux due to convection; q_r is the heat flux due to radiation and q_q is the solar radiation absorption at the boundary Γ_q; l, m and n are the direction cosines; and T_o is the temperature at time t_o. It is noteworthy that the convention in expression (3) is positive when heat flux flows outwards from the body [29].

The convective heat transfer between the structure's surface and the air is influenced by the wind speed and air temperature. The heat gain or loss from a surface due to convection is given as follows:

$$q_c = h_c (T - T_a) \tag{5}$$

where h_c is the convection heat transfer coefficient and T_a is the air temperature.

Due to the difference in temperature between the structure's surface and the surrounding air, the surface of the structure emits electromagnetic radiation known as thermal radiation. This radiation is measured by the Stefan–Boltzmann law as follows:

$$q_r = \varepsilon \, \sigma \left(T^4 - T_a^4 \right) \tag{6}$$

where ε is the emissivity of the surface and σ is the Stefan–Boltzmann constant given as 5.669×10^{-8} W(m^2 K). However, when T and T_a are close, which is the normal condition in civil engineering structures, it is possible to rewrite (6) in a quasi-linear form as follows:

$$q_r = h_r (T - T_a) \tag{7}$$

where h_r is the radiation linear coefficient defined as follows:

$$h_r = \varepsilon \, \sigma \left(T^2 - T_a^2 \right) (T - T_a) \tag{8}$$

As a result, the total heat transfer can be calculated by combining the contribution of both heat transfer mechanisms, convection and radiation, leading to the definition of a new coefficient called total thermal transmission coefficient, h_t. In essence, this new coefficient is a convection heat transfer coefficient that is updated to consider radiation.

Finally, the solar radiation boundary condition is given as follows:

$$q_q = a \, I_T \tag{9}$$

where a is the absorption coefficient and I_T is the solar irradiance.

6.2.2. Chemo-Thermal Model

Different mathematical formulations as a function of time, representing the evolution of heat of hydration in adiabatic conditions, have been proposed. In reality, however, hydration does not evolve adiabatically, and the heat source is greatly affected by the actual values of the temperatures that develop inside the concrete. As a result, the time parameter by itself is insufficient to accurately describe the progress of the hydration reaction. Due to the reciprocal relationship between the rate of hydration and the temperature of the concrete, it is necessary to explicitly model the effect of the actual temperature and the rate of reaction when calculating both the temperature development and the hardening process [30].

The evolution of the hydration reaction is represented by an Arrhenius-type equation that takes into consideration the thermo-activation and exothermic nature of the reaction, following the hydration kinetic model based on the thermodynamics of multiphase porous media proposed by Ulm and Coussy [31], in which the variation of the skeleton mass (reaction velocity) $\frac{dm}{dt}$ in [mol/s] is expressed as follows:

$$\frac{dm}{dt} = \frac{1}{\eta} A \exp\left(\frac{E_a}{RT}\right) \tag{10}$$

where η is a viscosity term representing the increase in physical barrier of calcium silicate hydrates (CSH), which tends to isolate the cement grain from the free water, and depends on the state of the hydration reaction; A is the affinity of the chemical reaction or, in other words, the thermodynamic force associated to the rate of hydration formation, which also depends on the state of the hydration reaction; E_a is the apparent thermal activation energy, which is considered to be constant with relation to the hydration degree; R is the universal constant of gases [8.314 J/(mol K)]; and T is the temperature in K.

For practical reasons, it is helpful to rewrite the model in terms of the normalized variable called hydration degree, defined as the relation between the mass of the skeleton at time t normalized by the mass of the skeleton when hydration is complete, i.e., $\xi(t) = m(t)/m_\infty$

$$\dot{\xi} = \widetilde{A}(\xi) \exp\left(-\frac{E_a}{RT}\right) \tag{11}$$

where $\dot{\xi}$ is the time derivative of ξ, and the function $\widetilde{A}(\xi)$ is the normalized affinity which completely characterizes the macroscopic hydration kinetics for a given concrete mixture.

Then, the problem of heat transfer during concrete hydration is obtained from Equation (1), substituting the internally generated heat per unit of volume and time G by the term $L\dot{\xi}$, where L represents the latent heat of hydration of the material:

$$\frac{\partial}{\partial x}\left[k_x \frac{\partial T}{\partial x}\right] + \frac{\partial}{\partial y}\left[k_y \frac{\partial T}{\partial y}\right] + \frac{\partial}{\partial z}\left[k_z \frac{\partial T}{\partial z}\right] + L\dot{\xi} = \rho\, c \frac{\partial T}{\partial t} \tag{12}$$

The simultaneous solution of the two last equation represents the thermochemical coupling, which is a nonlinear problem in T and ξ.

Among the different empirical relationships used to represent the normalized affinity $\widetilde{A}(\xi)$, the empirical relationship reported by Cervera et al. [32] was adopted in this work:

$$\widetilde{A}(\xi) = \frac{k_\xi}{\eta_{\xi o}}\left(\frac{A_{\xi o}}{k_\xi \xi_\infty} + \xi\right)(\xi_\infty - \xi) \exp\left(-\bar{\eta}\frac{\xi}{\xi_\infty}\right) \tag{13}$$

where ξ_∞ is the asymptotic degree of hydration, and k_ξ, $A_{\xi o}$, $\eta_{\xi o}$ and $\bar{\eta}$ are material properties.

6.2.3. Thermal Properties

The thermal properties of the concrete were determined using the method reported by the U.S. Bureau of Reclamation in [33]. This method is based on the mix proportions and petrographic composition of aggregates. It assumes that each material composing the concrete contributes to the conductivity and specific heat in proportion to the amount of the material present in the concrete. Taking into account that the average composition of the concrete used in the dam, expressed in part by weight, was 1:1.66:6.66 (cement:sand:granite) with a water/cement ratio of 0.52 [20], and assuming a reference temperature of 20 °C, we found that k = 2.65 W/(m °C) and c = 866 J/(kg °C).

Table 1 lists the concrete and rock mass foundation parameters that were employed in the thermal analysis.

Table 1. Thermal material properties.

Properties	Rock Mass Foundation	Concrete
Density ρ [kg/m^3]	2657	2460
Specific heat c [J/(kg °C)]	715	866
Thermal conductivity k [W/(m °C)]]	4.91	2.65

The computation of the S-shape function representing the heat of hydration of the concrete was based on the data given in [15], which refers to three samples of cement tested at the ages of 3, 7 and 28 days. Additionally, the heat at the ages of 90 and 365 days was estimated applying a percentage of increase based on the 28 days' value given in [20], resulting in the following:

$$Q = A\exp\left(\frac{B}{t}\right) \tag{14}$$

where Q is the heat of hydration in [kJ/m^3], t is the time in [h], A = kJ/m^3 and $B = -65$ h, as shown in Figure 9.

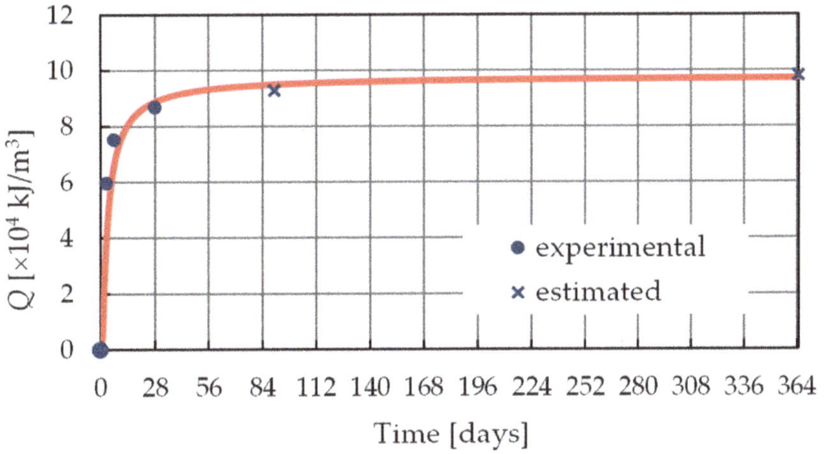

Figure 9. Concrete heat of hydration development.

After applying the approach presented in [34] and assuming $\xi_\infty = 0.75$ and $E_a/R = 4000$ K, the parameters of Equation (13) result in the following: $k_\xi/\eta_{\xi o} = 34.17\ \text{s}^{-1}$; $A_{\xi o}/k_\xi = 1.34 \times 10^{-2}$; and $\bar{\eta} = 6.26$.

6.2.4. Boundary Conditions

Convection/radiation heat transfer and solar radiation flux absorption boundary conditions were used for the dam's air-exposed boundaries. Fixed reservoir water temperature boundary conditions were used for the dam's submerged boundaries. Convection/radiation heat transfer boundary conditions were used for the air-exposed boundaries. Furthermore, adiabatic boundary conditions at the lateral faces and a fixed temperature boundary condition of 12 °C at the bottom were applied to the artificial boundaries of the rock mass foundation.

The average monthly temperature recorded at the dam site during the construction, initial impoundment and first period of operation, along with a daily variation extrapolated from more recent readings, were used to estimate the daily air temperature variation. The solar irradiance was represented by an exponential function derived from the graphics presented in [20] for the different Portuguese regions:

$$\frac{I_b}{\cos Z} = I_o \exp(-0.968 + 0.760 \cos Z) \tag{15}$$

where I_b is the beam component of the solar radiation, Z is the solar zenith angle and I_o is the solar constant (1367 W/m^2).

The total heat transfer coefficient h_t was set to a constant value of 24 W/(m^2 K) for the whole model, with the exception of the formwork-insulated surfaces. Due to the lack of information about the formwork characteristics, an empirical value of 0.10 times the total heat transfer coefficient of the concrete was adopted, yielding h_t= 2.4 W/(m^2 K). The concrete absorption coefficient was assumed as 0.5.

As mentioned before, the reservoir water temperature was introduced as a prescribed temperature. For the initial period, from the start of the impoundment to 31 December 1956, a constant value over the completed depth of 11 °C was adopted. After 31 December 1956, the approximation given by Bofang [35] was adopted as follows:

$$T^{water}(y,d) = T_m^{water}(y) - T_a^{water}(y) \cos\left\{\frac{2\pi}{365}[d - d_o(y)]\right\} \ [^\circ\mathrm{C}] \tag{16}$$

with

$$T_m^{water}(y) = 11 + 5\exp(-0.16y) \ [^\circ\mathrm{C}] \tag{17}$$

$$T_a^{water}(y) = -6\exp(-0.067y) \ [^\circ\mathrm{C}] \tag{18}$$

$$d_o(y) = 4450[1 - \exp(-0.00038y)] + 24.6 \ [\mathrm{days}] \tag{19}$$

where y is the depth of the water; d is the fractional day of the year; and T_m^{water}, T_a^{water} and d_o are the annual mean temperature, the amplitude of annual variation and the phase difference of water temperature at depth y, respectively.

More details about how to apply dam boundary conditions can be seen in [36].

6.2.5. Concrete Placement Schedule

Due to the lack of sufficient data on the concrete placement schedule, it was determined using the data available in the original LNEC technical reports by trial-and-error processes. In particular, the quarterly construction progress (Figure 10a), the volume of concrete pouring per month (Figure 10b) and the installation date for embedded monitoring devices were used.

The concrete placement of the dam was performed with a lift thickness of 2 m and an interval of placement of 3 days. However, the lift thickness and the intervals of placement adopted in the simulation have to be adapted to the 5 m height of the finite element used to represent the dam, resulting in one-week interval of placement. Therefore, the "birth" of a new element in height occurred, at least, one week after the "birth" of the underlying element. Figure 10c shows the final one-week interval placement schedule adopted for the simulation.

6.2.6. Analysis and Results

The in-house code PATQ [34], which uses a fully implicit backward Euler finite difference scheme for the time discretization and a finite element scheme for the spatial discretization, was used to perform the transient thermal analysis. Due to the dependence of the hydration rate on temperature, it uses a two-level iterative procedure. At the structural level, the iteration is caused by the nonlinear dependence of the "thermal body force" on the temperature. At the local level (that is, at each integration point), the iteration is a result of the nonlinear dependence of the degree of hydration (internal variable) on the temperature (free variable).

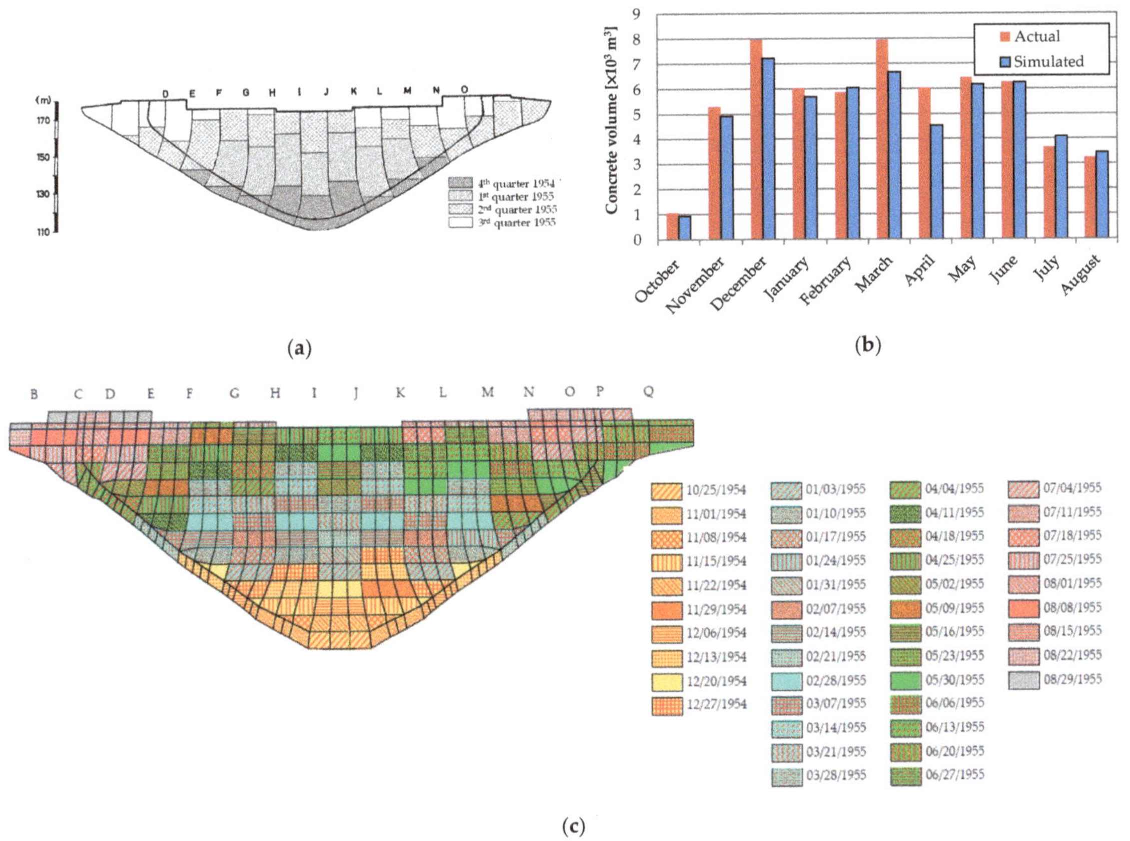

Figure 10. Concrete placement schedule: (**a**) quarterly construction progress; (**b**) comparison of the actual and simulated volume of concrete pouring per month; (**c**) concrete placement schedule adopted for the simulation.

The birth and death approach was used to update the finite element model at each step of construction. The analysis was carried out using an incremental time stepping of 1 h.

To evaluate the performance of the model, the simulation was performed over an extended period from the beginning of the construction, in October 1954, to December 1958. In this way, the simulation encompasses the construction phase, the first filling of the reservoir and the two first times the reservoir was emptied.

A comparison of the predicted and measured temperatures was performed to validate the model. Figure 11 compares the temperatures measured by thermometers T12, T14, T17, T20 and T22 located at an elevation of 150 m in monolith I–J to those obtained with the numerical model at the same position.

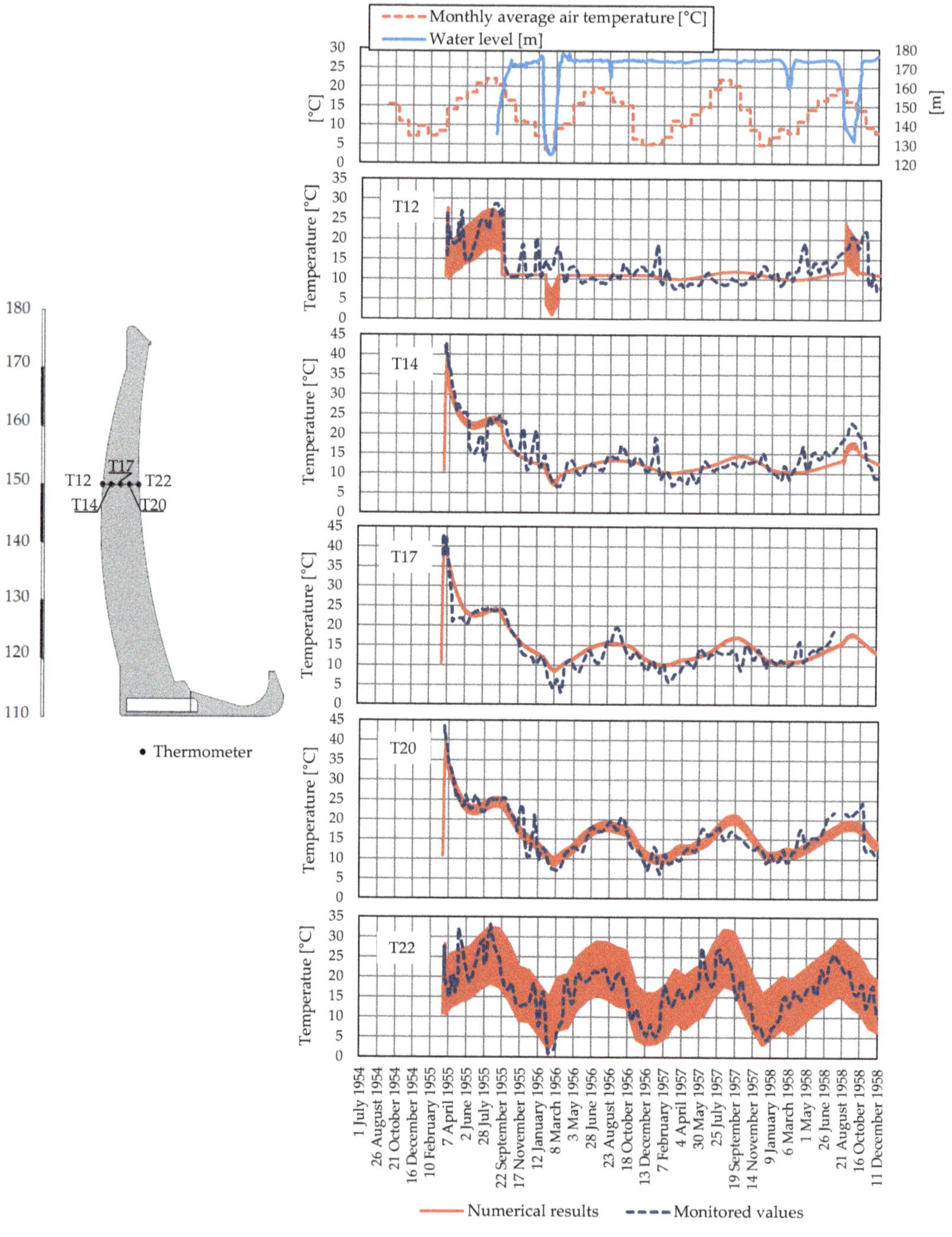

Figure 11. Monthly average air temperature, reservoir levels and comparison of predicted and monitored temperatures at thermometers in monoliths I–J at an elevation of 150 m.

It is important to note that monitoring data measured before 1970 were stored in the form of hand-drawn lines charts. Therefore, WebPlotDigitalizer [37] was used in order to extract the corresponding numerical data for the analysis.

Overall, between the numerically calculated temperatures and the monitored data, a very good agreement was observed. Consequently, we had confidence that the thermal model could accurately describe the temperature loads.

6.3. Mechanical Analysis

Lombardi's paper, which was presented at the Dam Fracture Workshop in Locarno in September 1990, pointed out that a dam:

> ... is not a simple, ideal, elastic body. Vertical contraction joints subdivide the dam in blocks. Even after grouting they will represent a discontinuity in as much as they cannot be grouted to the dam faces but only to the water-stops. Additionally, the grouting is seldom carried out in a completely satisfactory manner. Furthermore, the concrete is placed in layers and lifts making weakness planes possible at frequent elevations, and inhomogeneities and anisotropies – which can hardly be detected – are likely to exist in the concrete mass (Lombardi [38], p. 4).

Therefore, as the above quotation suggests, a model considering elastic, isotropic, and homogeneous materials which are free from discontinuities is not a true representation of the concrete of the dam or the rock mass foundation. Nevertheless, it will be always necessary to adopt some degree of simplification. For one, it is impossible to know and more so to represent all the inhomogeneities, anisotropies and discontinuities present in the dam and its foundation. Moreover, an idealized simple model can better explain and characterize the dominant features of the physical phenomena studied.

Therefore, in order to characterize the actual behavior of the dam, two different finite element models were adopted. The two models were selected in order to demarcate the bandwidth within which the behavior of the dam is expected. The first one is a linear continuous model, which is the standard model used in a dam analysis. In the second model, the nonlinear model, the contraction joints and the dam–foundation interface are represented explicitly by no-tension interface elements.

6.3.1. Mechanical Properties

In order to take into account the creep effect, the double power law [39] was adopted for the concrete as follows:

$$J(t-t') = \frac{1}{E_o} + \frac{\varphi_1}{E_o}\left(t'^{-m} + \alpha\right)(t-t')^n \tag{20}$$

where $J(t-t')$ is the compliance function (or the creep function), i.e., the strain at age t caused by a unit of uniaxial constant stress acting since age t'; E_o is the asymptotic modulus; and n, m, α and φ_1 are the material parameters.

Actually, as the finite element code uses the library MATPAR given by Bažant in [40], the input data for expression (20) are expressed in terms of the following five parameters: E_{28}, E_o/E_{28}, n, m and α. By calibrating the model with the recently recorded displacements, E_o = 40.5 GPa was obtained. Based on this value and using the standard relationship E_o/E_{28} = 1.5, E_{28} was fixed in 27 GPa. The remaining three parameters n, m and α were estimated using the empirical formulas given in [39] for w/c = 0.52 and f'_c = 27.5 MPa.

The rock mass foundation was considered as a linear elastic material. According to the rock field tests carried out by LNEC, the rock mass foundation was divided in three different zones, reflecting that near-surface rock is more weathered and fractured, as shown in Figure 8.

Zero-thickness interface elements were used to represent the contraction joints and the dam–foundation interface. In this formulation, the contact constraint is enforced by the penalty method, where the normal stiffness k_n and the tangential stiffnesses k_s and k_t play the role of penalty coefficients. This means that they have to be set as high as possible to guarantee that no penetration takes place while the joints are closed, but not so high to avoid ill-posed problems. Hence, the normal and tangential stiffness have no physical meaning, eliminating the need for extra experimental investigations. Moreover, the assumptions of no-tension and no-sliding conditions to characterize the normal and tangential behavior, respectively, also help us avoid the need for additional material parameters.

The material parameters utilized in the mechanical analysis are listed in Table 2.

Table 2. Mechanical material properties.

Material	Properties	Values
Concrete	Double power law	
	E_o [GPa]	40.50
	n	0.12
	m	0.34
	α	0.048
	φ_1	1.78
	Poisson's ratio ν	0.20
	Coefficient of thermal expansion α [1/°C]	10^{-5}
Rock mass foundation	Young's modulus E [GPa]	15.00, 5.00 or 1.80
	Poisson's ratio ν	0.20
	Coefficient of thermal expansion α [1/°C]	0.00
Joints	$k_s = k_t$ [GPa/m]	2000.00
	k_n [GPa/m]	2000.00
	f_t	0.00

6.3.2. Loads

The analysis was carried out considering the dead load, normal water load and internal strains caused by temperature changes.

The dead load corresponds to the weight of the concrete, and it was applied staggered at the "birth" of each element. As the contraction joints are open during the construction phase, the corresponding interface elements were considered inactive at this stage in order to simulate the cantilever behavior of the monoliths during construction.

The normal water load corresponds to the hydrostatic pressures acting on the dam's upstream face resulting from the reservoir. For the computation of the normal water load, the pressure is considered to vary linearly with depth and to act normally on the dam surface.

The thermal load was computed from the temperature variation obtained in the chemo-thermal analysis.

The uplift load at the dam–foundation interface was ignored because the dam is very thin.

6.3.3. Analysis and Results

The phases of the construction, the initial impoundment and the first period of operation were solved incrementally through time. The in-house code PAVK [28] was used for this analysis.

During the construction phase, the time intervals were constrained by the concrete placement and formwork striking dates. After that, until 30 April 1956, a smaller time interval of 1 or 2 days was used in order to follow the effect of the reservoir rise on the upstream face of the dam. Finally, the analysis was completed with a two-week interval, except when a geodetic survey was carried out.

Figure 12 compares the radial displacements obtained with the numerical model with the corresponding values measured by geodetic survey triangulation. This figure shows, from top to bottom, in the first graph, the monthly average temperature and the rising reservoir water level; and in the following three graphs, the comparison of the radial displacements at elevations of 170 m, 150 m and 130 m in block I–J. The negative direction indicates that the dam moved downstream.

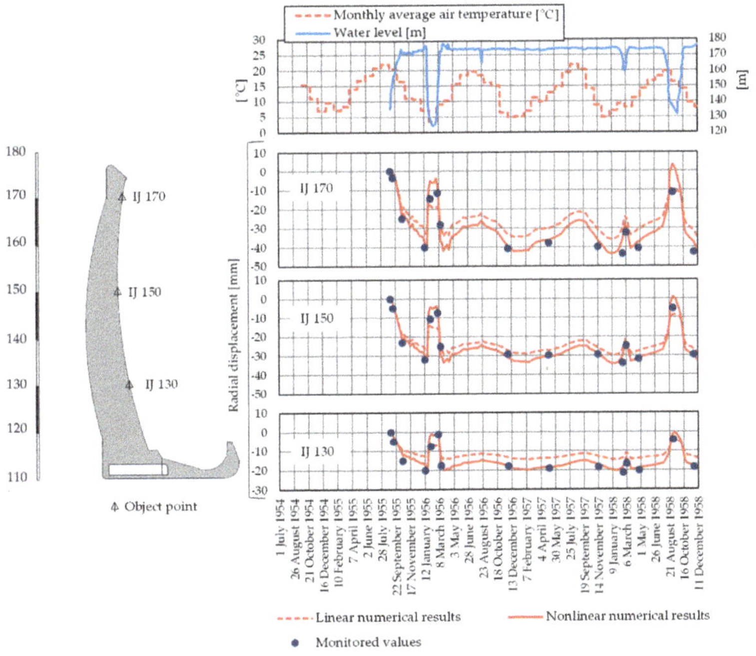

Figure 12. Monthly average air temperature, reservoir levels and comparison of predicted and monitored radial displacements in monoliths I–J at elevations of 170 m, 150 m and 130 m.

As can be observed, the nonlinear model shows a better agreement with the monitored radial displacements than the linear one. The comparison of the models shows that during impoundment and emptying periods, the nonlinear model reflects larger displacements than the linear one. This apparent greater flexibility of the nonlinear model is due to the opening/closing of the contraction joints and the dam–foundation interface in response to the stress changes.

In contrast, when the reservoir is full, leading to the closing of the contraction joints, both models show almost similar displacement fluctuations. This fact corroborates that the greater flexibility in the response of the dam for periods of low reservoir levels is due to the movements of the discontinuities and not to a lower concrete modulus of elasticity.

The importance of considering the discontinuities in the model is also confirmed by Figure 13, which shows the vertical stresses estimated and measured near the upstream heel and the downstream toe of monoliths I–J.

The measured stresses were recorded by stress meters located at 1 m from the faces of the dam. It is worth noting that the stress meter is a device that allows the direct measurement of compressive stress. In contrast with strain meters, the stress meter is fully responsive to compressive stress and very closely indicates the true stress at all times without any further analysis, i.e., no elasticity modulus needs to be adopted, and without regard to deformation due to causes other than stress [41].

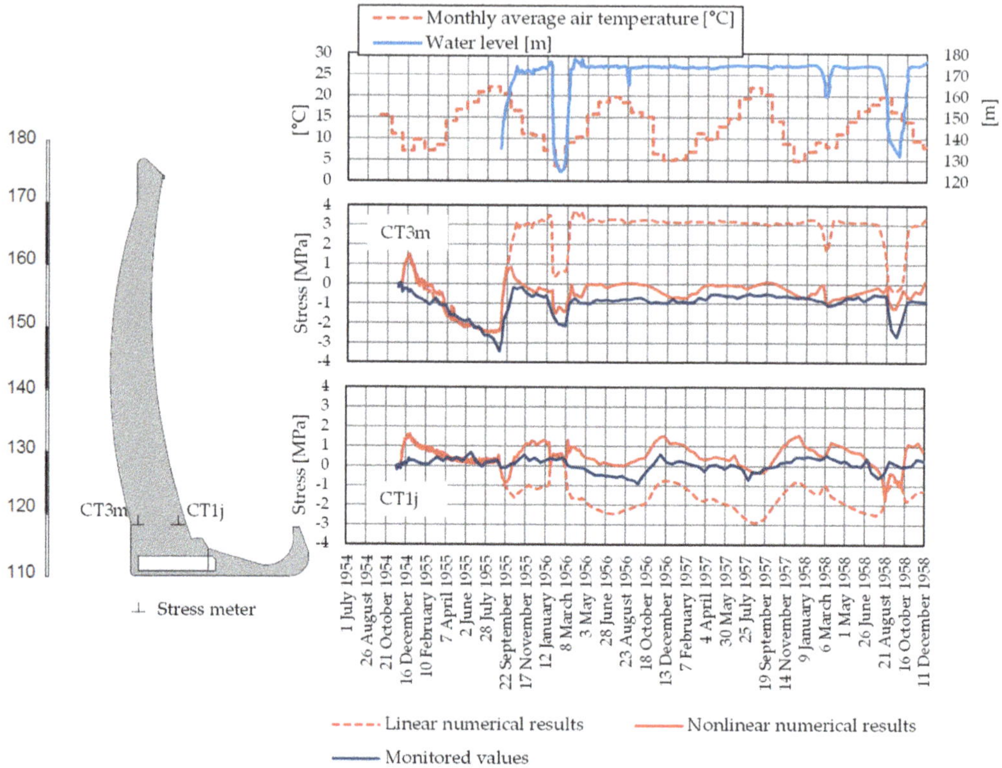

Figure 13. Monthly average air temperature, reservoir levels and comparison of predicted and monitored vertical stresses in monoliths I–J at elevations of 118 m.

Analyzing Figure 13 (stress meter CT3m), it can be observed that when the reservoir reaches approximately the level of 160 m, the two models no longer stay together. This is because tensile stresses appear in the upstream heel of the dam, inducing the opening of the corresponding dam–foundation interface element. This behavior is fully validated by the stresses measured in stress meter CT3m.

The opening of the interface element also modifies the stresses at the downstream toe of the dam, as shown by Figure 13 (stress meter CT1j). This is due to the redistribution of the stresses, which causes the decrease in the cantilever stresses at the expense of an increase of the arch stresses. This fact is validated by the stress meter CT1j.

7. Validation of the Diagnosis Hypothesis

According to the previous analysis, the nonlinear model is the most representative model of the behavior of the dam. Therefore, this model was used to study the cracking of the upstream face of the dam.

Figures 14 and 15 show the principal stress plots obtained during the initial impoundment over the upstream face of the dam. The surface stresses were computed in the center of each external face by extrapolating the stresses obtained in the integration points.

The length and direction of the arrows show the relative magnitude and direction of compression (blue) and tension (red).

The initial impoundment started on 3 September 1955 with an empty reservoir. The rise in the reservoir level took place at a very high rate, reaching a level of 160 m on 26 September 1955.

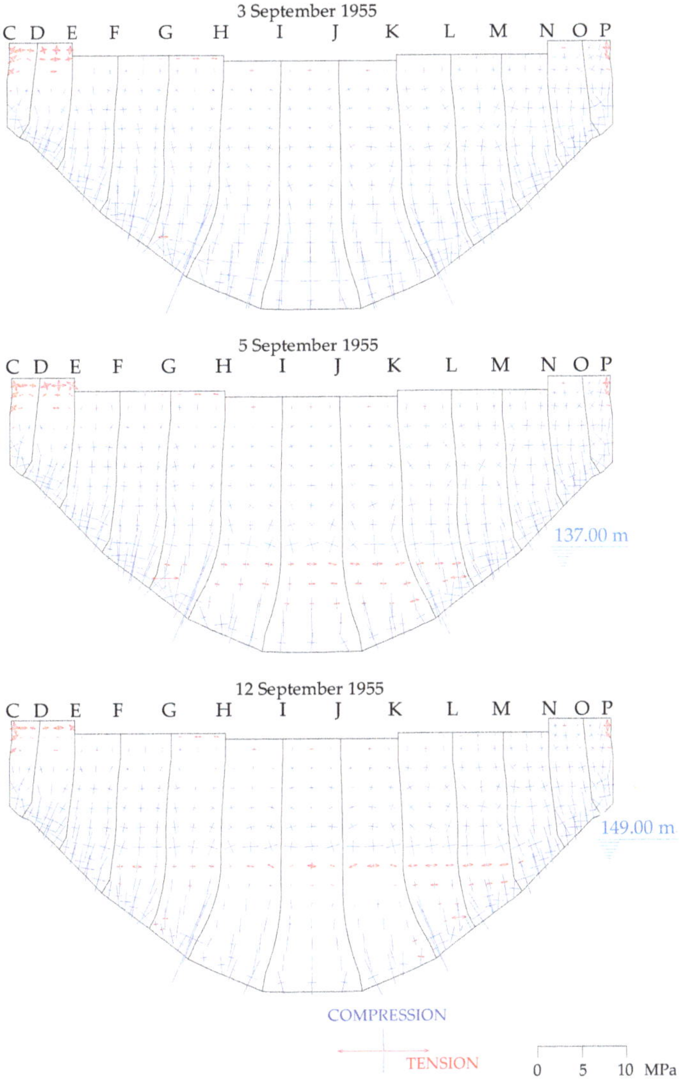

Figure 14. Distribution of principal stresses over the upstream surface for the empty reservoir with a water level of 137 m and 149 m.

As the water rose, the surface in contact with it cooled rapidly. The surface contraction due to the cooling was restrained by the hotter interior concrete, which did not contract as rapidly as the surface, and by the concrete located above the water level. As a result, an equilibrium deformation was obtained, inducing thermal compressive stresses in the hotter concrete and tensile stresses in the colder concrete, as can be observed in the principal stress distributions represented in the figures.

Up to the water level of 149 m, the self-weight counterbalanced the vertical (cantilever) stresses generated by the temperature variation. However, when the water level reached a level of 153 m, the thermal vertical stresses started to be predominant, with a maximum expression at a water level of 155 m. The nine days between the water levels of 155 m and

160 m were enough time to cool the emerged concrete and lower the vertical tensile stresses.

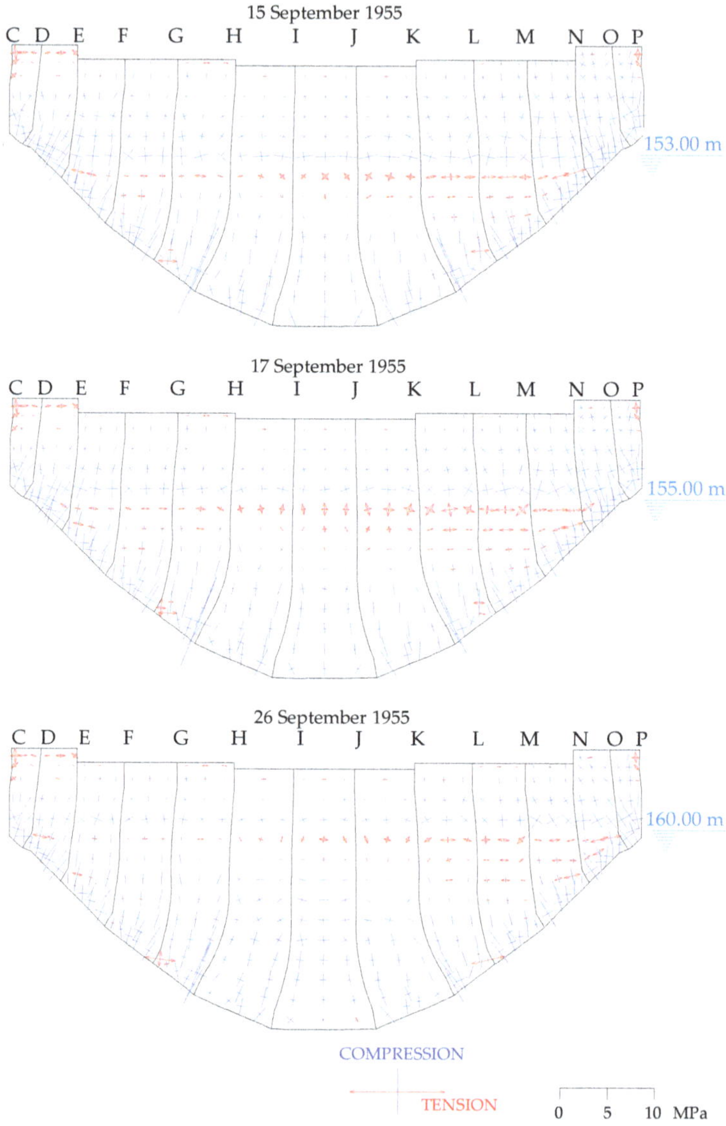

Figure 15. Distribution of principal stresses over the upstream surface for water levels of 153 m, 155 m and 160 m.

8. Prediction of Future Behavior

In order to corroborate that the presence of cracks has no influence on the static behavior of the dam, a comparison of the radial displacements measured by geodetic triangulation and those obtained with the finite element model are represented in Figure 16 for the period of 2001–2016.

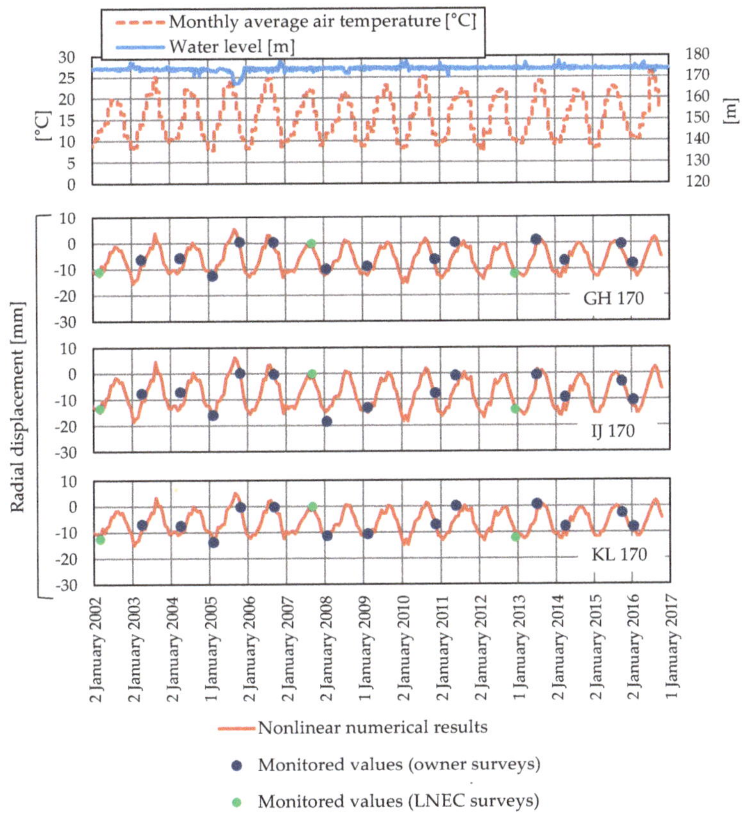

Figure 16. Monthly average air temperature, reservoir levels and monitored radial displacement between 2002 and 2016.

As for its potential seismic performance, an explanation can be found in the U.S. Army Corps of Engineering Manual EM 1110-2-6053, illustrated in Figure 17:

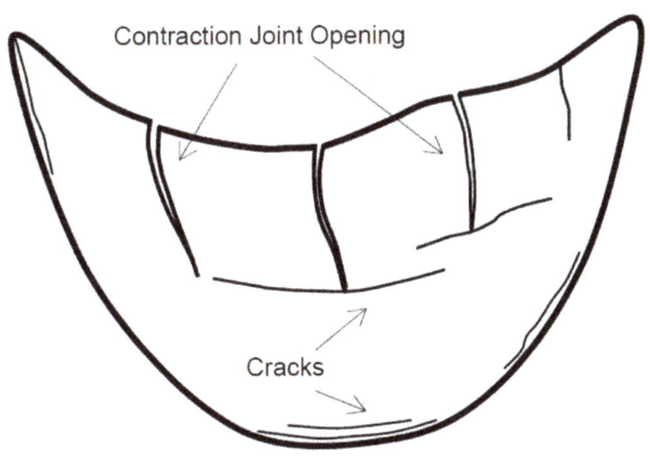

Figure 17. Response of arch dams to major earthquakes (adapted from [42]).

In arch dams, potentially opened contraction joints and cracked lift lines may subdivide the monolithic arch structure into partially free cantilever blocks, capable of transmitting only compressive or frictional forces. In this situation, any failure mode of the arch structure would more likely involve sliding stability of the partially free cantilevers. For small and moderate joint openings, the partially free cantilever blocks, bounded by opened joints, may remain stable through interlocking (wedging) with adjacent blocks. The extent of interlocking depends on the depth and type of shear keys and the amount of joint opening (EM 1110-2-6053 [42], p. 2–9).

9. Discussion and Conclusions

This paper presented the study, characterization and numerical simulation of the diagnosis procedure of the pathology affecting an old, thin dam. The cause of the main damage observed in the dam was attributed to the thermal cracking of weak construction joints during the initial impoundment. This hypothesis was validated through a chemo-thermo-mechanical finite element model.

For the thermal analysis of its construction, initial impoundment and first period of operation, a chemo-thermal model based on the chemical affinity concept was used.

Sequentially, a nonlinear viscoelastic analysis was performed in order to obtain the structural response of the dam. In this analysis, the contraction joints and the dam–foundation interface were modelled using zero-thickness interface elements under no-tension and no-sliding conditions, which avoided the need to determine new material parameters.

The conclusion of the analysis was that tensile vertical stresses developed on the upstream face of the dam during the initial impoundment together with weak construction joints caused the appearance of cracks in the upstream face of the dam. Over time, water penetration expanded these cracks, reaching the downstream face.

Author Contributions: Conceptualization, N.S.L.; methodology, N.S.L.; software, N.S.L. and E.C.; validation, N.S.L. and E.C.; formal analysis, N.S.L.; investigation, N.S.L. and E.C.; writing—original draft preparation, N.S.L.; writing—review and editing, N.S.L.; visualization, N.S.L.; supervision, N.S.L. All authors have read and agreed to the published version of the manuscript.

Funding: This research received no external funding.

Data Availability Statement: The data presented in this study are available in the article.

Conflicts of Interest: The authors declare no conflict of interest.

References

1. Law, S.; Troja, N. Blog: Hydropower Growth and Development through the Decades. Available online: https://www.hydropower.org/blog/blog-hydropower-growth-and-development-through-the-decades (accessed on 14 February 2023).
2. White, W.R. *World Water: Resources, Usage and the Role of Man-Made Reservoirs*, 3rd ed.; Foundation for Water Research: Marlow, Bucks, UK, 2019.
3. Perry, D.M.; Praskievicz, S.J. A New Era of Big Infrastructure? (Re)developing Water Storage in the U.S. West in the Context of Climate Change an Environmental Regulation. *Water Altern.* **2017**, *10*, 437–454.
4. Wieland, M. Hydropower over the Next Decade. *Int. Water Power Dam Constr.* 2020. Available online: https://www.waterpowermagazine.com/features/featurehydropower-over-the-next-decade-7874055/ (accessed on 14 February 2023).
5. Godart, B.; de Rooij, M.; Wood, J.M.G. *Guide to Diagnosis and Appraisal of AAR Damage to Concrete Structures. Part 1 Diagnosis (AAR 6.1)*; Springer: Dordrecht, The Netherlands, 2013.
6. Saouma, V.E. *Diagnosis & Prognosis of AAR Affected Structures. State-o-the-Art Report of the Rilem Technical Committee 259-ISR*; Springer: Cham, Switzerland, 2021.
7. von Fay, K.F. *Guide to Concrete Repair*; U.S. Department of the Interior, Bureau of Reclamation, Technical Service Center: Denver, CO, USA, 2015.
8. Wieland, M. Life-span of storage dams. *Int. Water Power Dam. Constr.* **2010**, 32–35. Available online: https://www.waterpowermagazine.com/features/featurelife-span-of-storage-dams (accessed on 14 February 2023).
9. FIP. *Repair and Strengthening of Concrete Structures: FIP Guide to Good Practice*; Fédération Internationale de la Précontrainte, Thomas Telford Services Ltd.: London, UK, 1991.

10. Pardo-Bosch, F.; Aguado, A. Damage Diagnosis in Concrete Dams with Presented Expansive Damage Based on Medical Propaedeutics. *J. Perform. Constr. Facil.* **2017**, *31*, 04017048. [CrossRef]
11. Blanco, A.; Pardo-Bosch, F.; Cavalaro, S.; Aguado, A. Lessons Learned about the Diagnosis of Pathologies in Concrete Dams: 30 Year of Research and Practice. *Constr. Build. Mater.* **2019**, *197*, 356–368. [CrossRef]
12. Campos, A.; López, C.M.; Blanco, A.; Aguado, A. Structural Diagnosis of a Concrete Dam with Cracking and High Nonrecoverable Displacements. *J. Perform. Constr. Facil.* **2016**, *30*, 04016021. [CrossRef]
13. Blanco, A.; Cavalaro, S.H.P.; Segura, I.; Segura-Castillo, L.; Aguado, A. Expansions with different origins in a concrete dam with bridge over spillway. *Constr. Build. Mater.* **2018**, *163*, 861–874. [CrossRef]
14. Campos, A.; López, C.M.; Blanco, A.; Aguado, A. Effects of an internal sulfate attack and an alkali-aggregate reaction in a concrete dam. *Constr. Build. Mater.* **2018**, *166*, 668–683. [CrossRef]
15. Pereira, J.A.; Silva, J.C. O aproveitamento hidroelétrico da Bouçã (1ª parte). In *Electricidade*; Empresa Editorial Electrotécnica (EDEL Lda): Lisbon, Portugal, 1957; Volume 1, pp. 71–94. Available online: https://www.colecoesfundacaoedp.edp.pt/Nyron/Library/Catalog/winlibimg.aspx?skey=77941E702E3949049D14F9EA163CE945&doc=162919&img=152155 (accessed on 14 February 2023). (In Portuguese)
16. Rocha, M.; Serafim, J.L. Analysis of Concrete Dams by Model Tests. In Proceedings of the 5th Congress on Large Dams, Paris, France, May–June 1955; Volume V, pp. 1307–1344.
17. Rocha, M.; Serafim, J.L.; Silveira, A.F.; Matos, M.E. Model tests and observation of Bouçã Dam. In Proceedings of the 6th International Congress on Large Dams, New York, NY, USA, September 1958; Volume II, pp. 927–960.
18. Rocha, M.; Serafim, J.L.; Ferreira, M.J.E. The Determination of the Safety Factors of Arch Dams by Means of Models. In Proceedings of the International Colloquium on Models of Structures, Madrid, Spain, June 1959; Rilem Bulletin Nº 7. 1960; pp. 68–78.
19. Rocha, M.; Silveira, A.F.; Azevedo, M.C.; Ferreira, M.J.E. The Use of Model Tests in the Determination of the Safety Factor in Arch Dams. *Water Power* **1969**, *21*, 461–465.
20. Silveira, A. *Temperatures Variations in Dams*; Memória nº 177; LNEC: Lisbon, Portugal, 1961. (In Portuguese)
21. Rouse, G.C.; Lockman, W.T.; Stubbs, J.A. *Structural Model Tests of Arch Dams—Glen Canyon and Morrow Point Dams*; U.S. Department of the Interior, Bureau of Reclamation: Denver, CO, USA, 1969.
22. Sarkaria, G.S. Lessons from Serious Incidents at Seven Arch Dams. In Proceedings of the 1997 Annual Conference of the Association of State Dam Safety Officials, 7–10 September 1997; Available online: https://damfailures.org/wp-content/uploads/2015/07/039_Lessons-from-Serious-Incidents-at-Seven-Arch-Dams.pdf (accessed on 14 February 2023).
23. Lemos, J.V.; Lamas, L. *Contribution of Manuel Rocha to Rock Mechanics and Dam Foundations*; LNEC: Lisbon, Portugal, 2013.
24. CNPGB. *Large Dams in Portugal*; Portuguese National Committee on Large Dams: Lisbon, Portugal, 1992.
25. Chopra, A.K. *Earthquake Engineering for Concrete Dams, Analysis, Design and Evaluation*; John Wiley & Sons Ltd.: Hoboken, NJ, USA, 2020.
26. Hendriks, M.A.N.; de Boer, A.; Belletti, B. *Guidelines for Nonlinear Finite Element Analysis of Concrete Structures*; Report RTD:1016-1:2017; Rijkswaterstaat Centre for Infrastructures: Utrecht, The Netherland, 2017.
27. Saouma, V.E.; Hariri-Ardebili, M.A. *Aging, Shaking, and Cracking of Infrastructures, From Mechanics to Concrete Dams and Nuclear Structures*; Springer: Cham, Switzerland, 2021.
28. Leitão, N.S.; Castilho, E.; Farinha, M.L.B. Towards a Better Understanding of Concrete Arch Dam Behavior during the First Filling of the Reservoir. *CivilEng* **2023**, *4*, 151–173. [CrossRef]
29. Lewis, R.W.; Nithiarasu, P.; Seetharamu, K.N. *Fundamentals of the Finite Element Method for Heat and Fluid Flow*; John Wiley & Sons, Ltd.: Chichester, West Sussex, UK, 2004.
30. van Breugel, K.; Braam, C.R.; Koenders, E.A.B. *Concrete Structures under Imposed Thermal and Shrinkage Deformations—Theory and Practice*; TUDelft: Delft, The Netherlands, 2013.
31. Ulm, F.J.; Coussy, O. Modeling of thermochemomechanical couplings of concrete at early ages. *J. Eng. Mech.* **1995**, *121*, 185–794. [CrossRef]
32. Cervera, M.; Oliver, J.; Prato, T. Thermo-chemo-mechanical model for concrete. I: Hydration and aging. *J. Eng. Mech.* **1999**, *125*, 1018–1027. [CrossRef]
33. U.S. Bureau of Reclamation. *Boulder Canyon Project Final Reports. Part VII—Cement and Concrete Investigation, Bulletin 1, Thermal Properties of Concrete*; U.S. Department of the Interior, Bureau of Reclamation: Denver, CO, USA, 1940. Available online: https://babel.hathitrust.org/cgi/pt?id=mdp.39015068182602 (accessed on 14 February 2023).
34. Castilho, E.; Schclar, N.; Tiago, C.; Farinha, M.L.B. FEA model for the simulation of the hydration process nd temperature evolution during the concreting of an arch dam. *Eng. Struct.* **2018**, *174*, 165–177. [CrossRef]
35. Bofang, Z. *Thermal Stresses and Temperature Control of Mass Concrete*, 1st ed.; Butterworth-Heinemann: Oxford, UK, 2014.
36. Leitão, N.S.; Castilho, E. Heat transfer analysis of infrastructures subjected to environmental actions: A finite element solver PAT. *Therm. Sci. Eng. Prog.* **2022**, *34*, 101447. [CrossRef]
37. Rohatgi, A. WebPlotDigitizer. Available online: https://automeris.io/WebPlotDigitizer/ (accessed on 14 February 2023).
38. Lombardi, G. Cracks in Arch Dams and Repair Works. In Proceedings of the Workshop on Application of Fracture Mechanics to Dam Engineering, Locarno, Switzerland, 17–18 September 1990; Available online: https://www.lombardi.ch/fr-fr/SiteAssets/Publications/997/Pubb-0176-L-Cracks%20in%20Ach%20Dams%20and%20repair%20works.pdf (accessed on 14 February 2023).

39. Bažant, Z.P.; Panula, L. Practical Prediction of Time-Dependent Deformation of Concrete. Part II: Basic creep. *Mater Struct.* **1978**, *11*, 317–328. [CrossRef]
40. Bažant, Z.P. Input of Creep and Shrinkage Characteristics for a Structural Analysis Program. *Mater Struct.* **1982**, *15*, 283–290. [CrossRef]
41. Raphel, J.M.; Carlson, R.W.; James, J. *Measurement of Structural Action in Dams*; James J. Gillick & CO.: Berkeley, CA, USA, 1954.
42. USACE. *Earthquake Design and Evaluation of Concrete Hydraulic Structures*; EM 1110-2-6053; U.S. Army Corps of Engineers: Washington, DC, USA, 2007.

MDPI AG
Grosspeteranlage 5
4052 Basel
Switzerland
Tel.: +41 61 683 77 34

Eng Editorial Office
E-mail: eng@mdpi.com
www.mdpi.com/journal/eng

www.ingramcontent.com/pod-product-compliance
Lightning Source LLC
Chambersburg PA
CBHW040313270926
51850CB00007BC/1226
9783725819621